PHYSICAL OCEANOGRAPHY

PHYSICAL OCEANOGRAPHY

Li Fang Liang
Min Li

Kruger Brentt
Publishers
2024

Kruger Brentt Publishers UK. LTD.
Company Number 9728962

Regd. Office: 68 St Margarets Road, Edgware, Middlesex HA8 9UU

ISBN: 978-1-78715-266-3

For information on all our publications visit our website at http://krugerbrentt.com/

PREFACE

Physical Oceanography, a branch of oceanography, is a scientific discipline that focuses on understanding the physical properties and processes of the Earth's oceans. It encompasses a wide range of phenomena, from the movements of ocean currents to the interactions between the atmosphere and the sea. This field plays a crucial role in unraveling the complexities of the oceans and their impact on the global climate system. Physical Oceanography is a multidisciplinary science that combines principles from physics, mathematics, and fluid dynamics to explore the intricate dynamics of the ocean. This i serves as a gateway into the fascinating world of oceanic processes, emphasizing the interconnected nature of the physical aspects that govern the behavior of the seas. Introduction to Physical Oceanography is designed to cater to a wide audience, ranging from students and researchers in oceanography to enthusiasts eager to grasp the fundamental principles governing our oceans. Physical oceanography, often regarded as the cornerstone of ocean sciences, investigates the physical properties and processes governing the oceans. This multidisciplinary field delves into the fundamental dynamics of seawater, exploring the intricate interplay of forces that drive currents, regulate temperatures, and sculpt the underwater landscapes. The Backdrop of Historical Circumstances comprehend the present, one must first glance into the past. This chapter provides a historical panorama, tracing the evolution of human understanding of the oceans. From the early navigational challenges faced by seafarers to the technological advancements that have propelled oceanography into the 21st century, the historical backdrop sets the stage for our contemporary exploration.

The physical setting of the oceans follows, revealing their expansive reach, diverse topography, and pivotal role in regulating the Earth's climate. This section lays the groundwork for a deeper exploration of the factors that influence oceanic behavior. As the ocean and atmosphere engage in a perpetual dance, this chapter uncovers the profound impact of atmospheric conditions on oceanic processes. From the gentle caress of breezes to the ferocity of storms, the intricate relationship between the oceans and the atmosphere comes to light. The ocean heat budget delving into the heart of climate regulation, we investigate the ocean's heat budget. How does the ocean absorb, store, and redistribute heat? This chapter unveils the pivotal role played by the oceans in shaping temperature variations and influencing climate patterns across the globe. The physical properties of seawater take center stage as we explore temperature, salinity, and density. These properties not only impact the movement of ocean currents but also play a fundamental role in shaping the distribution of marine life, fostering a delicate balance within the underwater ecosystems. The Equations of Motion - Some Mathematics introduces the equations of motion that govern the ebb and flow of seawater. Understanding these mathematical foundations is essential for predicting and interpreting the myriad oceanic phenomena that unfold beneath the surface. Equations of Motion with Viscosity adding a layer of complexity to our mathematical journey, we delve into the equations of motion with viscosity. This exploration sheds light on how viscosity influences the behavior of ocean currents, unraveling the intricate web of forces that shape underwater movement. Introduction to Physical Oceanography is designed to cater to a wide audience, ranging from students and researchers in oceanography to enthusiasts eager to grasp the fundamental principles governing our oceans.

We are grateful to all those persons as well as various books, manuals, periodicals, newsletters, magazines, journals etc. that helped in the preparation of this book. In spite of the best efforts, it is possible that some errors may have occurred into the compilation and editing of the book. Further queries, constructive suggestions and criticisms for the further improvement of the book are always welcome and shall be thankfully acknowledged.

Li Fang Liang

Min Li

CONTENTS

PREFACE V

1. PHYSICAL OCEANOGRAPHY: AN INTRODUCTION 1

1.1 Why study the Physics of the ocean? 1

1.2 Goals 2

1.3 Organization 3

1.4 The Big Picture 3

2. THE BACKDROP OF HISTORICAL CIRCUMSTANCES 7

2.1 Definitions 8

2.2 Eras of Oceanographic Exploration 8

2.3 Milestones in the Understanding of the Ocean 12

2.4 The Significance of Observations in the Field of Oceanography 16

2.5 Important Concepts: At a Glance 20

3. THE PHYSICAL SETTING: AN OVERVIEW 21

3.1 Ocean and Seas 22

3.2 Dimensions of the ocean 24

3.3 Sea-Floor Features 25

3.4 Measuring the Depth of the Ocean 29

3.5 Sea Floor Charts and Data Sets 33

3.6 Sound in the Ocean 35

4. INFLUENCES FROM THE ATMOSPHERE 39

4.1 The Earth in Space 39

4.2 Atmospheric Wind Systems 41

4.3 The Planetary Boundary Layer 42

4.4 Wind Measurement 43

4.5 Calculations of Wind 47

4.6 Wind Stress 48

4.7 Concise Summary 49

5. THE OCEAN HEAT BUDGET **51**

5.1 The Oceanic Heat Budget 51
5.2 Heat-Budget Terms 53
5.3 Direct Calculation of Fluxes 57
5.4 Indirect Calculation of Fluxes: Bulk Formulas 59
5.5 Global Data Sets for Fluxes 62
5.6 Geographic Distribution of Terms in the Heat Budget 66
5.7 Meridional Heat Transport 69
5.8 Variations in Solar Constant 71
5.9 Concise Summary 73

6. TEMPERATURE, SALINITY, AND DENSITY **75**

6.1 Salinity 75
6.2 Temperature 78
6.3 Geographical Distribution of Surface Temperature and Salinity 79
6.4 The Oceanic Mixed Layer and Thermocline 83
6.5 Density, Potential Temperature, and Neutral Density 85
6.6 Measurement of Temperature 90
6.7 Measurement of Conductivity or Salinity 95
6.8 Measurement of Pressure 96
6.9 Measurement of Temperature and Salinity with Depth 98
6.10 Light in the Ocean and Absorption of Light 100
6.11 Concise Summary 104

7. THE EQUATIONS OF MOTION: SOME MATHEMATICS **105**

7.1 Primary Factors Influencing Ocean Dynamics 105
7.2 Coordinate System 106
7.3 Types of Flow in the Ocean 107
7.4 Conservation of Mass and Salt 108
7.5 The Total Derivative (D/Dt) 109
7.6 Conservation of Mass: The Continuity Equation 114
7.7 Solutions to the Equations of Motion 116
7.8 Concise Summary 117

8. EQUATIONS OF MOTION WITH VISCOSITY **119**

8.1 The Influence of Viscosity 119
8.2 Turbulence 120

8.3 Calculation of Reynolds Stress 124
8.4 The Turbulent Boundary Layer Over a Flat Plate 125
8.5 Comments on Horizontal Mixing: 131
8.6 Stability 131
8.7 Concise Summary 136

9. THE RESPONSE AND BEHAVIOUR OF THE UPPER OCEAN TO THE INFLUENCE OF WIND 139

9.1 Inertial Motion 139
9.2 Ekman Layer at the Sea Surface 141
9.3 Ekman Mass Transport 149
9.4 Utilizing the Ekman Theory in Practical Scenarios 151
9.5 Langmuir Circulation 154
9.6 Concise Summary 155

10. MAINTAINING A GEOSTROPHIC BALANCE AND THE RESULTANT GEOSTROPHIC CURRENTS 157

10.1 Hydrostatic Equilibrium 157
10.2 Geostrophic Equations 159
10.3 Surface Geostrophic Currents from Altimetry 161
10.4 Geostrophic Currents from Hydrography 165
10.5 An Example using Hydrographic Data 170
10.6 Remarks on Geostrophic Currents 171
10.7 Currents from Hydrographic Sections 177
10.8 Lagrangian Measurements of Currents 179
10.9 Measurements Conducted using the Eulerian Method 185
10.10 Concise Summary 187

11. OCEAN CIRCULATION PROPELLED BY WIND FORCES 189

11.1 Sverdrup's Theory of the Oceanic Circulation 189
11.2 Stommel's Theory of Western Boundary Currents 195
11.3 Munk's Solution 196
11.4 Observed Surface Circulation in the Atlantic 199
11.5 Concise Summary 204

12. VORTICITY IN THE OCEAN: AT A GLANCE..205

12.2 Conservation of Vorticity 208
12.3 Influence of Vorticity 210

12.4 Vorticity and Ekman Pumping 211
12.5 Concise Summary 217
13. OCEANIC DEEP CIRCULATION..219
13.1 Deep Circulation 219
13.2 Importance of the Deep Circulation 220
13.3 Theory for the Deep Circulation 227
13.4 Observations of the Deep Circulation 231
13.5 Antarctic Circumpolar Current 238
13.6 Concise Summary 242
14. CONCEPTS OF EQUATORIAL PROCESSES..243
14.1 Introduction to Equatorial Processes 244
14.2 Variable Equatorial Circulation: El Niño/La Nina 249
14.3 El Niño Teleconnections 257
14.4 Observing El Niño 259
14.5 Forecasting El Niño 261
14.6 Concise Summary 263
15. NUMERICAL MODELS 265
15.1 Introduction: An Overview 265
15.2 Application of Numerical Models in Oceanography 267
15.3 Types of Global Ocean Models 269
15.4 Coastal Models 272
15.5 Assimilation Models 277
15.6 Coupled Numerical Models of the Atmosphere and Ocean 280
15.7 Concise Summary 283
16. INTRODUCTION TO OCEAN WAVES...285
16.1. Linear Theory of Ocean Surface Waves 285
16.2 Nonlinear Waves 290
16.3 Waves and the Concept of A Wave Spectrum 291
16.4 Ocean-Wave Spectra 297
16.5 Understanding of Wave Forecasting 301
16.7 Concise Summary 304
17. COASTAL PROCESSES AND TIDES 307
17.1 Waves in Shoaling and Coastal Processes 307
17.2 Tsunamis 312

17.3 Storm Surges 313
17.4 Principles Underlying Oceanic Tidal Phenomena 315
17.5 Forecasting Tidal Behavior 323
17.6 Concise Summary 327
REFERENCES 329
INDEX 357

01 PHYSICAL OCEANOGRAPHY: AN INTRODUCTION

Embarking on a journey into the depths of Earth's vast and mysterious oceans, physical oceanography unveils the intricate dynamics that govern the movement, properties, and behaviour of seawater. As we delve into the realm where the forces of nature converge with the fluidity of the sea, a fascinating exploration unfolds, offering insights into the fundamental processes shaping our planet's aquatic realms. Join us on this scientific odyssey as we navigate the currents, delve into the ocean's thermal structure, and unravel the complex interactions that define the physical aspects of our planet's vast and dynamic oceans.

The role of the ocean on weather and climate is often discussed in the news. Who has not heard of El Niño and changing weather patterns, the Atlantic hurricane season and storm surges? Yet, what exactly is the role of the ocean? And, why do we care?

1.1 WHY STUDY THE PHYSICS OF THE OCEAN?

The answer depends on our interests, which devolve from our use of the ocean. Three broad themes are important:

- We get food from the ocean. Hence we may be interested in processes which influence the sea just as farmers are interested in the weather and climate. The ocean not only has weather such as temperature changes and currents, but the oceanic weather fertilizes the sea. The atmospheric weather seldom fertilizes fields except for the small amount of nitrogen fixed by lightning.
- We use the ocean. We build structures on the shore or just offshore. We use the ocean for transport. We obtain oil and gas below the ocean. And, we use the ocean for recreation, swimming, boating, fishing, surfing, and diving. Hence we are interested in processes that influence these activities, especially waves, winds, currents, and temperature.
- The ocean influence the atmospheric weather and climate. The ocean influence the distribution of rainfall, droughts, floods, regional climate, and the development of storms, hurricanes, and typhoons. Hence we are interested in air-sea interactions, especially the fluxes of heat and water across the sea surface, the transport of heat by the ocean, and the influence of the ocean on climate and weather patterns.

These themes influence our selection of topics to study. The topics then deter- mine what we measure, how the measurements are made, and the geographic areas of interest. Some processes are local, such as the breaking of waves on a beach, some are regional, such as the influence of the North Pacific on Alaskan weather, and some are global, such as the influence of the ocean on changing climate and global warming.

If indeed, these reasons for the study of the ocean are important, let's begin a voyage of discovery. Any voyage needs a destination. What is ours?

1.2 GOALS

At the most basic level, I hope you, the students who are reading this text, will become aware of some of the major conceptual schemes (or theories) that form the foundation of physical oceanography, how they were arrived at, and why they are widely accepted, how oceanographers achieve order out of a random ocean, and the role of experiment in oceanography (to paraphrase Shamos, 1995: p. 89).

More particularly, I expect you will be able to describe physical processes influencing the ocean and coastal regions: the interaction of the ocean with the atmosphere, and the distribution of oceanic winds, currents, heat fluxes, and water masses. The text emphasizes ideas rather than mathematical techniques. I will try to answer such questions as:

1. What is the basis of our understanding of physics of the ocean?
    - What are the physical properties of sea water?
    - What are the important thermodynamic and dynamic processes influencing the ocean?
    - What equations describe the processes and how were they derived?
    - What approximations were used in the derivation?
    - Do the equations have useful solutions?
    - How well do the solutions describe the process?
    - That is, what is the experimental basis for the theories?
    - Which processes are poorly understood? Which are well understood?
2. What are the sources of information about physical variables?
    - What instruments are used for measuring each variable?
    - What are their accuracy and limitations?
    - What historic data exist?
    - What platforms are used? Satellites, ships, drifters, moorings?
3. What processes are important? Some important process we will study include:
    - Heat storage and transport in the ocean.
    - The exchange of heat with the atmosphere and the role of the ocean in climate.

- Wind and thermal forcing of the surface mixed layer.
- The wind-driven circulation including the Ekman circulation, Ekman pumping of the deeper circulation, and upwelling.
- The dynamics of ocean currents, including geostrophic currents and the role of vorticity.
- The formation of water types and masses.
- The deep circulation of the ocean.
- Equatorial dynamics, El Niño, and the role of the ocean in weather.
- Numerical models of the circulation.
- Waves in the ocean including surface waves, inertial oscillations, tides, and tsunamis.
- Waves in shallow water, coastal processes, and tide predictions.
- What are a few of the major currents and water masses in the ocean, and what governs their distribution?

1.3 ORGANIZATION

Before beginning a voyage, we usually try to learn about the places we will visit. We look at maps and we consult travel guides. In this book, our guide will be the papers and books published by oceanographers. We begin with a brief overview of what is known about the ocean. We then proceed to a description of the ocean basins, for the shape of the seas influences the physical processes in the water. Next, we study the external forces, wind and heat, acting on the ocean, and the ocean's response. As we proceed, I bring in theory and observations as necessary.

By the time we reach chapter 7, we will need to understand the equations describing dynamic response of the ocean. So we consider the equations of motion, the influence of earth's rotation, and viscosity. This leads to a study of wind-driven ocean currents, the geostrophic approximation, and the usefulness of conservation of vorticity.

Toward the end, we consider some particular examples: the deep circulation, the equatorial ocean and El Niño, and the circulation of particular areas of the ocean. Next we look at the role of numerical models in describing the ocean. At the end, we study coastal processes, waves, tides, wave and tidal forecasting, tsunamis, and storm surges.

1.4 THE BIG PICTURE

The ocean is one part of the earth system. It mediates processes in the atmosphere by the transfers of mass, momentum, and energy through the sea surface. It receives water and dissolved substances from the land. And, it lays down sediments that eventually become rocks on land. Hence an understanding of the ocean is important for understanding the earth as a system, especially for understanding important problems such as global change or global warming. At a lower level, physical oceanography and meteorology are merging. The ocean provides the feedback leading to slow changes in the atmosphere.

As we study the ocean, I hope you will notice that we use theory, observations, and numerical models to describe ocean dynamics. *None is sufficient by itself.*

- Ocean processes are nonlinear and turbulent. Yet we don't really understand the theory of non-linear, turbulent flow in complex basins. Theories used to describe the ocean are much simplified approximations to reality.
- Observations are sparse in time and space. They provide a rough description of the time-averaged flow, but many processes in many regions are poorly observed.
- Numerical models include much-more-realistic theoretical ideas, they can help interpolate oceanic observations in time and space, and they are used to forecast climate change, currents, and waves. Nonetheless, the numerical equations are approximations to the continuous analytic equations that describe fluid flow, they contain no information about flow between grid points, and they cannot yet be used to describe fully the turbulent flow seen in the ocean.

By combining theory and observations in numerical models we avoid some of the difficulties associated with each approach used separately (figure 1.1). Continued refinements of the combined approach are leading to ever-more-precise descriptions of the ocean. The ultimate goal is to know the ocean well enough to predict the future changes in the environment, including climate change or the response of fisheries to over fishing.

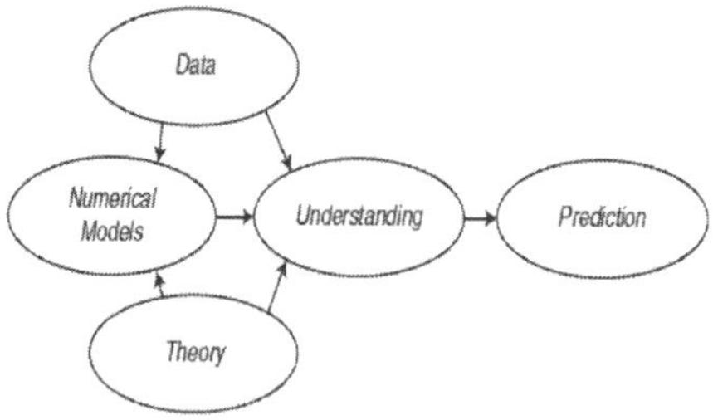

Figure 1.1 Data, numerical models, and theory are all necessary to understand the ocean. Eventually, an understanding of the ocean-atmosphere-land system will lead to predictions of future states of the system.

The combination of theory, observations, and computer models is relatively new. Four decades of exponential growth in computing power has made avail- able desktop computers capable of simulating important physical processes and oceanic dynamics.

All of us who are involved in the sciences know that the computer has be- come an essential tool for research . . . scientific computation has reached the point where it is on a par with laboratory experiment and mathematical theory as a tool for research in science and engineering—Langer (1999).

The combination of theory, observations, and computer models also implies a new way of doing oceanography. In the past, an oceanographer would devisea theory, collect data to test the theory, and publish the results. Now, the tasks have become so specialized that few can do it all. Few excel in theory, collecting data, and numerical simulations. Instead, the work is done more and more by teams of scientists and engineers.

02 THE BACKDROP OF HISTORICAL CIRCUMSTANCES

Our knowledge of oceanic currents, winds, waves, and tides goes back thousands of years. Polynesian navigators traded over long distances in the Pacific as early as 4000 bc (Service, 1996). Pytheas explored the Atlantic from Italy to Norway in 325 bc. Arabic traders used their knowledge of the reversing winds and currents in the Indian Ocean to establish trade routes to China in the Middle Ages and later to Zanzibar on the African coast. And, the connection between tides and the sun and moon was described in the Samaveda of the Indian Vedic period extending from 2000 to 1400 bc (Pugh, 1987). Those oceanographers who tend to accept as true only that which has been measured by instruments, have much to learn from those who earned their living on the ocean.

Modern European knowledge of the ocean began with voyages of discovery by Bartholomew Dias (1487–1488), Christopher Columbus (1492–1494), Vasco de Gama (1497–1499), Ferdinand Magellan (1519–1522), and many others. They laid the foundation for global trade routes stretching from Spain to the Philip- pines in the early 16th century. The routes were based on a good working knowledge of trade winds, the westerlies, and western boundary currents in the Atlantic and Pacific (Couper, 1983: 192–193).

The early European explorers were soon followed by scientific voyages of discovery led by (among many others) James Cook (1728–1779) on the *Endeavour*, *Resolution*, and *Adventure*, Charles Darwin (1809–1882) on the *Beagle*, Sir James Clark Ross and Sir John Ross who surveyed the Arctic and Antarctic regions from the *Victory*, the *Isabella*, and the *Erebus*, and Edward Forbes(1815–1854) who studied the vertical distribution of life in the ocean. Others collected oceanic observations and produced useful charts, including Edmond Halley who charted the trade winds and monsoons and Benjamin Franklin who charted the Gulf Stream.

Slow ships of the 19th and 20th centuries gave way to satellites, drifters, and autonomous instruments toward the end of the 20th century. Satellites now observe the ocean, air, and land. Thousands of drifters observe the upper two kilometres of the ocean. Data from these systems, when fed into numerical models allows the study of earth as a system. For the first time, we can study how biological, chemical, and physical systems interact to influence our environment.

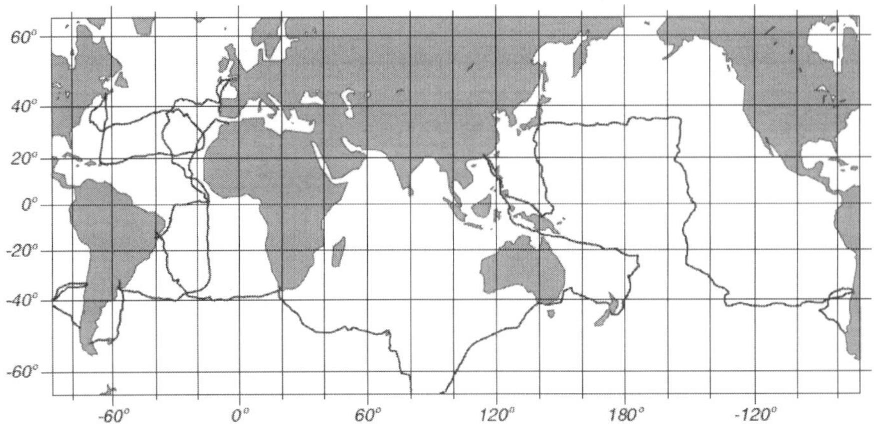

Figure 2.1 Example from the era of deep-sea exploration: Track of H.M.S. *Challenger* during the British Challenger Expedition 1872–1876. After Wust (1964).

2.1 DEFINITIONS

The long history of the study of the ocean has led to the development of various, specialized disciplines each with its own interests and vocabulary. The more important disciplines include:

- *Oceanography* is the study of the ocean, with emphasis on its character as an environment. The goal is to obtain a description sufficiently quantitative to be used for predicting the future with some certainty. *Geophysics* is the study of the physics of the earth.
- *Physical Oceanography* is the study of physical properties and dynamics of the ocean. The primary interests are the interaction of the ocean with the atmosphere, the oceanic heat budget, water mass formation, currents, and coastal dynamics. Physical Oceanography is considered by many to be a sub-discipline of geophysics.
- *Geophysical Fluid Dynamics* is the study of the dynamics of fluid motion on scales influenced by the rotation of the earth. Meteorology and oceanography use geophysical fluid dynamics to calculate planetary flow fields.
- *Hydrography* is the preparation of nautical charts, including charts of ocean depths, currents, internal density field of the ocean, and tides.
- *Earth-system Science* is the study of earth as a single system comprising many interacting subsystems including the ocean, atmosphere, cryosphere, and biosphere, and changes in these systems due to human activity.

2.2 ERAS OF OCEANOGRAPHIC EXPLORATION

The exploration of the sea can be divided, somewhat arbitrarily, into various eras (Wust, 1964). I have extended his divisions through the end of the 20th century.

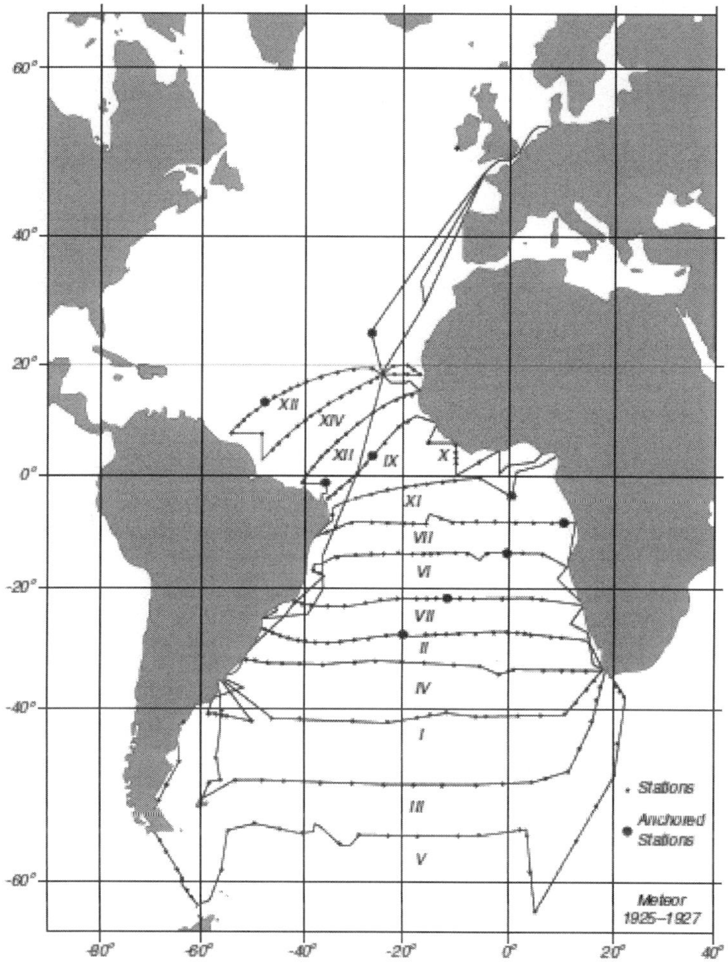

Figure 2.2 Example of a survey from the era of national systematic surveys. Track of the R/V *Meteor* during the German Meteor Expedition. Redrawn from Wust (1964).

1. **Era of Surface Oceanography:** Earliest times to 1873. The era is characterized by systematic collection of mariners' observations of winds, currents, waves, temperature, and other phenomena observable from the deck of sailing ships. Notable examples include Halley's charts of the trade winds, Franklin's map of the Gulf Stream, and Matthew Fontaine Maury's *Physical Geography of the Sea.*

2. **Era of Deep-Sea Exploration:** 1873–1914. Characterized by a few, wide- ranging oceanographic expeditions to survey surface and subsurface conditions, especially near colonial claims. The major example is the *Challenger*

Expedition (figure 2.1), but also the *Gazelle* and *Fram* Expeditions.

1. **Era of National Systematic Surveys:** 1925–1940. Characterized by detailed surveys of colonial areas. Examples include *Meteor* surveys of the Atlantic (figure 2.2), and the *Discovery* Expeditions.

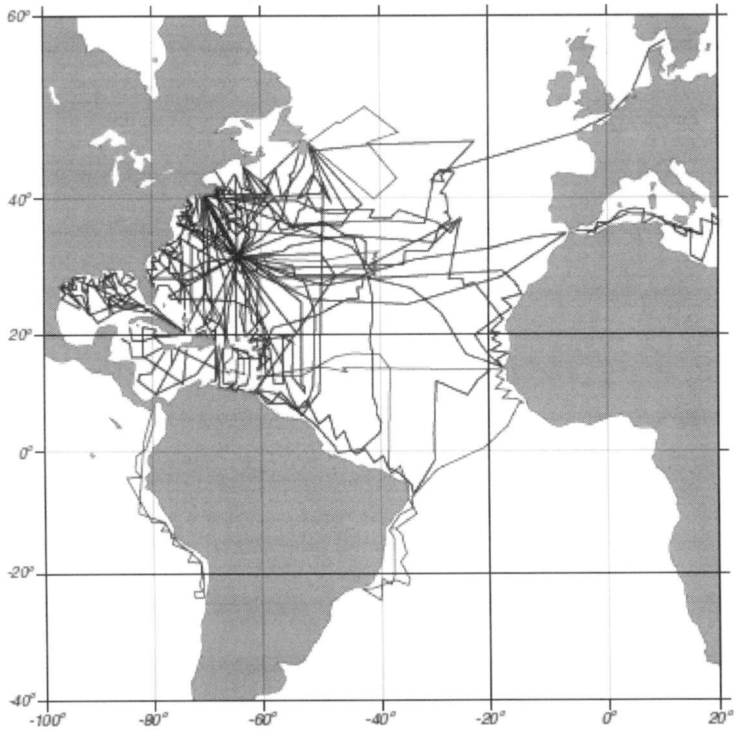

Figure 2.3 Example from the era of new methods. The cruises of the R/V *Atlantis* out of Woods Hole Oceanographic Institution. After Wust (1964).

2. **Era of New Methods:** 1947–1956. Characterized by long surveys using new instruments (figure 2.3). Examples include seismic surveys of the Atlantic by *Vema* leading to Heezen's maps of the sea floor.

3. **Era of International Cooperation:** 1957–1978. Characterized by multinational surveys of ocean and studies of oceanic processes. Examples include the Atlantic Polar Front Program, the norpac cruises, the International Geophysical Year cruises, and the International Decade of Ocean Exploration (figure 2.4). Multiship studies of oceanic processes include mode, polymode, norpax, and jasin experiments.

4. **Era of Satellites:** 1978–1995. Characterized by global surveys of oceanic processes from space. Examples include Seasat, NOAA 6–10, nimbus–7, Geosat, Topex/Poseidon, and ers–1 & 2. Era of Earth System Science: 1995–Characterized by global studies of the interaction of biological, chemical, and physical processes in the ocean and atmosphere and on land using *in situ* (which means from measurements made in the water) and space data in numerical models. Oceanic examples include the World Ocean Circulation Experiment (woce) (figure 2.5) and Topex/Poseidon (figure 2.6), the Joint Global Ocean Flux Study (jgofs), the Global Ocean Data Assimilation Experiment (GODAE), and the SeaWiFS, Aqua, and Terra satellites.

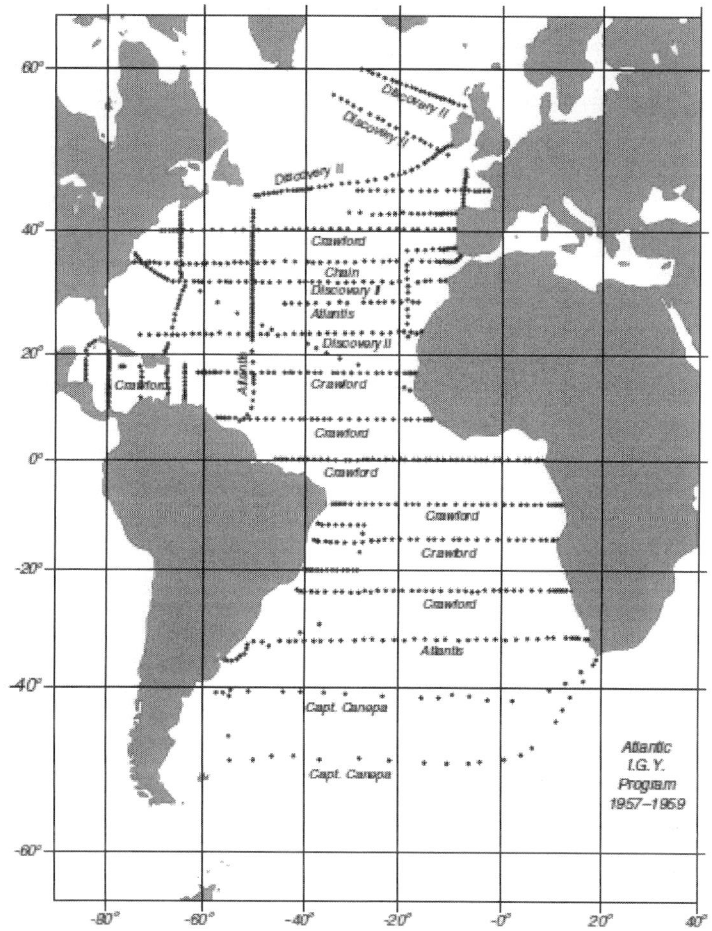

Figure 2.4 Example from the era of international cooperation . Sections measured by the International Geophysical Year Atlantic Program 1957-1959. After Wust (1964).

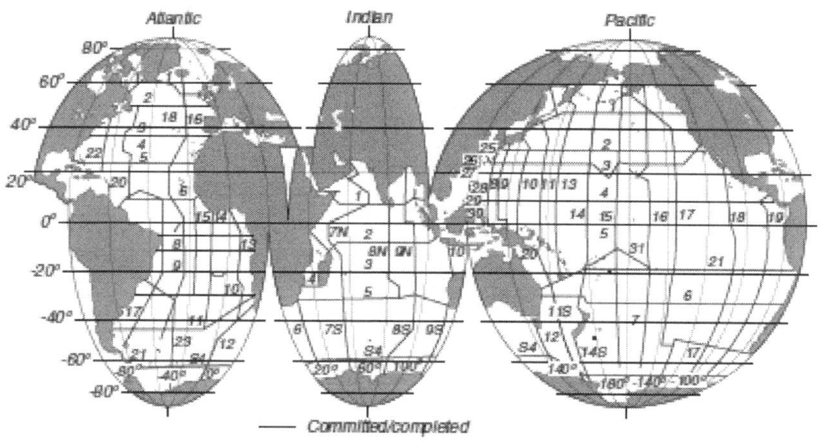

Figure 2.5 World Ocean Circulation Experiment: Tracks of research ships making a one-time global survey of the ocean of the world. From World Ocean Circulation Experiment.

2.3 MILESTONES IN THE UNDERSTANDING OF THE OCEAN

What have all these programs and expeditions taught us about the ocean? Let's look at some milestones in our ever increasing understanding of the ocean beginning with the first scientific investigations of the 17th century. Initially progress was slow. First came very simple observations of far reaching importance by scientists who probably did not consider themselves oceanographers, if the term even existed. Later came more detailed descriptions and oceanographic experiments by scientists who specialized in the study of the ocean.

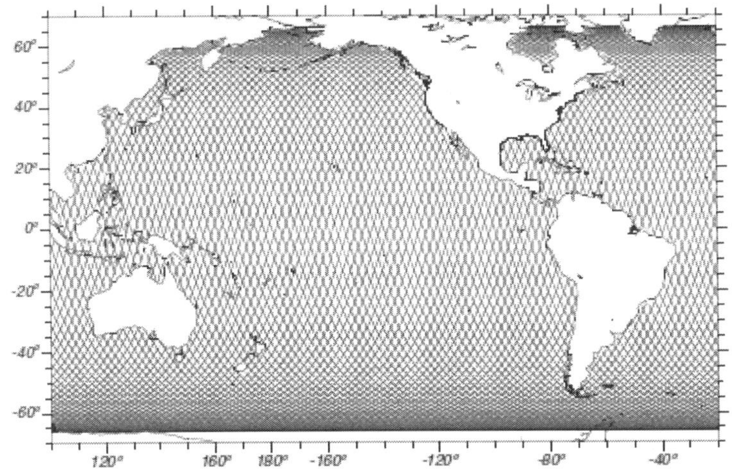

Figure 2.6 Example from the era of satellites. Topex/Poseidon tracks in the Pacific Ocean during a 10-day repeat of the orbit. From Topex/Poseidon Project.

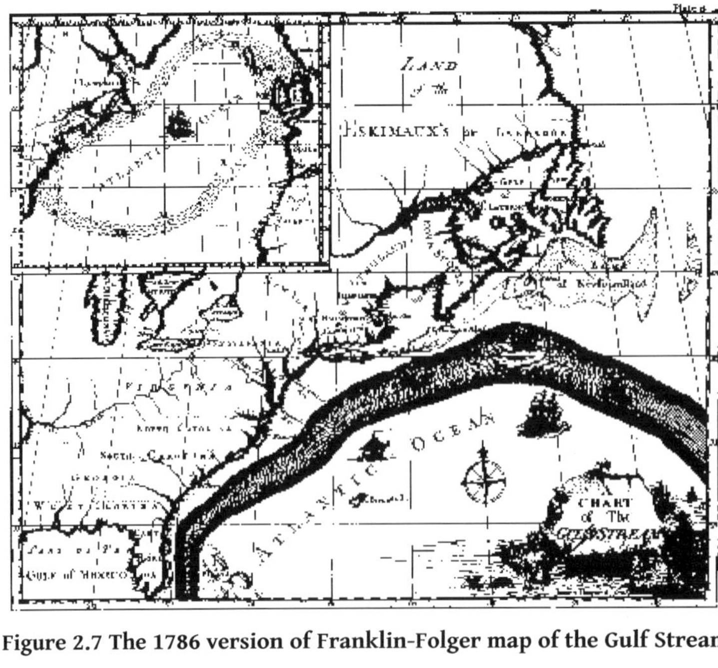

Figure 2.7 The 1786 version of Franklin-Folger map of the Gulf Stream.

1685 Edmond Halley, investigating the oceanic wind systems and currents, published "An Historical Account of the Trade Winds, and Monsoons, observable in the Seas between and near the Tropics, with an attempt to assign the Physical cause of the said Winds" *Philosophical Transactions.*

1735 George Hadley published his theory for the trade winds based on conservation of angular momentum in "Concerning the Cause of the General Trade-Winds" *Philosophical Transactions*, 39: 58-62.

1751 Henri Ellis made the first deep soundings of temperature in the tropics, finding cold water below a warm surface layer, indicating the water came from the polar regions.

1769 Benjamin Franklin, as postmaster, made the first map of the Gulf Stream using information from mail ships sailing between New England and England collected by his cousin Timothy Folger (figure 2.7).

1775 Laplace's published his theory of tides.

1800 Count Rumford proposed a meridional circulation of the ocean with water sinking near the poles and rising near the Equator.

1847 Matthew Fontaine Maury published his first chart of winds and currents based on ships logs. Maury established the practice of international exchange of environmental data, trading logbooks for maps and charts de- rived from the data.

1872–1876 Challenger Expedition marks the beginning of the systematic study of the biology, chemistry, and physics of the ocean of the world.

1885 Pillsbury made direct measurements of the Florida Current using current meters deployed from a ship moored in the stream.

1903 Founding of the Marine Biological Laboratory of the University of California. It later became the Scripps Institution of Oceanography.

1910–1913 Vilhelm Jerkens published *Dynamic Meteorology and Hydrography* which laid the foundation of geophysical fluid dynamics. In it he developed the idea of fronts, the dynamic meter, geostrophic flow, air-sea interaction, and cyclones.

1930 Founding of the Woods Hole Oceanographic Institution.

1942 Publication of *The ocean* by Sverdrup, Johnson, and Fleming, a comprehensive survey of oceanographic knowledge up to that time.

Post WW 2 The need to detect submarines led the navies of the world to greatly expand their studies of the sea. This led to the founding of oceanography departments at state universities, including Oregon State, Texas A&M

University, University of Miami, and University of Rhode Island, and the founding of national ocean laboratories such as the various Institutes of Oceanographic Science.

1947–1950 Sverdrup, Stommel, and Munk publish their theories of the wind-driven circulation of the ocean. Together the three papers lay the foundation for our understanding of the ocean's circulation.

1949 Start of California Cooperative Fisheries Investigation of the California Current. The most complete study ever undertaken of a coastal current.

1952 Cromwell and Montgomery rediscover the Equatorial Undercurrent in the Pacific.

1955 Bruce Hamon and Neil Brown develop the CTD for measuring conductivity and temperature as a function of depth in the ocean.

1958 Stommel publishes his theory for the deep circulation of the ocean.

1963 Sippican Corporation (Tim Francis, William Van Allen Clark, Graham Campbell, and Sam Francis) invents the Expendable BathyThermograph xbt now perhaps the most widely used oceanographic instrument deployed from ships.

1969 Kirk Bryan and Michael Cox develop the first numerical model of the oceanic circulation.

1978 NASA launches the first oceanographic satellite, Seasat. The project developed techniques used by generations of remotes sensing satellites.

1979–1981 Terry Joyce, Rob Pinkel, Lloyd Regier, F. Rowe and J. W. Young develop techniques leading to the acoustic doppler current profiler for measuring ocean-surface currents from moving ships, an instrument widely used in oceanography.

1988 nasa Earth System Science Committee headed by Francis Bretherton outlines how all earth systems are interconnected, thus breaking down the barriers separating traditional sciences of astrophysics, ecology, geology, meteorology, and oceanography.

1991 Wally Broecker proposes that changes in the deep circulation of the ocean modulate the ice ages, and that the deep circulation in the Atlantic could collapse, plunging the northern hemisphere into a new ice age.

1992 Russ Davis and Doug Webb invent the autonomous, pop-up drifter that continuously measures currents at depths to 2 km.

1992 nasa and cnes develop and launch Topex/Poseidon, a satellite that maps ocean surface currents, waves, and tides every ten days, revolutionizing our understanding of ocean dynamics and tides.

1993 Topex/Poseidon science-team members publish first accurate global maps of the tides. More information on the history of physical oceanography can be found in Appendix A of W.S. von Arx (1962): *An Introduction to Physical Oceanography.*

Data collected from the centuries of oceanic expeditions have been used to describe the ocean. Most of the work went toward describing the steady state of the ocean, its currents from top to bottom, and its interaction with the atmosphere. The basic description was mostly complete by the early 1970s. Figure 2.8 shows an example from that time, the surface circulation of the ocean. More recent work has sought to document the variability of oceanic processes, to provide a description of the ocean sufficient to predict annual and internal variability, and to understand the role of the ocean in global processes. A theoretical understanding of oceanic processes is based on classical physics coupled with an evolving understanding of chaotic systems in mathematics and the application to the theory of turbulence.

19th Century Development of analytic hydrodynamics. Lamb's *Hydrodynamics* is the pinnacle of this work. Bjerknes develops geostrophic method widely used in meteorology and oceanography.

1925–40 Development of theories for turbulence based on aerodynamics and mixing-length ideas. Work of Prandtl and von Karman.

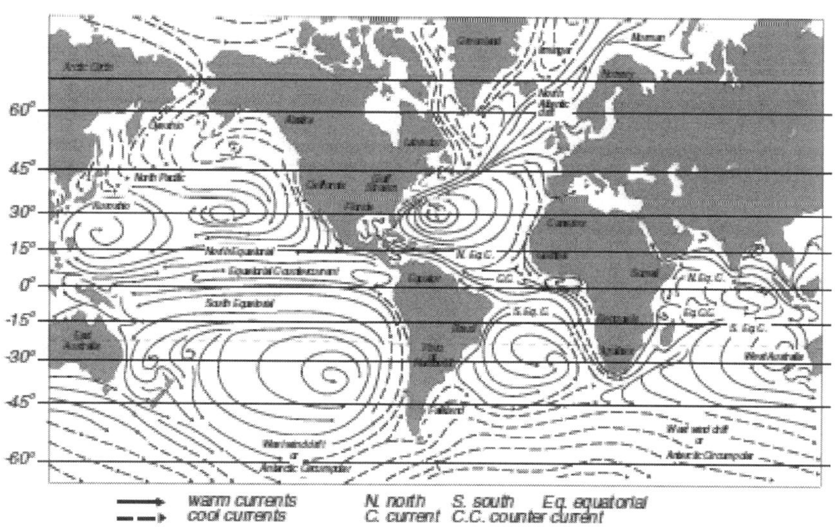

Figure 2.8 The time-averaged, surface circulation of the ocean during northern hemisphere winter deduced from a century of oceanographic expeditions. After Tomalin (1985: 16).

1940–1970 Refinement of theories for turbulence based on statistical correlations and the idea of isotropic homogeneous turbulence. Books by Batchelor (1967), Hinze (1975), and others.

1970– Numerical investigations of turbulent geophysical fluid dynamics based on high-speed digital computers.

1985– Mechanics of chaotic processes. The application to hydrodynamics is just beginning. Most motion in the atmosphere and ocean may be inherently unpredictable.

2.4 THE SIGNIFICANCE OF OBSERVATIONS IN THE FIELD OF OCEANOGRAPHY

The brief tour of theoretical ideas suggests that observations are essential for understanding the ocean. The theory describing a convicting, wind-forced, turbulent fluid in a rotating coordinate system has never been sufficiently well known that important features of the oceanic circulation could be predicted before they were observed. In almost all cases, oceanographers resort to observations to understand oceanic processes.

At first glance, we might think that the numerous expeditions mounted since 1873 would give a good description of the ocean. The results are indeed impressive. Hundreds of expeditions have extended into all ocean. Yet, much of the ocean is poorly explored.

By the year 2000, most areas of the ocean will have been sampled from top to bottom only once. Some areas, such as the Atlantic, will have been sparsely sampled three times: during the International Geophysical Year in 1959, during the Geochemical Sections cruises in the early 1970s, and during the World Ocean Circulation Experiment from 1991 to 1996. All areas will be vastly under sampled. This is the sampling problem.

Our samples of the ocean are insufficient to describe the ocean well enough to predict its variability and its response to changing forcing. *Lack of sufficient samples is the largest source of error in our understanding of the ocean.* The lack of observations has led to a very important and widespread conceptual error:

"The absence of evidence was taken as evidence of absence." The great difficulty of observing the ocean meant that when a phenomenon was not observed, it was assumed it was not present. The more one is able to observe the ocean, the more the complexity and subtlety that appears— Wunsch (2002a). As a result, our understanding of the ocean is often too simple to be correct.

Selecting Oceanic Data Sets : Much of the existing oceanic data have been organized into large data sets. For example, satellite data are processed and distributed by groups working with NASA. Data from ships have been collected and organized by other groups. Oceanographers now rely more and more on such collections of data produced by others.

The use of data produced by others introduces problems:

- How accurate are the data in the set?
- What are the limitations of the data set? And,
- How does the set compare with other similar sets? Anyone who uses public or private data sets is wise to obtain answers to such questions.

If you plan to use data from others, here are some guidelines.

1. *Use well documented data sets.* Does the documentation completely de- scribe the sources of the original measurements, all steps used to process the data, and all criteria used to exclude data? Does the data set include version numbers to identify changes to the set?

2. *Use validated data.* Has accuracy of data been well documented? Was accuracy determined by comparing with different measurements of the same variable? Was validation global or regional?

3. *Use sets that have been used by others and referenced in scientific papers.* Some data sets are widely used for good reason. Those who produced the sets used them in their own published work and others trust the data.

4. *Conversely, don't use a data set just because it is handy.* Can you document the source of the set? For example, many versions of the digital, 5-minute maps of the sea floor are widely available. Some date back to the first sets produced by the U.S. Defence Mapping Agency, others are from the etopo-5 set. Don't rely on a colleague's statement about the source. Find the documentation. If it is missing, find another data set.

Designing Oceanic Experiments: Observations are exceedingly important for oceanography, yet observations are expensive because ship time and satellites are expensive. As a result, oceanographic experiments must be carefully planned. While the design of experiments may not fit well within an historical chapter, perhaps the topic merits a few brief comments because it is seldom mentioned in oceanographic textbooks, although it is prominently described in texts for other scientific fields. The design of experiments is particularly important because poorly planned experiments lead to ambiguous results, they may measure the wrong variables, or they may produce completely useless data.

Sampling Error: Sampling error is the largest source of error in the geosciences. It is caused by a set of samples not representing the population of the variable being measured. A population is the set of all possible measurements, and a sample is the sampled subset of the population. We assume each measurement is perfectly accurate.

To determine if your measurement has a sampling error, you must first completely specify the problem you wish to study. This defines the population. Then, you must determine if the samples represent the population. Both steps are necessary.

Suppose your problem is to measure the annual-mean sea-surface temperature of the ocean to determine if global warming is occurring. For this problem, the population is the set of all possible measurements of surface temperature, in all regions in all months. If the sample mean is to equal the true mean, the samples must be uniformly distributed throughout the year and over all the area of the ocean, and sufficiently dense to include all important variability in time and space. This is impossible. Ships avoid stormy regions such as high latitudes in winter, so ship samples tend not to represent the population of surface temperatures. Satellites may not sample uniformly throughout the daily cycle, and they may not observe temperature at high latitudes in winter because of persistent clouds, although they tend to sample uniformly in space and throughout the year in most regions. If daily variability is small, the satellite samples will be more representative of the population than the ship samples.

From the above, it should be clear that oceanic samples rarely represent the population we wish to study. We always have sampling errors.

In defining sampling error, we must clearly distinguish between instruent errors and sampling errors. Instrument errors are due to the inaccuracy of the instrument. Sampling errors are due to a failure to make a measurement. Consider the example above: the determination of mean sea-surface temperature. If the measurements are made by thermometers on ships, each measurement has a small error because thermometers are not perfect. This is an instrument error. If the ships avoids high latitudes in winter, the absence of measurements at high latitude in winter is a sampling error.

Meteorologists designing the Tropical Rainfall Mapping Mission have been investigating the sampling error in measurements of rain. Their results are general and may be applied to other variables. For a general description of the problem see North & Nakamoto (1989).

The first and most important aspect of the design of any experiment is to determine *why* you wish to make a measurement before deciding how you will make the measurement or what you will measure.

- What is the purpose of the observations? Do you wish to test hypotheses or describe processes?
- What accuracy is required of the observation?
- What resolution in time and space is required? What is the duration of measurements?
- Consider, for example, how the purpose of the measurement changes how you might measure salinity or temperature as a function of depth:
- If the purpose is to describe water masses in an ocean basin, then measurements with 20–50 m vertical spacing and 50–300 km horizontal spacing, repeated once per 20–50 years in deep water are required.

- If the purpose is to describe vertical mixing in the open equatorial Pacific, then 0.5–1.0 mm vertical spacing and 50–1000 km spacing between locations repeated once per hour for many days may be required.

Accuracy, Precision, and Linearity: While we are on the topic of experiments, now is a good time to introduce three concepts needed throughout the book when we discuss experiments: precision, accuracy, and linearity of a measurement.

Accuracy is the difference between the measured value and the true value. *Precision* is the difference among repeated measurements.

The distinction between accuracy and precision is usually illustrated by the simple example of firing a rifle at a target. Accuracy is the average distance from the center of the target to the hits on the target. Precision is the average distance between the hits. Thus, ten rifle shots could be clustered within a circle 10 cm in diameter with the center of the cluster located 20 cm from the center of the target. The accuracy is then 20 cm, and the precision is roughly 5 cm.

Linearity requires that the output of an instrument be a linear function of the input. Nonlinear devices rectify variability to a constant value. So a nonlinear response leads to wrong mean values. Non-linearity can be as important as accuracy. For example, let

$$Output = Input + 0.1(Input)^2$$
$$Input = a \sin \omega t$$

then

$$Output = a \sin \omega t + 0.1\,(a \sin \omega t)^2$$
$$Output = Input + \frac{0.1}{2}a^2 - \frac{0.1}{2}a^2 \cos 2\omega t$$

Note that the mean value of the input is zero, yet the output of this nonlinear instrument has a mean value of $0.05a^2$ plus an equally large term at twice the input frequency. In general, if *input* has frequencies ω1 and ω2, then *output* of a non-linear instrument has frequencies ω1 ± ω2. Linearity of an instrument is especially important when the instrument must measure the mean value of a turbulent variable. For example, we require linear current meters when measuring currents near the sea surface where wind and waves produce a large variability in the current.

Sensitivity to other variables of interest: Errors may be correlated with other variables of the problem. For example, measurements of conductivity are sensitive to temperature. So, errors in the measurement of temperature in salinometers leads to errors in the measured values of conductivity or salinity.

2.5 IMPORTANT CONCEPTS: AT A GLANCE

From the above, I hope you have learned:

- The ocean is not well known. What we know is based on data collected from only a little more than a century of oceanographic expeditions supplemented with satellite data collected since 1978.
- The basic description of the ocean is sufficient for describing the time-averaged mean circulation of the ocean, and recent work is beginning to describe the variability.
- Observations are essential for understanding the ocean. Few processes have been predicted from theory before they were observed.
- Lack of observations has led to conceptual pictures of oceanic processes that are often too simplified and often misleading.
- Oceanographers rely more and more on large data sets produced by others. The sets have errors and limitations which you must understand before using them.
- The planning of experiments is at least as important as conducting the experiment.
- Sampling errors arise when the observations, the samples, are not representative of the process being studied. Sampling errors are the largest source of error in oceanography.
- Almost all our observations of the ocean now come from satellites, drifters, and autonomous instruments. Fewer and fewer observations come from ships at sea.

03 THE PHYSICAL SETTING: AN OVERVIEW

Surveying the physical setting provides a comprehensive glimpse into the environmental foundation of a particular area. This encompasses an array of elements, including topography, climate, geology, and hydrology. Such an overview offers essential insights into the natural characteristics that shape the landscape. By delving into the physical setting, one can grasp the intricacies of the surroundings, understanding how various factors interconnect and influence the overall dynamics of the region.

Earth is an oblate ellipsoid, an ellipse rotated about its minor axis, with an equatorial radius of R_e = 6, 378.1349 km (West, 1982) slightly greater than the polar radius of R_p = 6, 356.7497 km. The small equatorial bulge is due to earth's rotation.

Distances on earth are measured in many different units, the most common are degrees of latitude or longitude, meters, miles, and nautical miles. *Latitude* is the angle between the local vertical and the equatorial plane. A meridian is the inter-section at earth's surface of a plane perpendicular to the equatorial plane and passing through earth's axis of rotation. *Longitude* is the angle between the standard meridian and any other meridian, where the standard meridian is the one that passes through a point at the Royal Observatory at Greenwich, England. Thus longitude is measured east or west of Greenwich.

A degree of latitude is not the same length as a degree of longitude except at the equator. Latitude is measured along great circles with radius R, where R is the mean radius of earth. Longitude is measured along circles with radius$R$ cos ϕ, where ϕ is latitude. Thus 1° latitude = 111 km, and 1° longitude= 111 cos ϕ km. Because distance in degrees of longitude is not constant, oceanographers measure distance on maps using degrees of latitude.

Nautical miles and meters are connected historically to the size of earth. Gabriel Mouton proposed in 1670 a decimal system of measurement based on the length of an arc that is one minute of a great circle of earth. This eventually became the nautical mile. Mouton's decimal system eventually became the metric system based on a different unit of length, the meter, which was originally intended to be one ten-millionth the distance from the Equator to the pole along the Paris meridian. Although the tie between nautical miles, meters, and earth's radius was soon abandoned because it was not practical, the approximations are very good.

For example, earth's polar circumference is approximately 40,008 km. Therefore one ten-millionth of a quadrant is 1.0002 m. Similarly, a nautical mile should be 1.8522 km, which is very close to the official definition of the *international nautical mile*: 1 nm ≡ 1.8520 km.

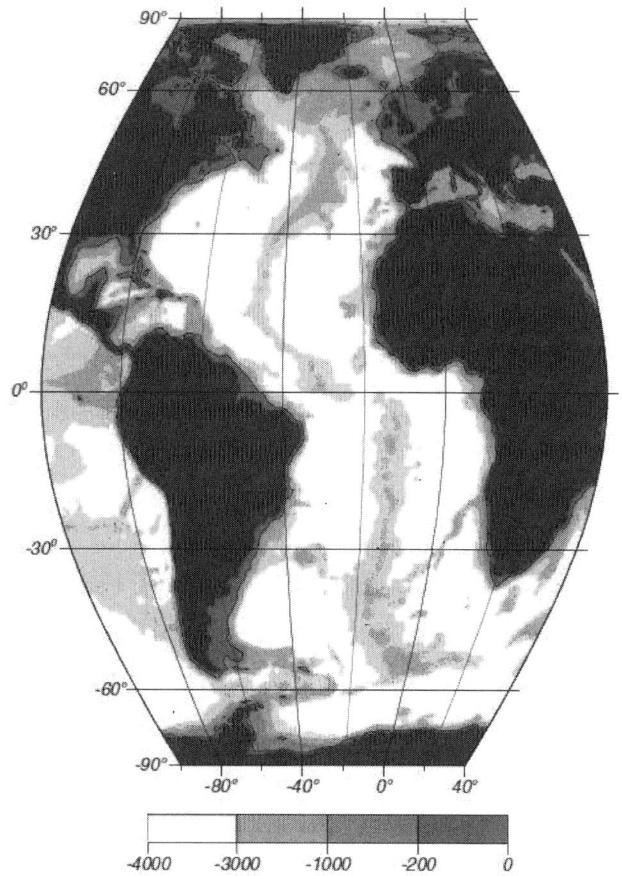

Figure 3.1 The Atlantic Ocean viewed with an Eckert VI equal-area projection. Depths, in meters, are from the etopo 30[1] data set. The 200 m contour outlines continental shelves.

3.1 OCEAN AND SEAS

There is only one ocean. It is divided into three named parts by international agreement: the Atlantic, Pacific, and Indian ocean (International Hydrographic Bureau, 1953). Seas, which are part of the ocean, are defined in several ways. I consider two.

- **The Atlantic Ocean** extends northward from Antarctica and includes all of the Arctic Sea, the European Mediterranean, and the American Mediterranean more commonly known as the Caribbean sea (figure 3.1). The boundary between the Atlantic and Indian Ocean is the meridian of Cape Agulhas (20°E).

- The boundary between the Atlantic and Pacific is the line forming the shortest distance from Cape Horn to the South Shetland Islands. In the north, the Arctic Sea is part of the Atlantic Ocean, and the Bering Strait is the boundary between the Atlantic and Pacific.
- **The Pacific Ocean** extends northward from Antarctica to the Bering Strait (figure 3.2). The boundary between the Pacific and Indian Ocean follows the line from the Malay Peninsula through Sumatra, Java, Timor, Australia at Cape Londonderry, and Tasmania. From Tasmania to Antarctica it is the meridian of South East Cape on Tasmania 147°E.

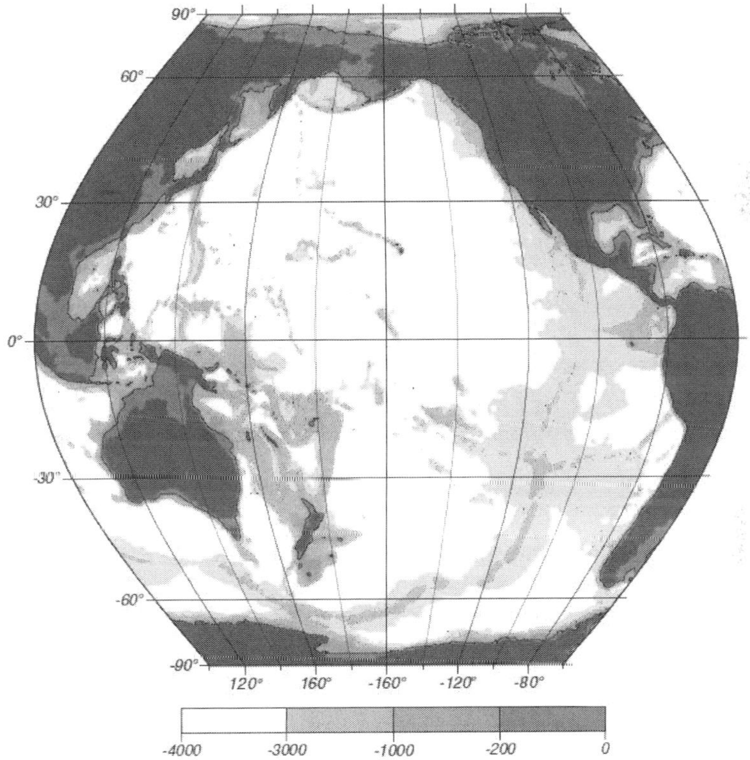

Figure 3.2 The Pacific Ocean viewed with an Eckert VI equal-area projection. Depths, in meters, are from the etopo 30[1] data set. The 200 m contour outlines continental shelves.

The Indian Ocean extends from Antarctica to the continent of Asia including the Red Sea and Persian Gulf (figure 3.3). Some authors use the name Southern Ocean to describe the ocean surrounding Antarctica.

Mediterranean Seas are mostly surrounded by land. By this definition, the Arctic and Caribbean Seas are both Mediterranean Seas, the Arctic Mediterranean and the Caribbean Mediterranean.

Marginal Seas are defined by only an indentation in the coast. The Arabian Sea and South China Sea are marginal seas.

3.2 DIMENSIONS OF THE OCEAN

The ocean and seas cover 70.8% of the surface of earth, which amounts to 361,254,000 km^2. The areas of the named parts vary considerably (table 3.1).

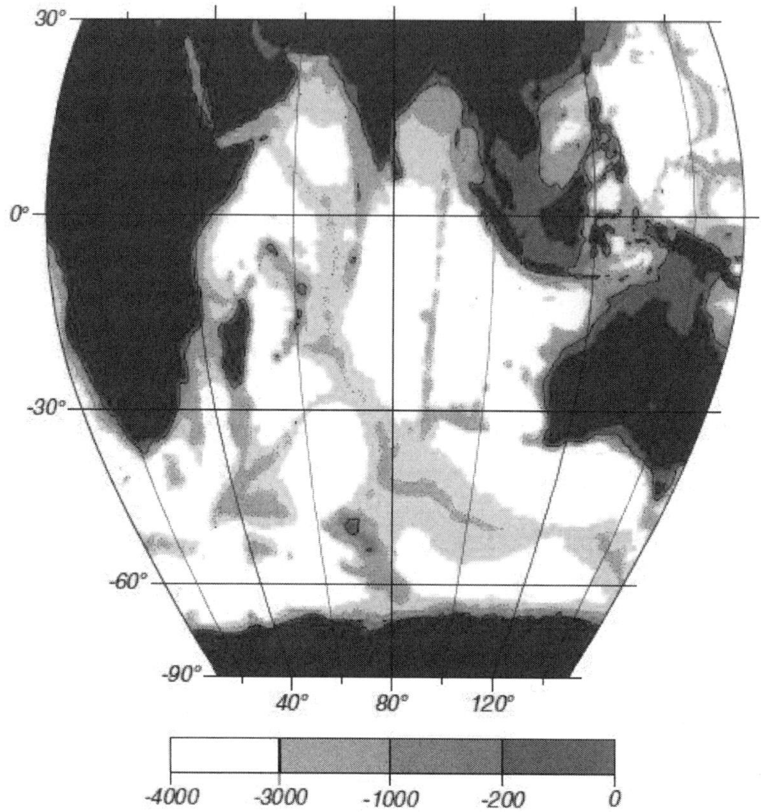

Figure 3.3 The Indian Ocean viewed with an Eckert VI equal-area projection. Depths, in meters, are from the etopo 30[1] data set. The 200 m contour outlines continental shelves.

Oceanic dimensions range from around 1500 km for the minimum width of the Atlantic to more than 13,000 km for the north-south extent of the Atlantic and the width of the Pacific. Typical depths are only 3–4 km. So horizontal dimensions of ocean basins are 1,000 times greater than the vertical dimension. A scale model of the Pacific, the size of an 8.5 × 11 in sheet of paper, would have dimensions similar to the paper: a width of 10,000 km scales to 10 in, and a depth of 3 km scales to 0.003 in, the typical thickness of a piece of paper.

Because the ocean is so thin, cross-sectional plots of ocean basins must have a greatly exaggerated vertical scale to be useful. Typical plots have a vertical scale that is 200 times the horizontal scale (figure 3.4). This exaggeration distorts our view of the ocean. The edges of the ocean basins, the continental slopes, are not steep cliffs as shown in the figure at 41°W and 12°E. Rather, they are gentle slopes dropping down 1 meter for every 20 meters in the horizontal.

The small ratio of depth to width of the ocean basins is very important for understanding ocean currents. Vertical velocities must be much smaller than horizontal velocities. Even over distances of a few hundred kilometers, the vertical velocity must be less than 1% of the horizontal velocity. I will use this information later to simplify the equations of motion.

Table 3.1 Surface Area of the ocean †

Pacific Ocean	181.34×10^6 km^2
Atlantic Ocean	106.57×10^6 km^2
Indian Ocean	74.12×10^6 km^2

† From Menard and Smith (1966)

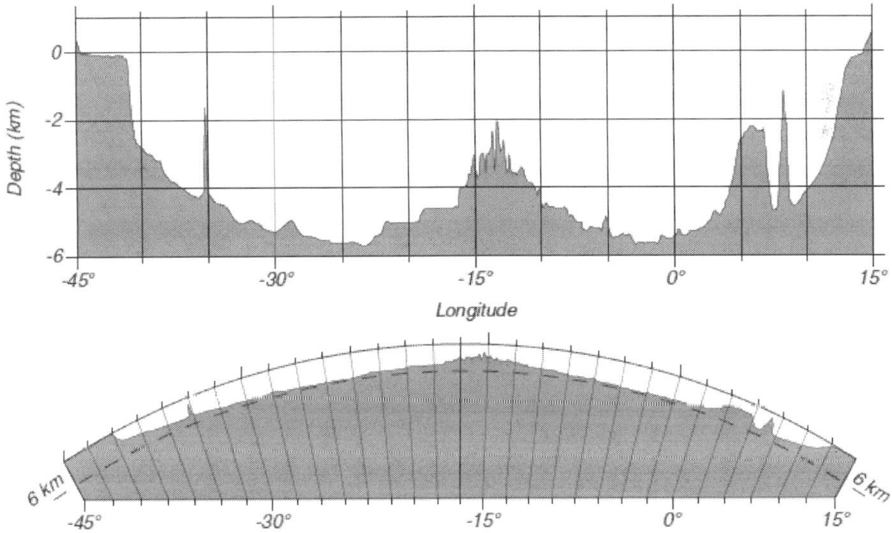

Figure 3.4 Cross-section of the south Atlantic along 25°S showing the continental shelf offshore of South America, a seamount near 35°W, the mid-Atlantic Ridge near 14°W, the Walvis Ridge near 6°E, and the narrow continental shelf off South Africa. Upper Vertical exaggeration of 180:1. Lower Vertical exaggeration of 30:1. If shown with true aspect ratio, the plot would be the thickness of the line at the sea surface in the lower plot.

The relatively small vertical velocities have great influence on turbulence. Three dimensional turbulence is fundamentally different than two-dimensional turbulence. In two dimensions, vortex lines must always be vertical, and there can be little vortex stretching. In three dimensions, vortex stretching plays a fundamental role in turbulence.

3.3 SEA-FLOOR FEATURES

Earth's rocky surface is divided into two types: oceanic, with a thin dense crust about 10 km thick, and continental, with a thick light crust about 40 km thick. The deep, lighter continental crust floats higher on the denser mantle than does the oceanic crust, and the mean height of the crust relative to sea level has two distinct values: continents have a mean elevation of 1100 m, the ocean has a mean depth of -3400 m (figure 3.5).

The volume of the water in the ocean exceeds the volume of the ocean basins, and some water spills over on to the low lying areas of the continents. These shallow seas are the continental shelves. Some, such as the South China Sea, are more than 1100 km wide. Most are relatively shallow, with typical depths of 50–100 m. A few of the more important shelves are: the East China Sea, the Bering Sea, the North Sea, the Grand Banks, the Patagonian Shelf, the Arafura Sea and Gulf of Carpentaria, and the Siberian Shelf. The shallow seas help dissipate tides, they are often areas of high biological productivity, and they are usually included in the exclusive economic zone of adjacent countries.

The crust is broken into large plates that move relative to each other. New crust is created at the mid-ocean ridges, and old crust is lost at trenches. The relative motion of crust, due to plate tectonics, produces the distinctive features of the sea floor sketched in figure 3.6, including mid-ocean ridges, trenches, is- land arcs, and basins. The names of the sub-sea features have been defined by the International Hydrographic Organization (1953), and the following definitions are taken from Sverdrup, Johnson, and Fleming (1942), Shepard (1963), and Dietrich et al. (1980). *Basins* are deep depressions of the sea floor of more or less circular or oval form.

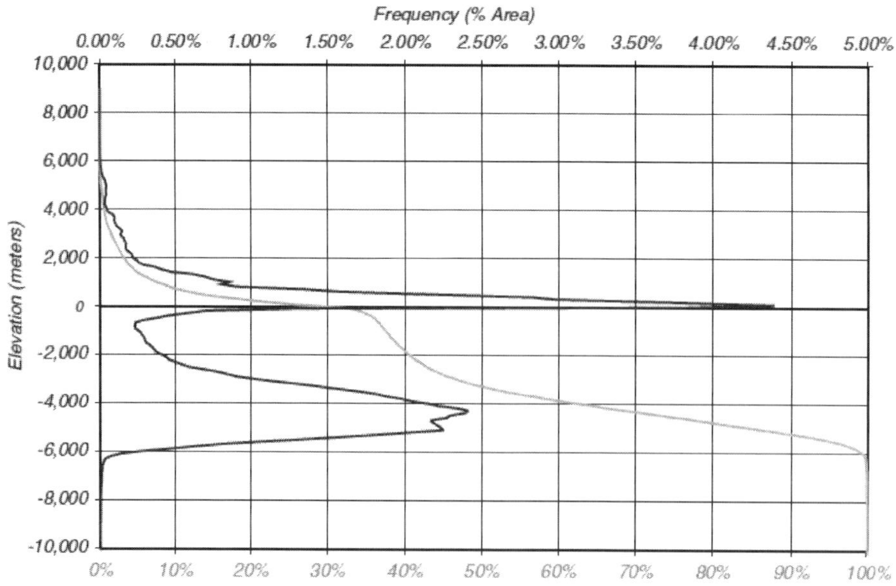

Figure 3.5 Histogram of height of land and depth of the sea as percentage of area of earth in 100 m intervals, showing the clear distinction between continents and sea floor. The cumulative frequency curve is the integral of the histogram. The curves are calculated from the etopo 2 data set by George Sharman of the NOAA National Geophysical Data Center.

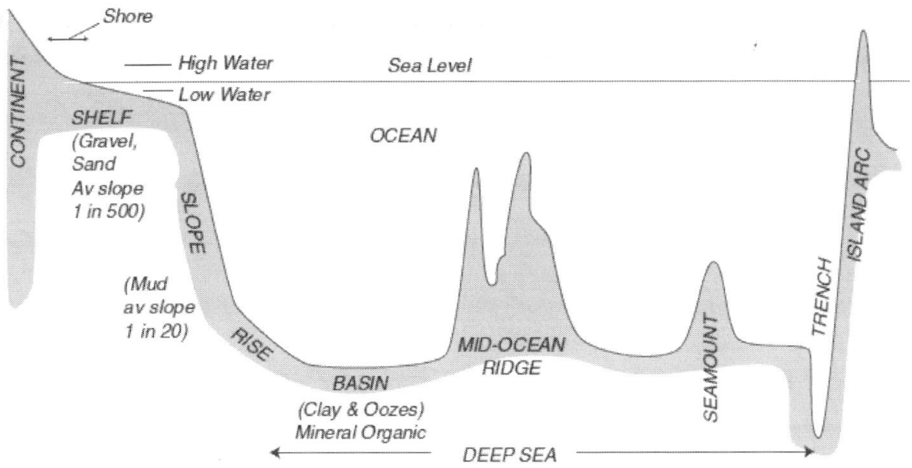

Figure 3.6 Schematic section through the ocean showing principal features of the sea floor. Note that the slope of the sea floor is greatly exaggerated in the figure.

Canyons are relatively narrow, deep furrows with steep slopes, cutting across the continental shelf and slope, with bottoms sloping continuously downward.

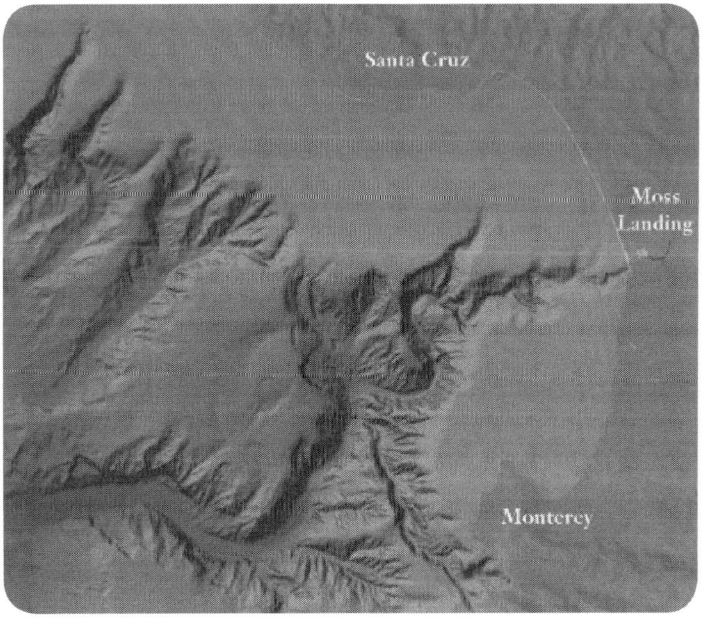

Figure 3.7 An example of a continental shelf, the shelf offshore of Monterey California showing the Monterey and other canyons. Canyons are common on shelves, often extending across the shelf and down the continental slope to deep water. Figure copyright Monterey Bay Aquarium Research Institute (mbari).

Continental shelves are zones adjacent to a continent (or around an island) and extending from the low-water line to the depth, usually about 120 m, where there is a marked or rather steep descent toward great depths. (figure 3.7)*Continental slopes* are the declivities seaward from the shelf edge into greater depth. *Plains* are very flat surfaces found in many deep ocean basins. *Ridges* are long, narrow elevations of the sea floor with steep sides and rough topography. *Seamounts* are isolated or comparatively isolated elevations rising 1000 m or more from the sea floor and with small summit area (figure 3.8).

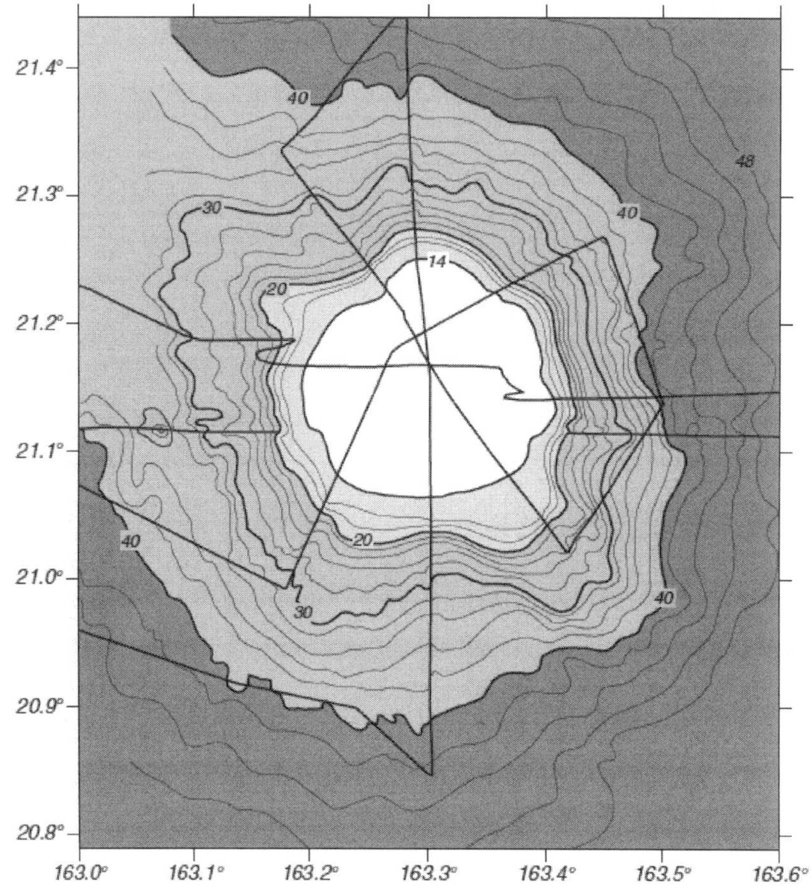

Figure 3.8 An example of a seamount, the Wilde Guyot. A guyot is a seamount with a flat top created by wave action when the seamount extended above sea level. As the seamount is carried by plate motion, it gradually sinks deeper below sea level. The depth was contoured from echo sounder data collected along the ship track (thin straight lines) supplemented with side-scan sonar data. Depths are in units of 100 m. From William Sager, Texas A&M University.

Sills are the low parts of the ridges separating ocean basins from one another or from the adjacent sea floor. *Trenches* are long, narrow, and deep depressions of the sea floor, with relatively steep sides (figure 3.9).

Sub-sea features strongly influences the ocean circulation. Ridges separate deep waters of the ocean into distinct basins. Water deeper than the sill between two basins cannot move from one to the other. Tens of thousands of seamounts are scattered throughout the ocean basins. They interrupt ocean currents, and produce turbulence leading to vertical mixing in the ocean.

3.4 MEASURING THE DEPTH OF THE OCEAN

The depth of the ocean is usually measured two ways:

- Using acoustic echo-sounders on ships, or
- Using data from satellite altimeters.

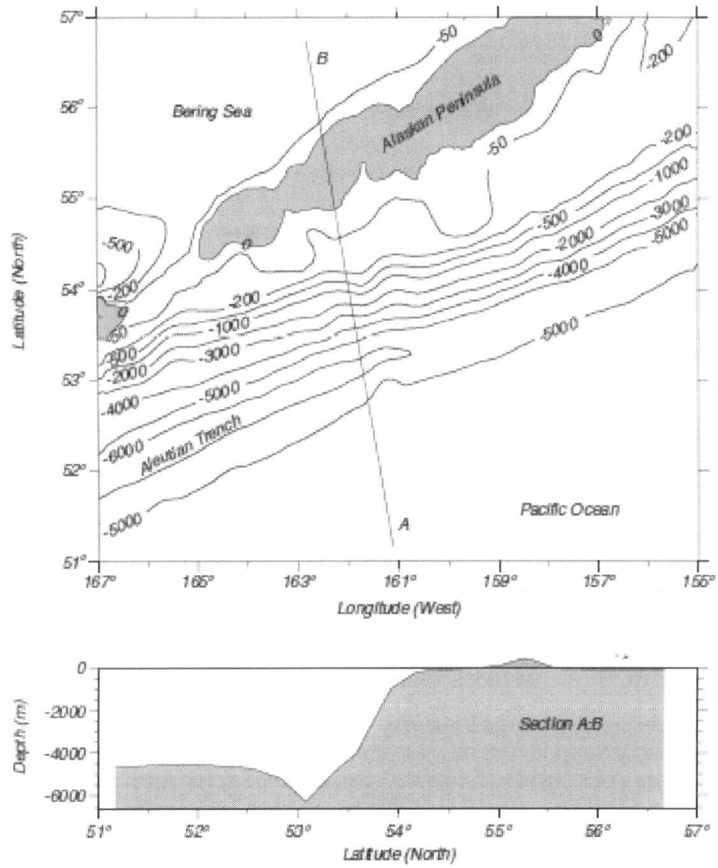

Figure 3.9 An example of a trench, the Aleutian Trench; an island arc, the Alaskan Peninsula; and a continental shelf, the Bering Sea. The island arc is composed of volcanoes produced when oceanic crust carried deep into a trench melts and rises to the surface. Top: Map of the Aleutian region of the North Pacific. Bottom: Cross-section through the region.

Echo Sounders : Most maps of the ocean are based on measurements made by echo sounders. The instrument transmits a burst of 10–30 kHz sound and listens for the echo from the sea floor. The time interval between transmission of the pulse and reception of the echo, when multiplied by the velocity of sound, gives twice the depth of the ocean (figure 3.10).

The first transatlantic echo soundings were made by the U.S. Navy Destroyer *Stewart* in 1922. This was quickly followed by the first systematic survey of an ocean basin, made by the German research and survey ship *Meteor* during its expedition to the south Atlantic from 1925 to 1927. Since then, oceanographic and naval ships have operated echo sounders almost continuously while at sea. Millions of miles of ship-track data recorded on paper have been digitized to produce data bases used to make maps. The tracks are not well distributed.

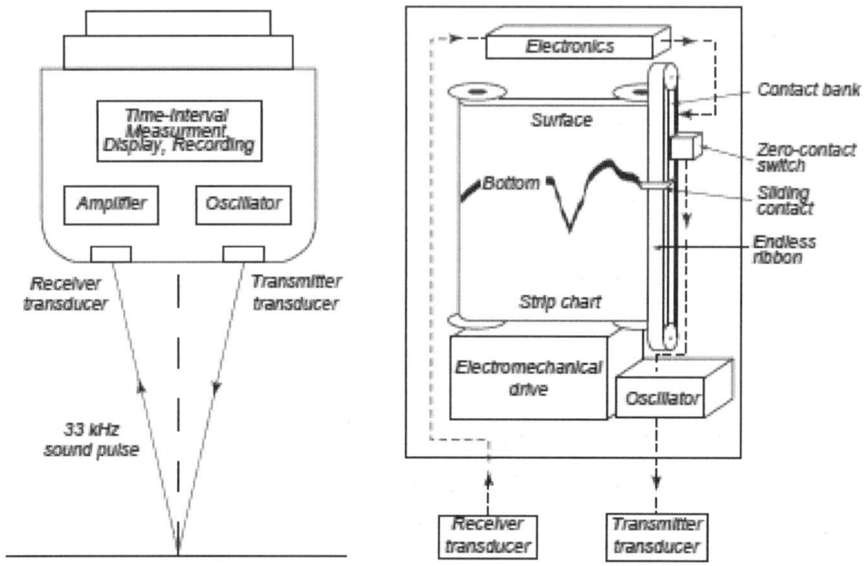

Figure 3.10 Left: Echo sounders measure depth of the ocean by transmitting pulses of sound and observing the time required to receive the echo from the bottom. Right: The time is recorded by a spark burning a mark on a slowly moving roll of paper. After Dietrich et al. (1980: 124).

Tracks tend to be far apart in the southern hemisphere, even near Australia (figure 3.11) and closer together in well mapped areas such as the North Atlantic. Echo sounders make the most accurate measurements of ocean depth. Their accuracy is ±1%.

Satellite Altimetry : Gaps in our knowledge of ocean depths between ship tracks have now been filled by satellite-altimeter data. Altimeters profile the shape of the sea surface, and its shape is very similar to the shape of the sea floor (Tapley and Kim, 2001; Cazenave and Royer, 2001; Sandwell and Smith, 2001). To see this, we must first consider how gravity influences sea level.

The Relationship Between Sea Level and the Ocean's Depth Excess mass at the sea floor, for example the mass of a seamount, increases local gravity because the mass of the seamount is larger than the mass of water it displaces. Rocks are more than three times denser than water. The excess mass increases local gravity, which attracts water toward the seamount. This changes the shape of the sea surface (figure 3.12).

Let's make the concept more exact. To a very good approximation, the sea surface is a particular *level surface* called the *geoid* (see box). By definition a level surface is a surface of constant gravitational potential, and it is everywhere perpendicular to gravity. In particular, it must be perpendicular to the local vertical determined by a plumb line, which is "a line or cord having at one end a metal weight for determining vertical direction" (Oxford English Dictionary). The excess mass of the seamount attracts the plumb line's weight, causing the plumb line to point a little toward the seamount instead of toward earth's center of mass. Because the sea surface must be perpendicular to gravity, it must have a slight bulge above a seamount as shown in figure 3.12. If there were no bulge, the sea surface would not be perpendicular to gravity. Typical seamounts produce a bulge that is 1–20 m high over distances of 100–200 kilometres. This bulge is far too small to be seen from a ship, but it is easily measured by satellite altimeters. Oceanic trenches have a deficit of mass, and they produce a depression of the sea surface.

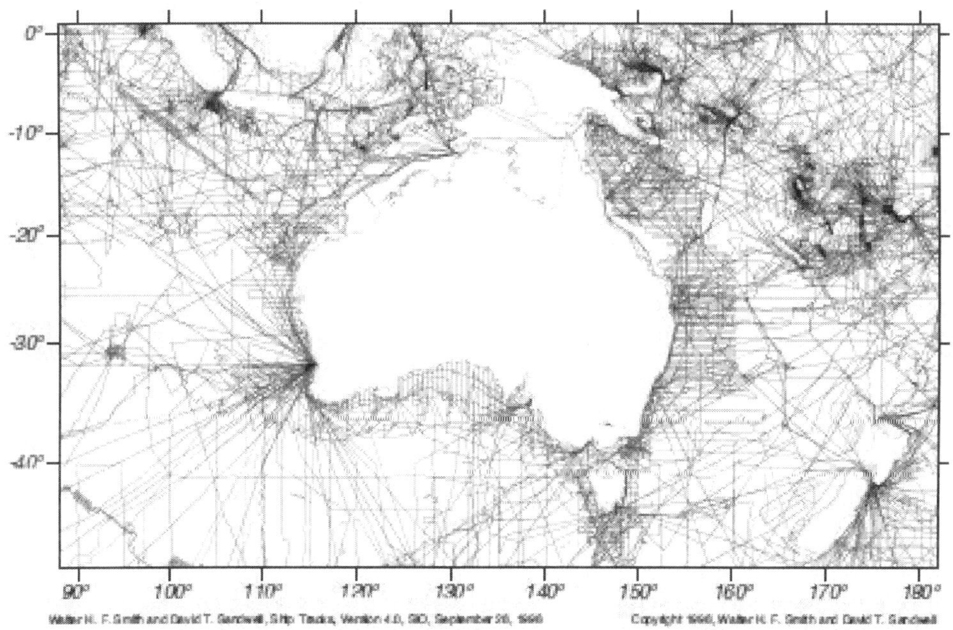

Figure 3.11 Locations of echo-sounder data used for mapping the ocean floor near Australia. Note the large areas where depths have not been measured from ships. From David Sandwell, Scripps Institution of Oceanography.

The correspondence between the shape of the sea surface and the depth of the water is not exact. It depends on the strength of the sea floor, the age of the sea-floor feature, and the thickness of sediments. If a seamount floats on the sea floor like ice on water, the gravitational signal is much weaker than it would be if the seamount rested on the sea floor like ice resting on a table top. As a result, the relationship between gravity and sea-floor topography varies from region to region.

Depths measured by acoustic echo sounders are used to determine the regional relationships. Hence, altimetry is used to interpolate between acoustic echo sounder measurements (Smith and Sandwell, 1994).

Satellite-altimeter systems Now let's see how altimeters measure the shape of the sea surface. Satellite altimeter systems include a radar to measure the height of the satellite above the sea surface and a tracking system to determine the height of the satellite in geocentric coordinates. The system measures the height of the sea surface relative to the center of mass of earth (figure 3.13).

The Geoid: The level surface that corresponds to the surface of an ocean at rest is a special surface, the *geoid*. To a first approximation, the geoid is an ellipsoid that corresponds to the surface of a rotating, homogeneous fluid in solid-body rotation, which means that the fluid has no internal flow. To a second approximation, the geoid differs from the ellipsoid because of local variations in gravity. The deviations are called *geoid undulations*. The maximum amplitude of the undulations is roughly ±60 m. To a third approximation, the geoid deviates from the sea surface because the ocean is not at rest. The deviation of sea level from the geoid is defined to be the *topography*. The definition is identical to the definition for land topography, for example the heights given on a topographic map.

The ocean's topography is caused by tides, heat content of the water, and ocean surface currents. I will return to their influence in chapters 10 and17. The maximum amplitude of the topography is roughly ±1 m, so it is small compared to the geoid undulations.

Geoid undulations are caused by local variations in gravity due to the uneven distribution of mass at the sea floor. Seamounts have an excess of mass because they are more dense than water. They produce an upward bulge in the geoid (see below). Trenches have a deficiency of mass. They produce a downward deflection of the geoid. Thus the geoid is closely related to sea-floor topography. Maps of the oceanic geoid have a remarkable resemblance to the sea-floor topography.

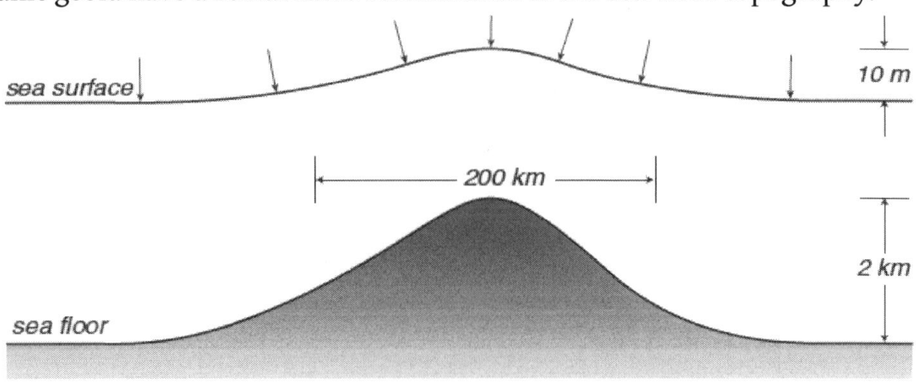

Figure 3.12 Seamounts are more dense than sea water. They increase local gravity, causing a plumb line at the sea surface (arrows) to be deflected toward the seamount. Because the surface of an ocean at rest must be perpendicular to gravity, the sea surface and the local geoid must have a slight bulge as shown. Such bulges are easily measured by satellite altimeters. As a result, satellite altimeter data can be used to map the sea floor. Note, the bulge at the sea surface is greatly exaggerated, a two-kilometre high seamount would produce a bulge of approximately 10 m.

This gives the shape of the sea surface. Many altimetric satellites have flown in space. All observed the marine geoid and the influence of sea-floor features on the geoid. The altimeters that produced the most useful data include Seasat (1978), geosat (1985–1988), ers–1 (1991– 1996), ers–2 (1995–), Topex/Poseidon (1992–2006), Jason (2002–), and Envisat (2002). Topex/Poseidon and Jason were specially designed to make extremely accurate measurements of sea-surface height. They measure sea-surface height with an accuracy of ±0.05 m.

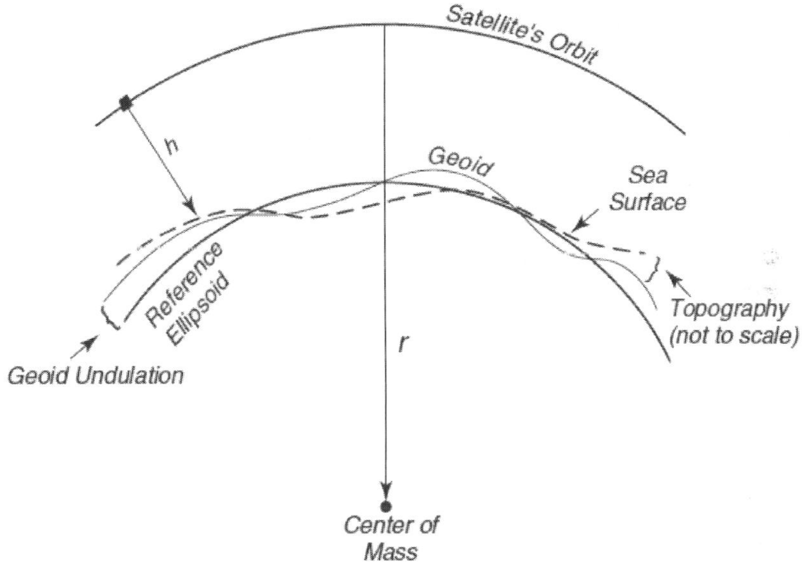

Figure 3.13 A satellite altimeter measures the height of the satellite above the sea surface. When this is subtracted from the height *r* of the satellite's orbit, the difference is sea level relative to the center of earth. The shape of the surface is due to variations in gravity, which produce the geoid undulations, and to ocean currents which produce the oceanic topography, the departure of the sea surface from the geoid. The reference ellipsoid is the best smooth approximation to the geoid. The variations in the geoid, geoid undulations, and topography are greatly exaggerated in the figure. From Stewart (1985).

Satellite Altimeter Maps of the Sea-floor Topography Seasat, geosat, ers– 1, and ers–2 were operated in orbits with ground tracks spaced 3–10 km apart, which was sufficient to map the geoid. By combining data from echo-sounders with data from geosat and ers–1 altimeter systems, Smith and Sandwell (1997) produced maps of the sea floor with horizontal resolution of 5–10 km and a global average depth accuracy of ±100 m.

3.5 SEA FLOOR CHARTS AND DATA SETS

Almost all echo-sounder data have been digitized and combined to make sea- floor charts. Data have been further processed and edited to produce digital data sets which are widely distributed in cd-rom format. These data have been supplemented with data from altimetric satellites to produce maps of the sea floor with horizontal resolution around 3 km.

The British Oceanographic Data Centre publishes the General Bathymetric Chart of the ocean (gebco) Digital Atlas on behalf of the Inter-governmental Oceanographic Commission of UNESCO and the International Hydrographic Organization. The atlas consists primarily of the location of depth contours, coastlines, and track lines from the gebco 5th Edition published at a scale of 1:10 million. The original contours were drawn by hand based on digitized echo-sounder data plotted on base maps.

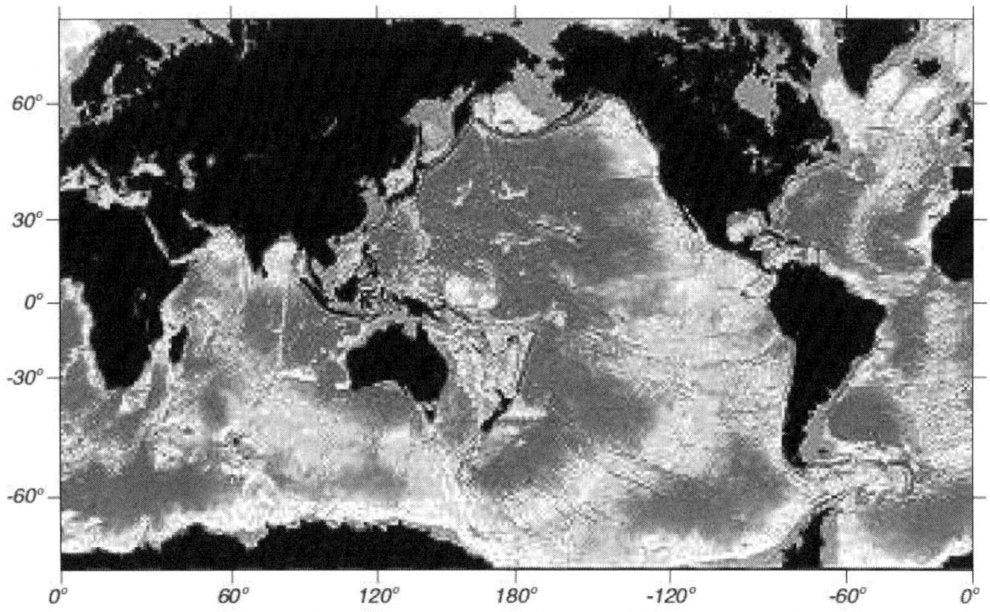

Figure 3.14 The sea-floor topography of the ocean with 3 km resolution produced from satellite altimeter observations of the shape of the sea surface. From Smith and Sandwell.

The U.S. National Geophysical Data Center publishes the etopo-2 cd-rom containing digital values of oceanic depths from echo sounders and altimetry and land heights from surveys. Data are interpolated to a 2-minute (2 nautical mile) grid. Ocean data between 64°N and 72°S are from the work of Smith and Sandwell (1997), who combined echo-sounder data with altimeter data from geosat and ers–1. Seafloor data northward of 64°N are from the International Bathymetric Chart of the Arctic Ocean. Seafloor data southward of 72°S are from are from the US Naval Oceanographic Office's Digital Bathymetric Data Base Variable Resolution. Land data are from the globe Project, that produced a digital elevation model with 0.5-minute (0.5 nautical mile) grid spacing using data from many nations.

National governments publish coastal and harbour maps. In the USA, the NOAA National Ocean Service publishes nautical charts useful for navigation of ships in harbours and offshore waters.

3.6 SOUND IN THE OCEAN

Sound provides the only convenient means for transmitting information over great distances in the ocean. Sound is used to measure the properties of the sea floor, the depth of the ocean, temperature, and currents. Whales and other ocean animals use sound to navigate, communicate over great distances, and find food.

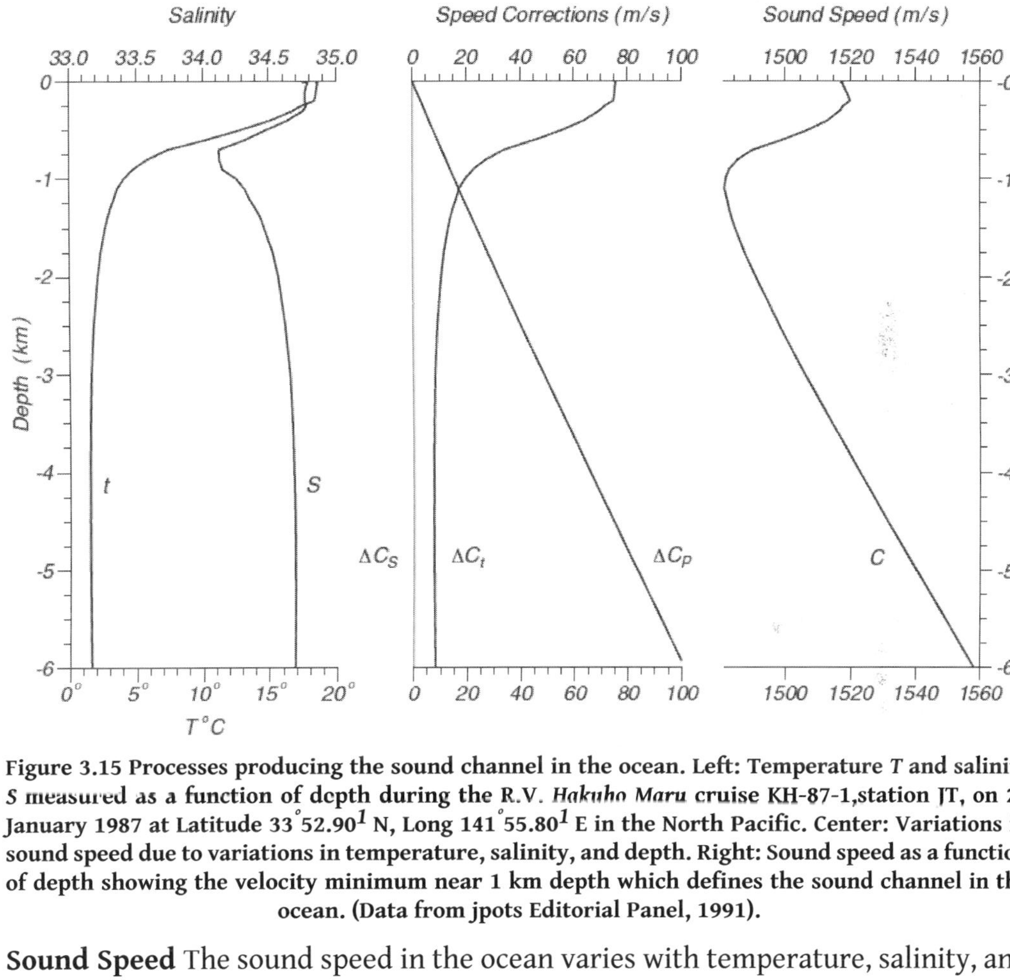

Figure 3.15 Processes producing the sound channel in the ocean. Left: Temperature *T* and salinity *S* measured as a function of depth during the R.V. *Hakuho Maru* cruise KH-87-1,station JT, on 28 January 1987 at Latitude 33°52.90' N, Long 141°55.80' E in the North Pacific. Center: Variations in sound speed due to variations in temperature, salinity, and depth. Right: Sound speed as a function of depth showing the velocity minimum near 1 km depth which defines the sound channel in the ocean. (Data from jpots Editorial Panel, 1991).

Sound Speed The sound speed in the ocean varies with temperature, salinity, and pressure (MacKenzie, 1981; Munk et al. 1995: 33):

$$C = 1448.96 + 4.591\,t - 0.05304\,t^2 + 0.0002374\,t^3 + 0.0160\,Z \qquad (3.1)$$
$$+ (1.340 - 0.01025\,t)(S - 35) + 1.675 \times 10^{-7}\,Z^2 - 7.139 \times 10^{-13}\,t\,Z^3$$

where C is speed in m/s, t is temperature in Celsius, S is salinity (see Chapter 6 for a definition of salinity), and Z is depth in meters. The equation has an accuracy of about 0.1 m/s (Dushaw et al. 1993). Other sound-speed equations have been widely used, especially an equation proposed by Wilson (1960) which has been widely used by the U.S. Navy.

For typical oceanic conditions, *C* is usually between 1450 m/s and 1550 m/s (figure 3.15). Using (3.1), we can calculate the sensitivity of *C* to changes of temperature, depth, and salinity typical of the ocean. The approximate values are: 40 m/s per 10°C rise of temperature, 16 m/s per 1000 m increase in depth, and 1.5 m/s per 1 increase in salinity. Thus the primary causes of variability of sound speed is temperature and depth (pressure). Variations of salinity are too small to have much influence.

If we plot sound speed as a function of depth, we find that the speed usually has a minimum at a depth around 1000 m (figure 3.16). The depth of minimum speed is called the *sound channel*. It occurs in all ocean, and it usually reaches the surface at very high latitudes.

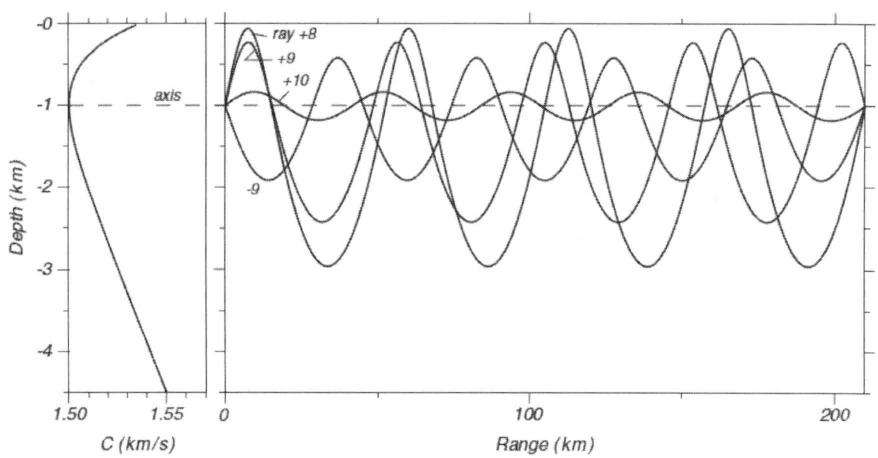

Figure 3.16 Ray paths of sound in the ocean for a source near the axis of the sound channel. After Munk et al. (1995).

The sound channel is important because sound in the channel can travel very far, sometimes half way around the earth. Here is how the channel works: Sound rays that begin to travel out of the channel are refracted back toward the center of the channel. Rays propagating upward at small angles to the horizontal are bent downward, and rays propagating downward at small angles to the horizontal are bent upward (figure 3.16). Typical depths of the channel vary from 10 m to 1200 m depending on geographical area.

Absorption of Sound: Absorption of sound per unit distance depends on the intensity *I* of the sound:

$$dI = -kI_0\,dx \qquad (3.2)$$

where *I0* is the intensity before absorption and *k* is an absorption coefficient which depends on frequency of the sound. The equation has the solution:

$$I = I_0 \exp(-kx) \qquad (3.3)$$

Typical values of k (in decibels dB per kilometer) are: 0.08 dB/km at 1000 Hz, and 50 dB/km at 100,000 Hz. Decibels are calculated from: $dB = 10 \log(I/I0)$, where $I0$ is the original acoustic power, I is the acoustic power after absorption. For example, at a range of 1 km a 1000 Hz signal is attenuated by only 1.8%:

$I = 0.982I0$. At a range of 1 km a 100,000 Hz signal is reduced to $I = 10^{-5}I0$.

The 30,000 Hz signal used by typical echo sounders to map the ocean's depths are little attenuated going from the surface to the bottom and back.

Very low frequency sounds in the sound channel, those with frequencies below 500 Hz have been detected at distances of megameters. In 1960 15-Hz sounds from explosions set off in the sound channel off Perth Australia were heard in the sound channel near Bermuda, nearly halfway around the world.

Later experiment showed that 57-Hz signals transmitted in the sound channel near Heard Island (75°E, 53°S) could be heard at Bermuda in the Atlantic and at Monterey, California in the Pacific (Munk et al. 1994).

Use of Sound: Because low frequency sound can be heard at great distances, the US Navy, in the 1950s, placed arrays of microphones on the sea floor in deep and shallow water and connected them to shore stations. The Sound Surveillance System sosus, although designed to track submarines, has found many other uses. It has been used to listen to and track whales up to 1,700 km away, and to find the location of sub-sea volcanic eruptions.

3.7 CONCISE SUMMARY

- If the ocean were scaled down to a width of 8 inches it would have depths about the same as the thickness of a piece of paper. As a result, the velocity field in the ocean is nearly 2-dimensional. Vertical velocities are much smaller than horizontal velocities.
- There are only three official ocean.
- The volume of ocean water exceeds the capacity of the ocean basins, and the ocean overflows on to the continents creating continental shelves.
- The depths of the ocean are mapped by echo sounders which measure the time required for a sound pulse to travel from the surface to the bottom and back. Depths measured by ship-based echo sounders have been used to produce maps of the sea floor. The maps have poor horizontal resolution in some regions because the regions were seldom visited by ships and ship tracks are far apart.
- The depths of the ocean are also measured by satellite altimeter systems which profile the shape of the sea surface. The local shape of the surface is influenced by changes in gravity due to sub-sea features. Recent maps based on satellite altimeter measurements of the shape of the sea surface

combined with ship data have depth accuracy of ±100 m and horizontal resolutions of ±3 km.

- Typical sound speed in the ocean is 1480 m/s. Speed depends primarily on temperature, less on pressure, and very little on salinity. The variability of sound speed as a function of pressure and temperature produces a horizontal sound channel in the ocean. Sound in the channel can travel great distances. Low-frequency sounds below 500 Hz can travel halfway around the world provided the path is not interrupted by land.

04 INFLUENCES FROM THE ATMOSPHERE

The realm above, where the vast expanse of Earth meets the boundless sky, is a theatre of constant interaction and influence – a domain shaped by atmospheric forces that extend their reach to the surface below. In the intricate dance between the Earth's surface and the atmosphere, a multitude of phenomena unfold, each leaving its mark on the environment we inhabit. Delving into the intricate web of atmospheric influences opens a gateway to understanding the complex dynamics that govern our weather patterns, climate variations, and the interconnected systems that contribute to the delicate balance of our planet

The sun and the atmosphere drive directly or indirectly almost all dynamical processes in the ocean. The dominant external sources and sinks of energy are sunlight, evaporation, infrared emissions from the sea surface, and sensible heating of the sea by warm or cold winds. Winds drive the ocean's surface circulation down to depths of around a kilometre. Wind and tidal mixing drive the deeper currents in the ocean.

The ocean, in turn, is the dominant source of heat that drives the atmospheric circulation. The uneven distribution of heat loss and gain by the ocean leads to winds in the atmosphere. Sunlight warms the tropical ocean, which evaporate, transferring heat in the form of water vapour to the atmosphere. The heat is released when the vapour condenses as rain. Winds and ocean currents carry heat poleward, where it is lost to space.

Because the atmosphere drives the ocean, and the ocean drives the atmosphere, we must consider the ocean and the atmosphere as a coupled dynamic system. In this chapter we will look mainly at the exchange of momentum between the atmosphere and the ocean. In the next chapter, we will look at heat exchanges. In chapter 14 we will look at how the ocean and the atmosphere interact in the Pacific to produce El Niño.

4.1 THE EARTH IN SPACE

Earth's orbit about the sun is nearly circular at a mean distance of 1.5×10^8 km. The eccentricity of the orbit is small, 0.0168. Thus earth is 3.4% further from the Sun at aphelion than at perihelion, the time of closest approach to the sun. Perihelion occurs every year in January, and the exact time changes by about 20 minutes per year. In 1995, it occurred on 3 January. Earth's axis of rotation is inclined 23.45° to the plane of earth's orbit around the sun (figure 4.1). The orientation is such that the sun is directly overhead at the Equator on the vernal and autumnal equinoxes, which occur on or about 21 March and 21 September each year.

The latitudes of 23.45° North and South are the Tropics of Cancer and Capricorn respectively. The tropics lie equator ward of these latitudes. As a 39result of the eccentricity of earth's orbit, maximum solar insolation averaged over the surface of the earth occurs in early January each year. As a result of the inclination of earth's axis of rotation, the maximum insolation at any location outside the tropics occurs around 21 June in the northern hemisphere, and around 21 December in the southern hemisphere.

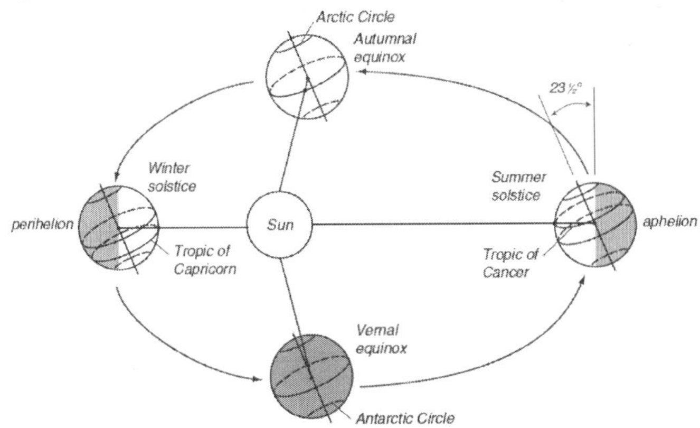

Figure 4.1 The earth in space. The ellipticity of earth's orbit around the sun and the tilt of earth's axis of rotation relative to the plane of earth orbit leads to an unequal distribution of heating and to the seasons. Earth is closest to the sun at perihelion.

If solar heat was rapidly redistributed over earth, maximum temperature would occur in January. Conversely, if heat were poorly redistributed, maximum temperature in the northern hemisphere would occur in summer. So it is clear that heat is not rapidly redistributed by winds and currents.

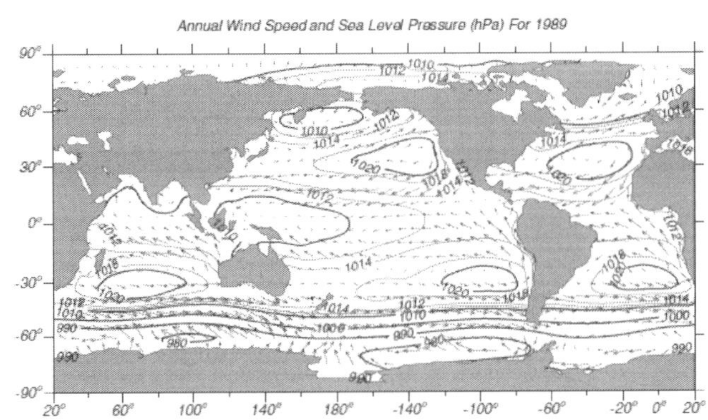

Figure 4.2 Map of mean annual wind velocity calculated from Trenberth et al. (1990) and sea-level pressure for 1989 from the nasa Goddard Space Flight Center's Data Assimilation Office (Schubert et al. 1993). The winds near 140°W in the equatorial Pacific are about 8m/s.

4.2 ATMOSPHERIC WIND SYSTEMS

Figure 4.2 shows the distribution of sea-level winds and pressure averaged over the year 1989. The map shows strong winds from the west between 40° to 60° latitude, the roaring forties, weak winds in the subtropics near 30° latitude, trade winds from the east in the tropics, and weaker winds from the east along the Equator. The strength and direction of winds in the atmosphere is the result of uneven distribution of solar heating and continental land masses and the circulation of winds in a vertical plane in the atmosphere.

A cartoon of the distribution of winds in the atmosphere (figure 4.3) shows that the surface winds are influenced by equatorial convection and other pro-cesses higher in the atmosphere. The mean value of winds over the ocean is (Wentz et al. 1984):

U_{10} - 7.4 m/s (4.1)

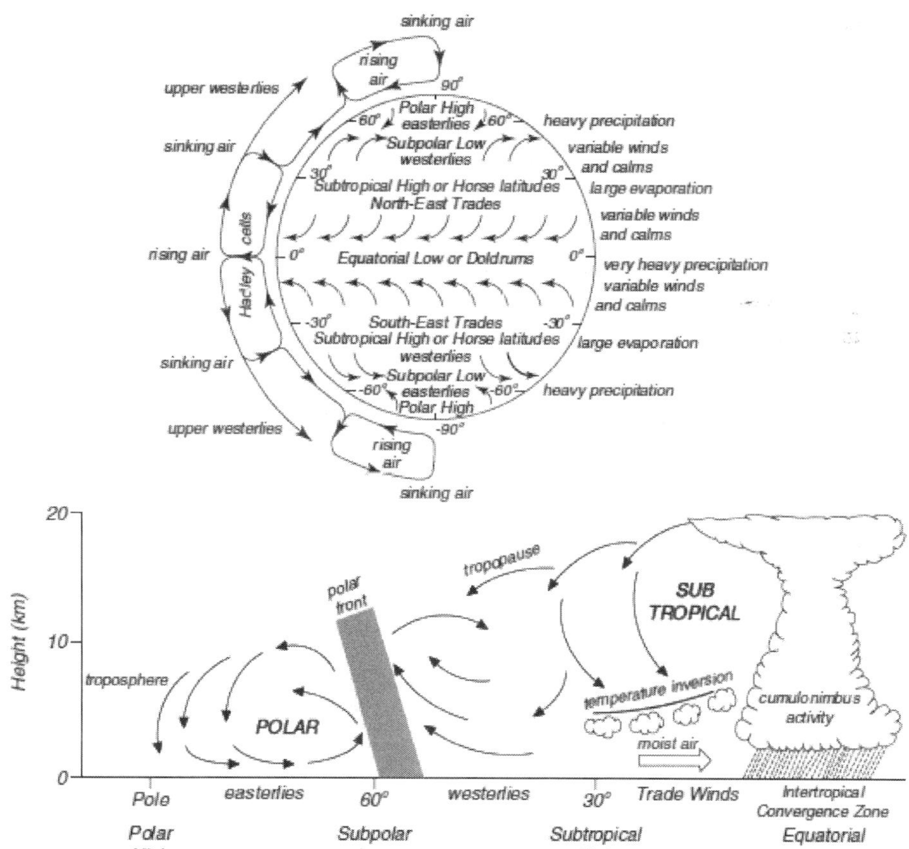

Figure 4.3 Sketch of earth's atmospheric circulation driven by solar heating in the tropics and cooling at high latitudes. Upper: The meridional cells in the atmosphere and the influence of earth's rotation on the winds. Bottom: Cross-section through the atmosphere showing the two major cells of meridional circulation. After The Open University (1989a: 14).

Maps of surface winds change somewhat with the seasons. The largest changes are in the Indian Ocean and the western Pacific Ocean (figure 4.4). Both regions are strongly influenced by the Asian monsoon. In winter, the cold air mass over Siberia creates a region of high pressure at the surface, and cold air blows south-eastward across Japan and on across the hot Kuroshio, extracting heat from the ocean. In summer, the thermal low over Tibet draws warm, moist air from the Indian Ocean leading to the rainy season over India.

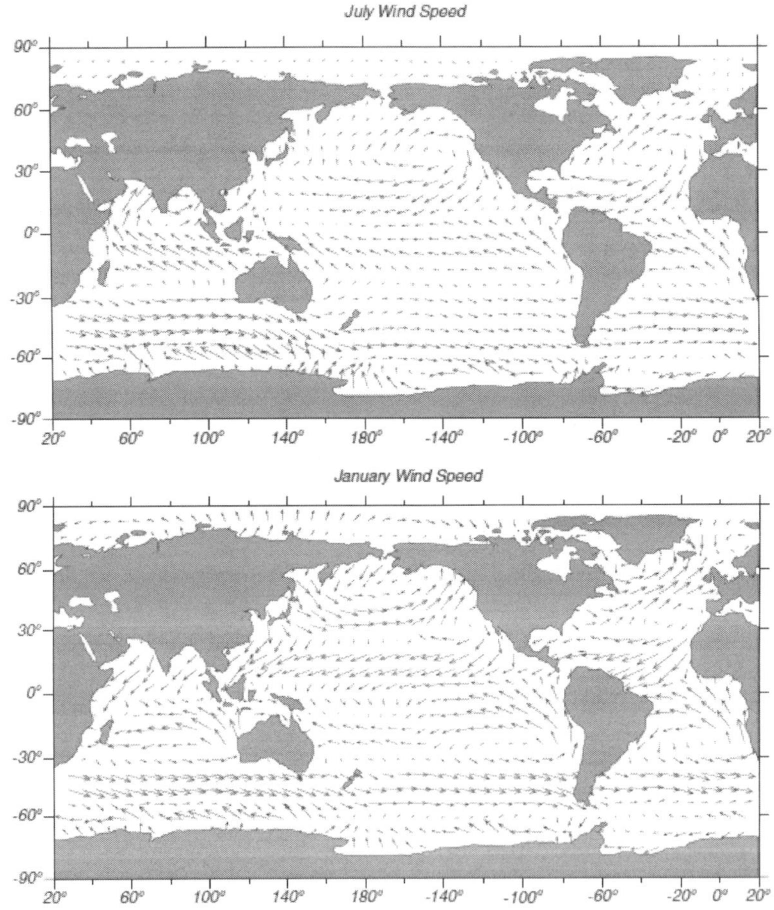

Figure 4.4 Mean, sea-surface winds for July and January calculated from the Trenberth et al. (1990) data set, which is based on the ECMWF reanalyses of weather data from 1980 to 1989.The winds near 140°W in the equatorial Pacific are about 8 m/s.

4.3 THE PLANETARY BOUNDARY LAYER

The atmosphere within 100 m of the sea surface is influenced by the turbulent drag of the wind on the sea and the fluxes of heat through the surface. This is the *atmospheric boundary layer*. It's thickness Z_i varies from a few tens of meters for weak winds blowing over water colder than the air to around a kilometre for stronger winds blowing over water warmer than the air.

The lowest part of the atmospheric boundary layer is the surface layer. Within this layer, which has thickness of ≈ $0.1Zi$, vertical fluxes of heat and momentum are nearly constant.

Wind speed varies as the logarithm of height within the surface layer for neutral stability. See "The Turbulent Boundary Layer Over a Flat Plate" in Chapter 8. Hence, the height of a wind measurement is important. Usually, winds are reported as the value of wind at a height 10 m above the sea $U10$.

4.4 WIND MEASUREMENT

Wind at sea has been measured for centuries. Maury (1855) was the first to systematically collect and map wind reports. Recently, the US National Atmospheric and Oceanic Administration NOAA has collected, edited, and digitized millions of observations going back over a century. The resulting *International Comprehensive Ocean, Atmosphere Data Set* icoads discussed in §5.5 is widely used for studying atmospheric forcing of the ocean.

Our knowledge of winds at the sea surface come from many sources. Here are the more important, listed in a crude order of relative importance:

Beaufort Scale : By far the most common source of wind data up to 1991 have been reports of speed based on the Beaufort scale. The scale is based on features, such as foam coverage and wave shape, seen by an observer on a ship (table 4.1).

The scale was originally proposed by Admiral Sir F. Beaufort in 1806 to give the force of the wind on a ship's sails. It was adopted by the British Admiralty in 1838 and it soon came into general use.

The International Meteorological Committee adopted the force scale for international use in 1874. In 1926 they adopted a revised scale giving the wind speed at a height of 6 meters corresponding to the Beaufort Number. The scale was revised again in 1946 to extend the scale to higher wind speeds and to give the equivalent wind speed at a height of 10 meters. The 1946 scale was based on the equation $U10 = 0.836B^3/2$, where B = Beaufort Number and $U10$ is the wind speed in meters per second at a height of 10 meters (List, 1966). More recently, various groups have revised the Beaufort scale by comparing Beaufort force with ship measurements of winds. Kent and Taylor (1997) compared the various revisions of the scale with winds measured by ships having anemometers at known heights. Their recommended values are given in table 4.1.

Observers on ships everywhere in the world usually report weather observations, including Beaufort force, at the same four times every day. The times are at 0000Z, 0600Z, 1200Z and 1800Z, where Z indicates Greenwich Mean Time. The reports are coded and reported by radio to national meteorological agencies. The biggest error in the reports is the sampling error. Ships are unevenly distributed over the ocean.

They tend to avoid high latitudes in winter and hurricanes in summer, and few ships cross the southern hemisphere (figure 4.5). Overall, the accuracy is around 10%.

Table 4.1 Beaufort Wind Scale and State of the Sea

Beaufort Number	Descriptive term	m/s	Appearance of the Sea
0	Calm	0	Sea like a mirror.
1	Light Air	1.2	Ripples with appearance of scales; no foam crests.
2	Light Breeze	2.8	Small wavelets; crests of glassy appearance, not breaking.
3.	Gentle breeze	4.9	Large wavelets; crests begin to break; scattered whitecaps.
4.	Moderate breeze	7.7	Small waves, becoming longer; numerous whitecaps.
5.	Fresh breeze	10.5	Moderate waves, taking longer to form; many whitecaps; some spray.
6.	Strong breeze	13.1	Large waves forming; whitecaps everywhere; more spray.
7.	Near gale	15.8	Sea heaps up; white foam from breaking waves begins to be blown into streaks.
8.	Gale	18.8	Moderately high waves of greater length; edges of crests begin to break into spindrift; foam is blown in well-marked streaks.
9.	Strong gale	22.1	High waves; sea begins to roll; dense streaks of foam; spray may reduce visibility.
10.	Storm	25.9	Very high waves with overhanging crests; sea takes white appearance as foam is blown in very dense streaks; rolling is heavy and visibility reduced.
11.	Violent storm	30.2	Exceptionally high waves; sea covered with white foam patches; visibility still more reduced.
12.	Hurricane	35.2	Air is filled with foam; sea completely white with driving spray; visibility greatly reduced.

From Kent and Taylor (1997)

Scatterometers: Observations of winds at sea now come mostly from scatterometers on satellites (Liu, 2002). The scatterometer is a instrument very much like a radar that measures the scatter of centimetre-wavelength radio waves from small, centimetre-wavelength waves on the sea surface. The area of the sea covered by small waves, their amplitude, and their orientation, depend on wind speed and direction. The scatterometer measures scatter from 2–4 directions, from which wind speed and direction are calculated.

The scatterometers on ers-1 and 2 have made global measurements of winds from space since 1991. The NASA scatterometer on adeos measured winds for a six-month period beginning November 1996 and ending with the premature failure of the satellite. It was replaced by another scatterometer on QuikScat, launched on 19 June 1999. Quikscat views 93% of the ocean every 24 hr with a resolution of 25 km.

Freilich and Dunbar (1999) report that, overall, the NASA scatterometer onadeos measured wind speed with an accuracy of ±1.3 m/s. The error in wind *90°*direction was ±17°. Spatial resolution was 25 km. Data from QuikScat has an accuracy of ±1 m/s.

Because scatterometers view a specific oceanic area only once a day, the data must be used with numerical weather models to obtain 6-hourly wind maps required for some studies.

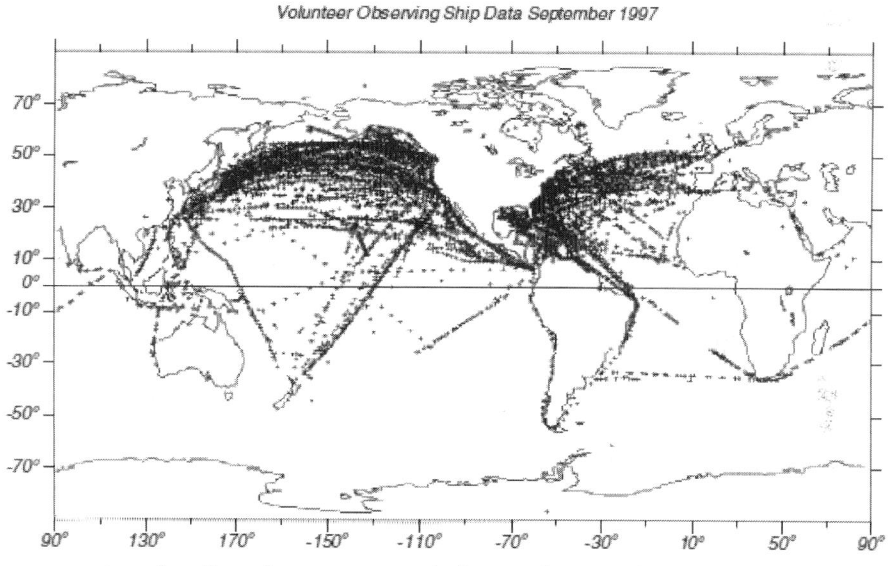

Figure 4.5 Location of surface observations made from volunteer observing ships and reported to national meteorological agencies. From NOAA, National Ocean Service.

Windsat: Windsat is an experimental, polarimetric, microwave radiometer developed by the US Navy that measures the amount and polarization of microwave radiation emitted from the sea at angles between 50° to 55° relative to the vertical and at five radio frequencies. It was launched on 6 January 2003 on the Coriolis satellite. The received radio signal is a function of wind speed, sea-surface temperature, water vapour in the atmosphere, rain rate, and the amount of water in cloud drops. By observing several frequencies simultaneously, data from the instrument are used for calculating the surface wind speed and direction, sea-surface temperature, total precipitable water, integrated cloud liquid water, and rain rate over the ocean regardless of time of day or cloudiness. Winds are calculated over most of the ocean on a 25-km grid once a day. Winds measured by

Windsat have an accuracy of ±2 m/s in speed and ±20° indirection over the range of 5–25 m/s.

Special Sensor Microwave SSM/I: Another satellite instrument that is used to measure wind speed is the Special-Sensor Microwave/Imager (ssm/i) carried since 1987 on the satellites of the U.S. Defence Meteorological Satellite Program in orbits similar to the NOAA polar-orbiting meteorological satellites. The instrument measures the microwave radiation emitted from the sea at an angle near 60° from the vertical. The radio signal is a function of wind speed, water vapour in the atmosphere, and the amount of water in cloud drops. By observing several frequencies simultaneously, data from the instrument are used for calculating the surface wind speed, water vapour, cloud water, and rain rate.

Winds measured by ssm/i have an accuracy of ± 2 m/s in speed. When combined with ECMWF 1000 mb wind analyses, wind direction can be calculated with an accuracy of ±22° (Atlas, Hoffman, and Bloom, 1993). Global, gridded data are available since July 1987 on a 0.25° grid every 6 hours. But remember, the instrument views a specific oceanic area only once a day, and the gridded, 6-hourly maps have big gaps.

Anemometers on Ships: Satellite observations are supplemented by winds reported to meteorological agencies by observers reading anemometers on ships. The anemometer is read four times a day at the standard Greenwich times and reported via radio to meteorological agencies.

Again, the biggest error is the sampling error. Very few ships carry calibrated anemometers. Those that do tend to be commercial ships participating in the Volunteer Observing Ship program (figure 4.5). These ships are met in port by scientists who check the instruments and replace them if necessary, and who collect the data measured at sea. The accuracy of wind measurements from these ships is about ±2 m/s.

Calibrated Anemometers on Weather Buoys : The most accurate measurements of winds at sea are made by calibrated anemometers on moored weather buoys. Unfortunately there are few such buoys, perhaps only a hundred scat- tered around the world. Some, such as Tropical Atmosphere Ocean tao array in the tropical Pacific (figure 14.14) provide data from remote areas rarely visited by ships, but most tend to be located just offshore of coastal areas. NOAA operates buoys offshore of the United States and the tao array in the Pacific. Data from the coastal buoys are averaged for eight minutes before the hour, and the observations are transmitted to shore via satellite links.

The best accuracy of anemometers on buoys operated by the us National Data Buoy Center is the greater of ±1 m/s or 10% for wind speed and ±10° for wind direction (Beardsley et al. 1997).

4.5 CALCULATIONS OF WIND

Satellites, ships, and buoys measure winds at various locations and times of the day. If you wish to use the observations to calculate monthly averaged winds over the sea, then the observations can be averaged and gridded. If you wish to use wind data in numerical models of the ocean's currents, then the data will be less useful. You are faced with a very common problem: How to take all observations made in a six-hour period and determine the winds over the ocean on a fixed grid?

One source of gridded winds over the ocean is the *surface analysis* calculated by numerical weather models. The strategy used to produce the six-hourly gridded winds is called *sequential estimation techniques* or *data assimilation.* "Measurements are used to prepare initial conditions for the model, which is then integrated forward in time until further measurements are available. The model is thereupon re-initialized" (Bennett, 1992: 67). The initial condition is called the *analysis.*

Usually, all available measurements are used in the analysis, including observations from weather stations on land, pressure and temperature reported by ships and buoys, winds from scatterometers in space, and data from meteorological satellites. The model interpolates the measurements to produce an analysis consistent with previous and present observations. Daley (1991) describes the techniques in considerable detail.

Surface Analysis from Numerical Weather Models: Perhaps the most widely used weather model is that run by the European Centre for Medium range Weather Forecasts ECMWF. It calculates a surface analysis, including surface winds and heat fluxes (see Chapter 5) every six hours on a $1^\circ \times 1^\circ$grid from an explicit boundary-layer model. Calculated values are archived on a 2.5°grid. Thus the wind maps from the numerical weather models lack the detail seen in maps from scatter meter data, which have a $1/4^\circ$ grid. ECMWF calculations of winds have relatively good accuracy. Freilich and Dunbar (1999) estimated that the accuracy for wind speed at 10 meters is ± 1.5 m/s, and $\pm 18^\circ$ for direction.

Accuracy in the southern hemisphere is probably as good as in the northern hemisphere because continents do not disrupt the flow as much as in the northern hemisphere, and because scatterometers give accurate positions of storms and fronts over the ocean.

The NOAA National Centers for Environmental Prediction and the US Navy also produces global analyses and forecasts every six hours.

Reanalyzed Data from Numerical Weather Models : Surface analyses of weather over some regions have been produced for more than a hundred years, and over the whole earth since about 1950. Surface analyses calculated by numerical models of the atmospheric circulation have been available for decades. Throughout this period, the methods for calculating surface analyses have constantly changed as meteorologists worked to make ever more accurate forecasts. Fluxes calculated from the analyses are therefore not consistent in time. The changes can be larger than the inter annual variability of the fluxes (White, 1996). To minimize this problem, meteorological agencies have taken all archived weather data and reanalyzed them using the best numerical models to produce a uniform, internally-consistent, surface analysis.

The reanalyzed data are used to study oceanic and atmospheric processes in the past. Surface analyses issued every six hours from weather agencies are used only for problems that require up-to-date information. For example, if you are designing an offshore structure, you will probably use decades of reanalyzed data. If you are operating an offshore structure, you will watch the surface analysis and forecasts put out every six hours by meteorological agencies.

Sources of Reanalyzed Data: Reanalyzed surface flux data are available from national meteorological centers operating numerical weather prediction models.

1. The U.S. National Centers for Environmental Predictions, working with the National Center for Atmospheric Research have produced the NCEP/ NCAR reanalysis based on 51 years of weather data from 1948 to 2005 using the 25 January 1995 version of their forecast model. The reanalysis period is being extended forward to include all date up to the present with about a three-day delay in producing data sets. The reanalysis uses surface and ship observations plus sounder data from satellites. Reanalysis products are available every six hours on a T62 grid having 192 × 94 grid points with a spatial resolution of 209 km and with 28 vertical levels. Important subsets of the reanalysis, including surface fluxes, are available on cd–rom (Kalnay et al. 1996; Kistler et al. 2000).

2. The European Centre for Medium-range Weather Forecasts ECMWF has reanalyzed 45 years of weather data from September 1957 to August 2002 (era-40) using their forecast model of 2001 (Uppala et al. 2005). The reanalysis uses mostly the same surface and ship data used by the NCEP/NCAR reanalysis plus data from the ers-1 and ers-2 satellites and ssm/i. The era-40 full-resolution products are available every six hours on a N80 grid having 160 × 320 grid points with a spatial resolution of1.125° and with 60 vertical levels. The era-40 basic-resolution products are available every six hours with a spatial resolution of 2.5° and with23 vertical levels. The reanalysis includes an ocean-wave model that calculates ocean wave heights and wave spectra every six hours on a 1.5°grid.

4.6 WIND STRESS

The wind, by itself, is usually not very interesting. Often we are much more interested in the force of the wind, or the work done by the wind. The horizontal force of the wind on the sea surface is called the *wind stress*. Put another way, it is the vertical transfer of horizontal momentum. Thus momentum is transferred from the atmosphere to the ocean by wind stress.

Wind stress *T* is calculated from:

$$T = \rho_a C_D U_{10}^2 \qquad (4.2)$$

where ρa = 1.3 kg/m^3 is the density of air, *U10* is wind speed at 10 meters, and *CD* is the *drag coefficient* . *CD* is measured using the techniques described in § 5.6. Fast response instruments measure wind fluctuations within 10–20 m of the sea surface, from which *T* is directly calculated. The correlation of *T* with U^2_{10} gives *CD* (figure 4.6).

Various measurements of *CD* have been published based on careful measurements of turbulence in the marine boundary layer. Trenberth et al. (1989) and Harrison (1989) discuss the accuracy of an effective drag coefficient relating wind stress to wind velocity on a global scale. Perhaps the best of the recently published values are those of Yelland and Taylor (1996) and Yelland et al. (1998) who give:

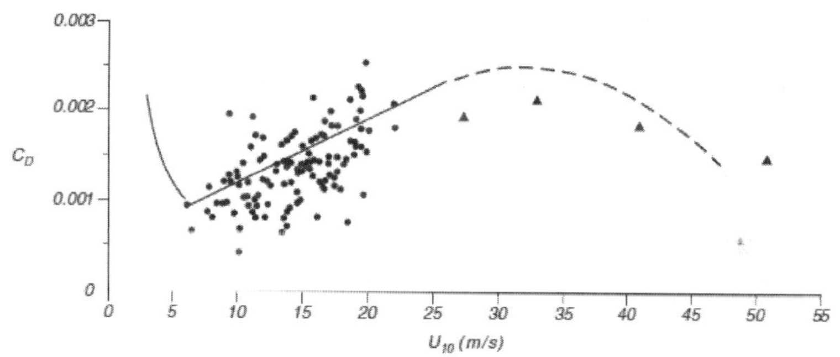

Figure 4.6 The drag coefficient as a function of wind speed U10 ten meters above the sea. Circles: Measured values from Smith (1980). Triangles: Measured values from Powell, Vickery, and Reinhold (2003). The solid line is from eq (4.3) proposed by Yelland and Taylor (1996). The dashed line is from Jarosz (2007).

$$1000\, C_D = 0.29 + \frac{3.1}{U_{10}} + \frac{7.7}{U_{10}^2} \qquad (3 \le U_{10} \le 6 \text{ m/s}) \qquad (4.3a)$$

$$1000\, C_D = 0.60 + 0.071\, U_{10} \qquad (6 \le U_{10} \le 26 \text{ m/s}) \qquad (4.3b)$$

for neutrally stable boundary layer. Other values are listed in their table 1 and in figure 4.6.

4.7 CONCISE SUMMARY

- Sunlight is the primary energy source driving the atmosphere and ocean.
- There is a boundary layer at the bottom of the atmosphere where wind speed decreases with as the boundary is approached, and in which fluxes of heat and momentum are constant in the lower 10–20 meters.

- Wind is measured many different ways. The most common until 1995 was from observations made at sea of the Beaufort force of the wind.
- Since 1995, the most important source of wind measurements is from scatter meters on satellites. They produce daily global maps with 25 km resolution.
- The surface analysis from numerical models of the atmosphere is the most useful source of global, gridded maps of wind velocity for dates before 1995. It also is a useful source for 6-hourly maps. Resolution is 100-250 km.
- The flux of momentum from the atmosphere to the ocean, the wind stress, is calculated from wind speed using a drag coefficient.

05 THE OCEAN HEAT BUDGET

The ocean heat budget refers to the intricate balance between the incoming and outgoing heat within the Earth's oceans. This dynamic equilibrium is crucial for understanding the ocean's role in the planet's overall energy exchange. Incoming solar radiation, primarily in the form of sunlight, penetrates the ocean surface, where it is absorbed by the water. This absorbed heat is then distributed and transported throughout the ocean layers through various processes such as conduction and ocean currents. On the other hand, the ocean releases heat back into the atmosphere through processes like evaporation and infrared radiation. Monitoring and studying the ocean heat budget is essential for comprehending climate patterns, sea level changes, and the overall impact of oceans on global weather systems.

About half the solar energy reaching earth is absorbed by the ocean and land, where it is temporarily stored near the surface. Only about a fifth of the available solar energy is directly absorbed by the atmosphere. Of the energy absorbed by the ocean, most is released locally to the atmosphere, mostly by evaporation and infrared radiation. The remainder is transported by currents to other areas especially mid latitudes.

Heat lost by the tropical ocean is the major source of heat needed to drive the atmospheric circulation. And, solar energy stored in the ocean from summer to winter helps ameliorate earth's climate. The thermal energy transported by ocean currents is not steady, and significant changes in the transport, particularly in the Atlantic, may have been important for the development of the ice ages. For these reasons, oceanic heat budgets and transports are important for understanding earth's climate and its short and long term variability.

5.1 THE OCEANIC HEAT BUDGET

Changes in energy stored in the upper ocean result from an imbalance between input and output of heat through the sea surface. This transfer of heat across or through a surface is called a *heat flux* . The flux of heat and water also changes the density of surface waters, and hence their buoyancy. As a result, the sum of the heat and water fluxes is often called the *buoyancy flux* .

The flux of energy to deeper layers is usually much smaller than the flux through the surface. And, the total flux of energy into and out of the ocean must be zero, otherwise the ocean as a whole would heat up or cool down. The sum of the heat fluxes into or out of a volume of water is the *heat budget.*

The major terms in the budget at the sea surface are:

- *Insolation QSW* , the flux of solar energy into the sea;
- *Net Infrared Radiation QLW* , net flux of infrared radiation from the sea;
- *Sensible Heat Flux QS* , the flux of heat out of the sea due to conduction;
- *Latent Heat Flux QL*, the flux of energy carried by evaporated water; and
- *Advection QV* , heat carried away by currents.

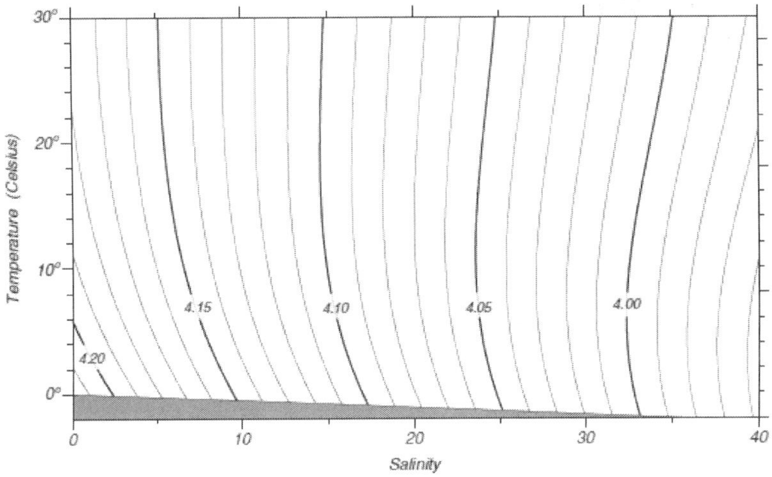

Figure 5.1 Specific heat of sea water at atmospheric pressure C_p in joules per gram per degree Celsius as a function of temperature in Celsius and salinity, calculated from the empirical formula given by Millero et al. (1973) using algorithms in Fofon off and Millard (1983). The lower line is the freezing point of salt water.

Conservation of heat requires:

$$Q = Q_{SW} + Q_{LW} + Q_S + Q_L + Q_V \tag{5.1}$$

where *Q* is the resultant heat gain or loss. Units for heat fluxes are watts/m^2. The product of flux times surface area times time is energy in joules. The change in temperature Δ*t* of the water is related to change in energy Δ*E* through:

$$\Delta E = C_p \, m \, \Delta t \tag{5.2}$$

where *m* is the mass of water being warmed or cooled, and *Cp* is the specific heat of sea water at constant pressure.

$$C_p \approx 4.0 \times 10^3 \text{ J} \cdot \text{kg}^{-1} \cdot {}^\circ\text{C}^{-1} \tag{5.3}$$

Thus, 4,000 joules of energy are required to heat 1.0 kilogram of sea water by 1.0°C (figure 5.1).

Importance of the Ocean in Earth's Heat Budget: To understand the importance of the ocean in earth's heat budget, let's make a comparison of the heat stored in the ocean with heat stored on land during an annual cycle. During the cycle, heat is stored in summer and released in the winter. The point is to show that the ocean store and release much more heat than the land.

To begin, use (5.3) and the heat capacity of soil and rocks

$$C_{p(rock)} = 800 \text{ J} \cdot \text{kg}^{-1} \cdot {}^{\circ}\text{C}^{-1} \qquad (5.4)$$

The volume of water which exchanges heat with the atmosphere on a seasonal cycle is 100 m^3 per square meter of surface, i.e. that mass from the surface to a depth of 100 meters. The density of water is 1000 kg/m^3, and the mass in contact with the atmosphere is density × volume = *mwater* = 100, 000 kg. The volume of land which exchanges heat with the atmosphere on a seasonal cycle is 1 m^3. Because the density of rock is 3,000 kg/m^3, the mass of the soil and rock in contact with the atmosphere is 3,000 kg.

The seasonal heat storage values for the ocean and land are therefore:

$$\begin{aligned}
\Delta E_{ocean} &= C_{p(water)}\, m_{water}\, \Delta t & \Delta t &= 10^{\circ}\text{C} \\
&= (4000)(10^5)(10^{\circ}) \text{ Joules} \\
&= 4.0 \times 10^9 \text{ Joules} \\
\Delta E_{land} &= C_{p(rock)}\, m_{rock}\, \Delta t & \Delta t &= 20^{\circ}\text{C} \\
&= (800)(3000)(20^{\circ}) \text{ Joules} \\
&= 4.8 \times 10^7 \text{ Joules}
\end{aligned}$$

$$\frac{\Delta E_{ocean}}{\Delta E_{land}} = 100$$

where Δt is the typical change in temperature from summer to winter.

The large storage of heat in the ocean compared with the land has important consequences. The seasonal range of air temperatures on land increases with distance from the ocean, and it can exceed 40°C in the center of continents, reaching 60°C in Siberia. Typical range of temperature over the ocean and along coasts is less than 10°C. The variability of water temperatures is still smaller.

5.2 HEAT-BUDGET TERMS

Let's look at the factors influencing each term in the heat budget.

Factors Influencing Insolation: Incoming solar radiation is primarily deter- mined by latitude, season, time of day, and cloudiness. The polar regions are heated less than the tropics, areas in winter are heated less than the same area in summer, areas in early morning are heated less than the same area at noon, and cloudy days have less sun than sunny days.

The following factors are important:

- The height of the sun above the horizon, which depends on latitude, season, and time of day. Don't forget, there is no insolation at night!
- The length of day, which depends on latitude and season.
- The cross-sectional area of the surface absorbing sunlight, which depends on height of the sun above the horizon.
- Attenuation, which depends on: i) Clouds, which absorb and scatter radiation. ii) Path length through the atmosphere, which varies as csc ϕ, where ϕ is angle of the sun above the horizon. iii) Gas molecules which absorb radiation in some bands (figure 5.2). H2O, O3, and CO2 are all important. iv) Aerosols which scatter and absorb radiation. Both volcanic and marine aerosols are important. And v) dust, which scatters radiation, especially Saharan dust over the Atlantic.

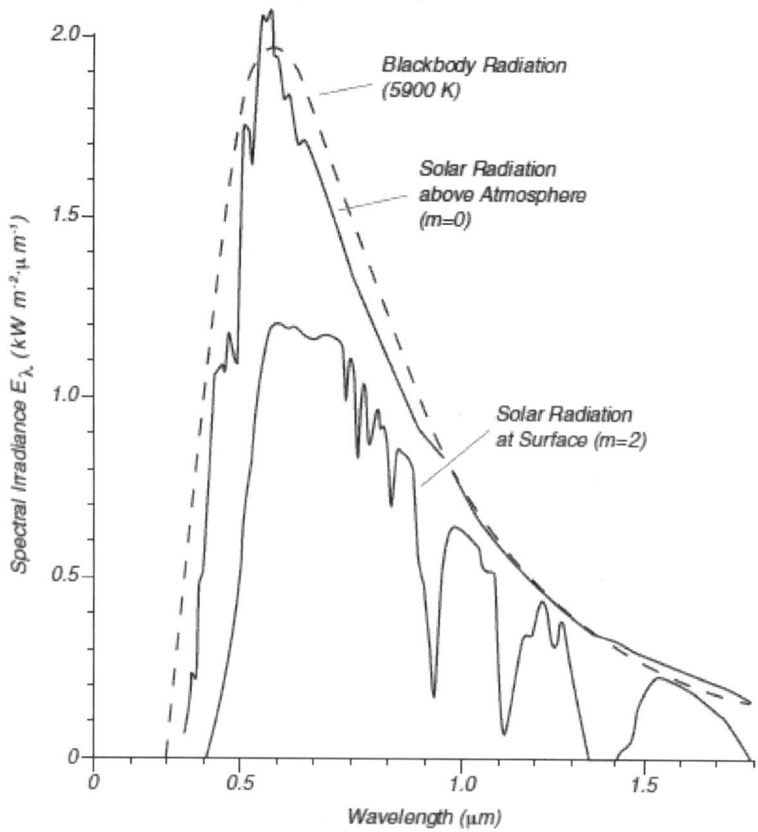

Figure 5.2 Insolation (spectral irradiance) of sunlight at top of the atmosphere and at the sea surface on a clear day. The dashed line is the best-fitting curve of blackbody radiation the size and distance of the sun. The number of standard atmospheric masses is designated by *m*. Thus *m* = 2 is applicable for sunlight when the sun is 30° above the horizon. After Stewart (1985: 43).

Reflectivity of the surface, which depends on solar elevation angle and roughness of sea surface. Solar inclination and cloudiness dominate. Absorption by ozone, water vapour, aerosols, and dust are much weaker.

The average annual value for insolation (figure 5.3) is in the range:

$$30\ \mathrm{W/m^2} < Q_{SW} < 260\ \mathrm{W/m^2} \qquad (5.5)$$

Factors Influencing Infrared Flux: The sea surface radiates as a blackbody having the same temperature as the water, which is roughly 290 K. The distribution of radiation as a function of wavelength is given by Planck's equation. Sea water at 290 K radiates most strongly at wavelengths near 10 μm. These wavelengths are strongly absorbed by clouds, and somewhat by water vapour. A plot of atmospheric transmittance as a function of wavelength for a clear atmosphere but with varying amounts of water vapour (figure 5.4) shows the atmosphere is nearly transparent in some wavelength bands called windows.

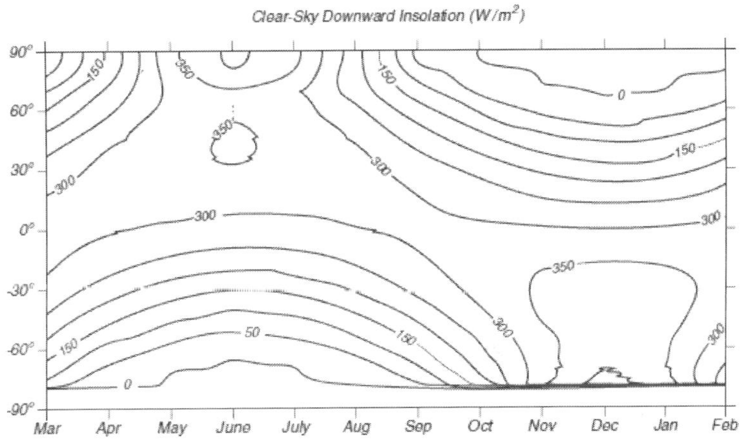

Figure 5.3 Monthly average of downward flux of sunlight through a cloud-free sky and into the ocean in W/m^2 during 1989 calculated by the Satellite Data Analysis Center at the NASA Langley Research Center (Darnell et al. 1992) using data from the International Satellite Cloud Climatology Project.

The transmittance on a cloud-free day through the window from 8 μm to 13μm is determined mostly by water vapour. Absorption in other bands, such as those at 3.5 μm to 4.0 μm depends on CO2 concentration in the atmosphere. As the concentration of CO2 increases, these windows close and more radiation is trapped by the atmosphere.

Because the atmosphere is mostly transparent to incoming sunlight, and somewhat opaque to outgoing infrared radiation, the atmosphere traps radiation. The trapped radiation, coupled with convection in the atmosphere, keepsearth's surface 33° warmer than it would be in the absence of a convecting, wet atmosphere but in thermal equilibrium with space. The atmosphere acts like the panes of glass on a greenhouse, and the effect is known as the *greenhouse effect* . See Hartmann

(1994: 24–26) for a simple discussion of the radiative balance of a planet. CO2, water vapour, methane, and ozone are all important greenhouse gasses.

The net infrared flux depends on:

- Clouds thickness. The thicker the cloud deck, the less heat escapes to space.
- Cloud height, which determines the temperature at which the cloud radiates heat back to the ocean. The rate is proportional to t^4, where t is the temperature of the radiating body in Kelvins. High clouds are colder than low clouds.
- Atmospheric water-vapour content. The more humid the atmosphere the less heat escapes to space.

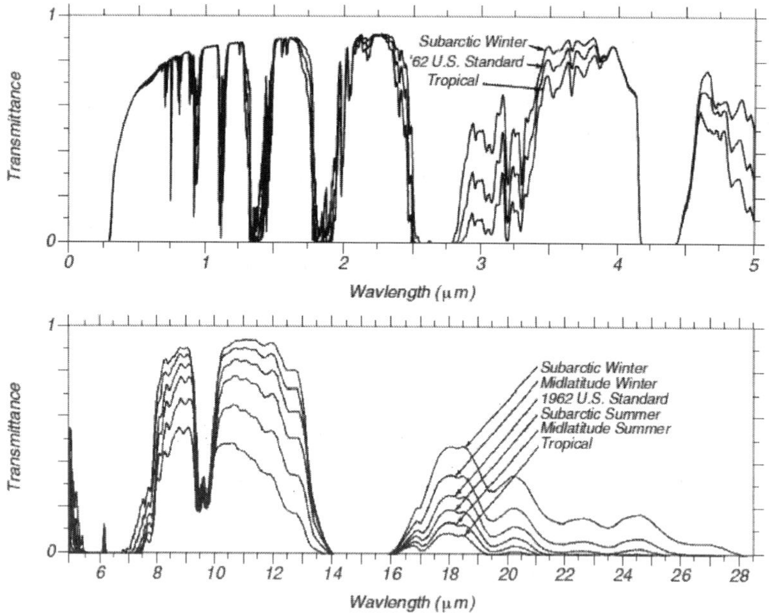

Figure 5.4 Atmospheric transmittance for a vertical path to space from sea level for six model atmospheres with very clear, 23 *km*, visibility, including the influence of molecular and aerosol scattering. Notice how water vapour modulates the transparency of the 10-14 μm atmospheric window, hence it modulates QLW , which is a maximum at these wavelengths. After Selby and McClatchey (1975).

1. Water Temperature. The hotter the water the more heat is radiated. Again, radiation depends of t^4.
2. Ice and snow cover. Ice emits as a black body, but it cools much faster than open water. Ice-covered seas are insulated from the atmosphere.

Water vapour and clouds influence the net loss of infrared radiation more than surface temperature. Hot tropical regions lose less heat than cold Polar Regions. The temperature range from poles to equator is $0^{\circ}C < t < 25^{\circ}C$ or273K < t < 298K, and the ratio of maximum to minimum temperature in Kelvins is 298/273 = 1.092.

Raised to the fourth power this is 1.42. Thus there is a 42% increase in emitted radiation from pole to equator. Over the same distance water vapour can change the net emitted radiance by 200%.

The average annual value for net infrared flux is in the narrow range:

$$-60 \text{ W/m}^2 < Q_{LW} < -30 \text{ W/m}^2 \tag{5.6}$$

Factors Influencing Latent-Heat Flux: Latent heat flux is influenced primarily by wind speed and relative humidity. High winds and dry air evaporate much more water than weak winds with relative humidity near 100%. In polar regions, evaporation from ice covered ocean is much less than from open water. In the arctic, most of the heat lost from the sea is through leads (ice-free areas). Hence the percent open water is very important for the arctic heat budget.

The average annual value for latent-heat flux is in the range:

$$-130 \text{ W/m}^2 < Q_L < -10 \text{ W/m}^2 \tag{5.7}$$

Factors Influencing Sensible-Heat Flux: Sensible heat flux is influenced by wind speed and air-sea temperature difference. High winds and large temperature differences cause high fluxes. Think of this as a wind-chill factor for the ocean.

The average annual value for sensible-heat flux is in the range:

$$-42 \text{ W/m}^2 < Q_S < -2 \text{ W/m}^2 \tag{5.8}$$

5.3 DIRECT CALCULATION OF FLUXES

Before we can describe the geographical distribution of fluxes into and out of the ocean, we need to know how they are measured or calculated.

Gust-Probe Measurements of Turbulent Fluxes: There is only one accurate method for calculating fluxes of sensible and latent heat and momentum at the sea surface: from direct measurement of turbulent quantities in the atmospheric boundary layer made by gust probes on low-flying aircraft or offshore platforms. Very few such measurements have been made. They are expensive, and they cannot be used to calculate heat fluxes averaged over many days or large areas. The gust-probe measurements are used only to calibrate other methods of calculating fluxes.

- Measurements must be made in the surface layer of the atmospheric boundary layer (See §4.3), usually within 30 m of the sea surface, because fluxes are independent of height in this layer.
- Measurements must be made by fast-response instruments (gust probes) able to make several observations per second on a tower, or every meter from a plane.
- Measurements include the horizontal and vertical components of the wind, the humidity, and the air temperature.

Fluxes are calculated from the correlation of vertical wind and horizontal wind, humidity, or temperature: Each type of flux is calculated from different measured variables, u', w', t', and q':

$$T = \langle \rho_a \, u' w' \rangle = \rho_a \, \langle u' w' \rangle \equiv \rho_a \, u_*^2 \tag{5.9a}$$

$$Q_S = C_p \, \langle \rho_a \, w' t' \rangle = \rho_a \, C_p \, \langle w' t' \rangle \tag{5.9b}$$

$$Q_L = L_E \, \langle w' q' \rangle \tag{5.9c}$$

where the brackets denotes time or space averages, and the notation is given in table 5.1. Note that *specific humidity* mentioned in the table is the mass of water vapour per unit mass of air.

Table 5.1 Notation Describing Fluxes

Symbol	Variable	Value and Units
C_p	Specific heat capacity of air	1030 J·kg^{-1}·K^{-1}
C_D	Drag coefficient (see 4.3)	$(0.50 + 0.071\ U10) \times 10^{-3}$
C_L	Latent heat transfer coefficient	1.2×10^{-3}
C_S	Sensible heat transfer coefficient	1.0×10^{-3}
L_E	Latent heat of evaporation	2.5×10^6 J/kg
q	Specific humidity of air	kg (water vapor)/kg (air) kg (water vapor)/kg (air)
q_a	Specific humidity of air 10 m above the sea	kg (water vapor)/kg (air)
q_s	Specific humidity of air at the sea surface	kg (water vapor)/kg (air)
Q_S	Sensible heat flux	W/m^2
Q_L	Latent heat flux	W/m^2
T	Wind stress	Pascals
t_a	Temperature of the air 10 m above the sea	K or °C
t_s	Sea-surface temperature	K or °C
t'	Temperature fluctuation	°C
u'	Horizontal component of fluctuation of wind	m/s
$u*$	Friction velocity	m/s
U_{10}	Wind speed at 10 m above the sea	m/s
w'	Vertical component of wind fluctuation	m/s
ρ_a	Density of air	1.3 kg/m^3
T	Vector wind stress	Pa

C_S and C_L from Smith (1988).

Radiometer Measurements of Radiative Fluxes : Radiometers on ships, offshore platforms, and even small islands are used to make direct measurements of radiative fluxes. Wideband radiometers sensitive to radiation from 0.3 μm to 50 μm can measure incoming solar and infrared radiation with an accuracy of around 3% provided they are well calibrated and maintained. Other, specialized radiometers can measure the incoming solar radiation, the downward infrared radiation, and the upward infrared radiation.

5.4 INDIRECT CALCULATION OF FLUXES: BULK FORMULAS

The use of gust-probes is very expensive, and radiometers must be carefully maintained. Neither can be used to obtain long-term, global values of fluxes. To calculate these fluxes from practical measurements, we use observed correlations between fluxes and variables that can be measured globally.

For fluxes of sensible and latent heat and momentum, the correlations are called *bulk formulas*. They are:

$$T = \rho_a C_D U_{10}^2 \quad (5.10a)$$

$$Q_S = \rho_a C_p C_S U_{10} (t_s - t_a) \quad (5.10b)$$

$$Q_L = \rho_a L_E C_L U_{10} (q_s - q_a) \quad (5.10c)$$

Air temperature t_a is measured using thermometers on ships. It cannot be measured from space using satellite instruments. *ts* is measured using thermometers on ships or from space using infrared radiometers such as the AVHRR.

Table 5.2 Accuracy of Wind and Fluxes Observed Globally From Space

Variable	Accuracy	Comments
Wind Speed	±1.5 m/s ±1.5 m/s	Instrument Error Sampling Error (Monthly Average)
Wind Stress	±10 % ±14 Pa	Drag Coefficient Error Assuming 10 m/s Wind Speed
Insolation	±5 % ±15 W/m^2 ±10 %	Monthly Average Monthly Average Daily Average
Rain Rate	±50 %	
Rainfall	±10 %	5° × 5° area for trmm
Net Long Wave Radiation Latent Heat Flux	±4–8 % ±15–27 W/m^2 ±35 W/m^2 ±15 W/m^2	Daily Average Daily Average Monthly Average

The specific humidity of air at 10 m above the sea surface *qa* is calculated from measurements of relative humidity made from ships. Gill (1982: pp: 39– 41, 43–44,

& 605–607) describes equations relating water vapour pressure, vapour density, and specific heat capacity of wet air. The specific humidity at the sea surface q_s is calculated from t_s assuming the air at the surface is saturated with water vapour. U_{10} is measured or calculated using the instruments or techniques described in Chapter 4. Note that wind stress is a vector with magnitude and direction. It is parallel to the surface in the direction of the wind.

The problem now becomes: How to calculate the fluxes across the sea surface required for studies of ocean dynamics? The fluxes include: 1) stress; 2) solar heating; 3) evaporation; 4) net infrared radiation; 5) rain; 5) sensible heat; and 6) others such as CO_2 and particles (which produce marine aerosols). Furthermore, the fluxes must be accurate. We need an accuracy of approximately$\pm$15 W/m^2. This is equivalent to the flux of heat which would warm or cool acolumn of water 100 m deep by roughly 1°C in one year. Table 5.2 lists typical accuracies of fluxes measured globally from space. Now, let's look at each variable.

Wind Speed and Stress: Stress is calculated from wind observations made from ships at sea and from scatterometers in space.

Insolation: is calculated from cloud observations made from ships and from visible-light radiometers on meteorological satellites. Satellite measurements are far more accurate than the ship data because it's very hard to measure cloudiness from below the clouds. Satellite measurements processed by the International Satellite Cloud Climatology Project ISCCP are the basis for maps of insolation and its variability from month to month (Darnell et al. 1988; Rossow and Schiffer 1991).

The basic idea behind the calculation of insolation is this. Sunlight at the top of the atmosphere is accurately known from the solar constant, latitude, longitude, and time. Sunlight is either reflected back to space by clouds, or it eventually reaches the sea surface. Only a small and nearly constant fraction is absorbed in the atmosphere. But, recent work by Cess et al. (1995) and Ramanathan et al. (1995) suggest that this basic idea may be incomplete, and that atmospheric absorption may be a function of cloudiness. Assuming atmospheric absorption is constant, insolation is calculated from:

$$\text{Insolation} = S\,(1 - A) - C$$

where S = 1365 W/m^2 is the solar constant, A is albedo, the ratio of incident to reflected sunlight, and C is a constant which includes absorption by ozone and other atmospheric gases and by cloud droplets. Insolation is calculated from cloud data (which also includes reflection from aerosols) collected from instruments such as the AVHRR on meteorological satellites. Ozone and gas absorption are calculated from known distributions of the gases in the atmosphere. Q_{SW} is calculated from satellite data with an accuracy of 5–7%.

Water Flux in (Rainfall): Rain rate is another variable that is very difficult to measure from ships. Rain collected from gauges at different locations on ships and from gauges on nearby docks all differ by more than a factor of two. Rain at sea falls mostly horizontally because of wind, and the ship's superstructure distorts the paths of raindrops. Rain in many areas falls mostly as drizzle, and it is difficult to detect and measure.

The most accurate measurements of rain rate in the tropics (±35°) are calculated from microwave radiometers and radar observations of rain at several frequencies using instruments on the Tropical Rain Measuring Mission TRMM launched in 1997. Rain for other times and latitudes can be calculated accurately by combining microwave data with infrared observations of the height of cloud tops and with rain gauge data (figure 5.5). Rain is also calculated from the reanalyses weather data by numerical models of the atmospheric circulation (Schubert, Rood, and Pfaendtner, 1993), and by combining ship and satellite observations with analyses from numerical weather-prediction models (Xie and Arkin, 1997).

The largest source of error is due to conversion of rain rate to cumulative rainfall, a sampling error. Rain is very rare, it is log-normally distributed, and most rain comes from a few storms. Satellites tend to miss storms, and data must be averaged over areas up to 5° on a side to obtain useful values of rainfall.

Net Long-Wave Radiation: Net Long-wave radiation is not easily calculated because it depends on the height and thickness of clouds, and the vertical distribution of water vapour in the atmosphere. It is calculated by numerical weather-prediction models or from observations of the vertical structure of the atmosphere from atmospheric sounders.

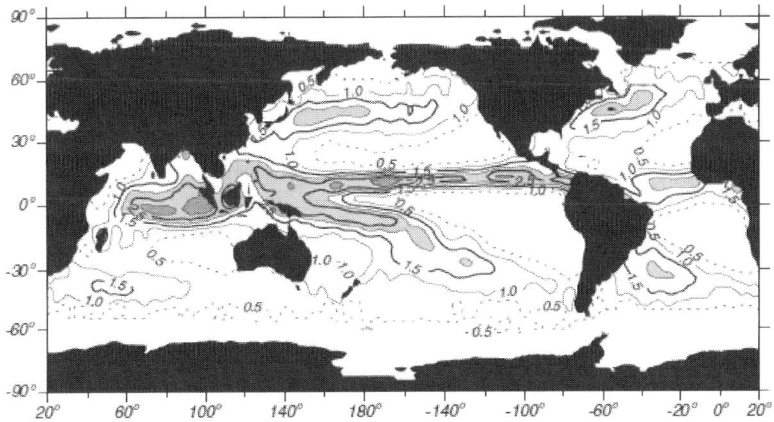

Figure 5.5 Rainfall in m/year calculated from data compiled by the Global Precipitation Climatology Project at NASA's Goddard Space Flight Center using data from rain gauges, infrared radiometers on geosynchronous meteorological satellites, and the ssm/i. Contour interval is 0.5 m/yr, light shaded areas exceed 2 m/yr, heavy shaded areas exceed 3 m/yr.

Water Flux Out (Latent Heat Flux): Latent heat flux is calculated from ship observations of relative humidity, water temperature, and wind speed using bulk formulas (5.10c) and ship data accumulated in the icoads described below.

The fluxes are not calculated from satellite data because satellite instruments are not very sensitive to water vapour close to the sea. Perhaps the best fluxes are those calculated from numerical weather models.

Sensible Heat Flux: Sensible heat flux is calculated from observations of air-sea temperature difference and wind speed made from ships, or by numerical weather models. Sensible fluxes are small almost everywhere except offshore of the east coasts of continents in winter when cold, Arctic air masses extract heat from warm, western, boundary currents. In these areas, numerical models give perhaps the best values of the fluxes. Historical ship report give the long-term mean values of the fluxes.

5.5 GLOBAL DATA SETS FOR FLUXES

Ship and satellite data have been processed to produce global maps of fluxes. Ship measurements made over the past 150 years yield maps of the long-term mean values of the fluxes, especially in the northern hemisphere. Ship data, however, are sparse in time and space, and they are being replaced more and more by fluxes calculated by numerical weather models and by satellite data.

The most useful maps are those made by combining level 3 and 4 satellite data sets with observations from ships, using numerical weather models. Let's look first at the sources of data, then at a few of the more widely used data sets.

International Comprehensive Ocean-Atmosphere Data Set: Data collected by observers on ships are the richest source of marine information. Slutz et al. (1985) describing their efforts to collect, edit, and publish all marine observations write:

Since 1854, ships of many countries have been taking regular observations of local weather, sea surface temperature, and many other characteristics near the boundary between the ocean and the atmosphere. The observations by one such ship-of-opportunity at one time and place, usually incidental to its voyage, make up a marine report. In later years fixed research vessels, buoys, and other devices have contributed data. Marine reports have been collected, often in machine-readable form, by various agencies and countries. That vast collection of data, spanning the ocean from the mid-nineteenth century to date, is the historical ocean- atmosphere record.

These marine reports have been edited and published as the *International Comprehensive Ocean-Atmosphere Data Set* ICOADS (Woodruff et al. 1987) available through the National Oceanic and Atmospheric Administration.

The ICOADS release 2.3 includes 213 million reports of marine surface conditions collected from 1784–2005 by buoys, other platform types, and by observers on merchant ships. The data set include fully quality-controlled (trimmed) reports and summaries. Each unique report contains 22 observed and derived variables, as well as flags indicating which observations were statistically trimmed or subjected to adaptive quality control. Here, statistically trimmed means outliers were removed from the data set. The summaries included in the data set give 14 statistics, such as the median and mean, for each of eight observed variables: air and sea surface temperatures, wind velocity, sea-level pressure, humidity, and cloudiness, plus 11 derived variables. The data set consists of an easily-used data base at three principal resolutions: 1) individual reports, 2) year-month summaries of the individual reports in 2°latitude by 2° longitude boxes from 1800 to 2005 and 1° latitude by 1° longitude boxes from 1960 to 2005, and 3) decade-month summaries. Note that data from 1784 through the early 1800s are extremely sparse–based on scattered ship voyages.

Duplicate reports judged inferior by a first quality control process designed by the National Climatic Data Center NCDC were eliminated or flagged, and "untrimmed" monthly and decadal summaries were computed for acceptable data within each 2° latitude by 2° longitude grid. Tighter, median-smoothed limits were used as criteria for statistical rejection of apparent outliers from the data used for separate sets of *trimmed* monthly and decadal summaries. Individual observations were retained in report form but flagged during this second quality control process if they fell outside 2.8 or 3.5 estimated standard- deviations about the smoothed median applicable to their 2° latitude by 2° longitude box, month, and 56–, 40–, or 30–year period (*i.e.*, 1854–1990, 1910–1949, or 1950–1979).

The data are most useful in the northern hemisphere, especially the North Atlantic. Data are sparse in the southern hemisphere and they are not reliable south of 30° S. Gleckler and Weare (1997) analyzed the accuracy of the ICOADS data for calculating global maps and zonal averages of the fluxes from 55°N to 40°S. They found that systematic errors dominated the zonal means. Zonal averages of insolation were uncertain by about 10%, ranging from ±10 W/m^{2}in high latitudes to ±25 W/m^2 in the tropics. Long wave fluxes were uncertain by about ±7 W/m^2. Latent heat flux uncertainties ranged from ±10 W/m^2 in some areas of the northern ocean to ±30 W/m^2 in the western tropical ocean to ±50 W/m^2 in western boundary currents. Sensible heat flux uncertainties tend to be around ±5 - 10 W/m^2.

Josey et al (1999) compared averaged fluxes calculated from ICOADS with fluxes calculated from observations made by carefully calibrated instruments on some ships and buoys. They found that mean flux into the ocean, when averaged over all the seas surface had errors of ±30 W/m^2. Errors vary seasonally and by region, and global maps of fluxes require corrections such as those proposed by DaSilva, Young, and Levitus (1995) shown in figure 5.7.

Satellite Data Raw data are available from satellite projects, but we need processed data. Various levels of processed data from satellite projects are produced (table 5.3):

Table 5.3 Levels of Processed Satellite Data

Level	Level of Processing
Level 1	Data from the satellite in engineering units (volts)
Level 2	Data processed into geophysical units (wind speed) at the time and place the satellite instrument made the observation
Level 3	Level 2 data interpolated to fixed coordinates in time and space Level 4 Level 3 data averaged in time and space or further processed

The operational meteorological satellites that observe the ocean include:

- NOAA series of polar-orbiting, meteorological satellites;
- U.S. Defence Meteorological Satellite Program dmsp polar-orbiting satellites, which carry the Special Sensor Microwave/ Imager (ssm/i);
- Geostationary meteorological satellites operated by NOAA (goes), Japan (gms) and the European Space Agency (meteosats).

Data are also available from instruments on experimental satellites such as:

- Nimbus-7, Earth Radiation Budget Instruments;
- Earth Radiation Budget Satellite, Earth Radiation Budget Experiment;
- The European Space Agency's ers–1 & 2;
- The Japanese Advanced Earth Observing System (JAEOS) and Midori;
- QuikScat;
- The Earth-Observing System satellites Terra, Aqua, and Envisat;
- The Tropical Rainfall Measuring Mission (TRMM); and,
- Topex/Poseidon and its replacement Jason-1.

Satellite data are collected, processed, and archived by government organizations. Archived data are further processed to produce useful flux data sets.

International Satellite Cloud Climatology Project: The International Satellite Cloud Climatology Project is an ambitious project to collect observations of clouds made by dozens of meteorological satellites from 1983 to 2000, to calibrate the satellite data, to calculate cloud cover using carefully verified techniques, and to calculate surface insolation and net surface infrared fluxes (Rossow and Schiffer, 1991). The clouds were observed with visible-light instruments on polar-orbiting and geostationary satellites.

Global Precipitation Climatology Project: This project uses three sources of data to calculate rain rate (Huffman, et al. 1995, 1997):

- Infrared observations of the height of cumulus clouds from goes satellites. The basic idea is that the more rain produced by cumulus clouds, the higher the cloud top, and the colder the top appears in the infrared. Thus rain rate at the base of the clouds is related to infrared temperature.
- Measurements by rain gauges on islands and land.
- Radio emissions from water drops in the atmosphere observed by ssm–i.

Accuracy is about 1 mm/day. Data from the project are available on a 2.5° latitude by 2.5° longitude grid from July 1987 to December 1995 from the Global Land Ocean Precipitation Analysis at the NASA Goddard Space Flight Center.

Xie and Arkin (1997) produced a 17-year data set based on seven types of satellite and rain-gauge data combined with the rain calculated from the NCEP/NCAR reanalyzed data from numerical weather models. The data set has the same spatial and temporal resolution as the Huffman data set.

Reanalyzed Output From Numerical Weather Models: Surface heat flux has been calculated from weather data using numerical weather models by various reanalysis projects described in §4.5. The fluxes are consistent with atmospheric dynamics, they are global, they are calculated every six hours, and they are available for many years on a uniform grid. For example, the NCAR/NCEP reanalysis, available on a cd-rom, include daily averages of wind stress, sensible and latent heat fluxes, net long and short wave fluxes, near-surface temperature, and precipitation.

Accuracy of Calculated Fluxes: Recent studies of the accuracy of fluxes computed by numerical weather models and reanalysis projects suggest:

- Heat fluxes from the NCEP and ECMWF reanalyses have similar global average values, but the fluxes have important regional differences. Fluxes from the Goddard Earth Observing System reanalysis are much less ac- curate (Taylor, 2000: 258). Chou et al (2004) finds large differences in fluxes calculated by different groups.
- The fluxes are biased because they were calculated using numerical models optimized to produce accurate weather forecasts. The time-mean values of the fluxes may not be as accurate as the time-mean values calculated directly from ship observations.
- The simulation of boundary-layer clouds is a significant source of error in calculated fluxes. The poor vertical resolution of the numerical models does not adequately resolve the low-level cloud structure (Taylor, 2001).
- The fluxes have zonal means that differ significantly from the same zonal means calculated from ICOADS data. The differences can exceed 40 W/m^2.
- The atmospheric models do not require that the net heat flux averaged over time and earth's surface be zero. The ECMWF data set averaged

over fifteen years gives a net flux of 3.7 /m^2 into the ocean. The NCEP reanalysis gives a net flux of 5.8 W/m^2 out of the ocean (Taylor, 2000: 206). ICOADS data give a net flux of 16 W/m^2 into the ocean (figure 5.7).

Thus reanalyzed fluxes are most useful for forcing climate models needing actual heat fluxes and wind stress. ICOADS data are most useful for calculating time- mean fluxes except perhaps in the southern hemisphere. Overall, Taylor (2000) notes that there are no ideal data sets, all have significant and unknown errors.

Output From Numerical Weather Models: Some projects require fluxes a few hours after observations are collected. The surface analysis from numerical weather models is a good source for this type of flux.

5.6 GEOGRAPHIC DISTRIBUTION OF TERMS IN THE HEAT BUDGET

Various groups have used ship and satellite data in numerical weather models to calculate globally averaged values of the terms for earth's heat budget. The values give an overall view of the importance of the various terms (figure 5.6). Notice that insolation balances infrared radiation at the top of the atmosphere. At the surface, latent heat flux and net infrared radiation tend to balance insolation, and sensible heat flux is small.

Note that only 20% of insolation reaching earth is absorbed directly by the atmosphere while 49% is absorbed by the ocean and land. What then warms the atmosphere and drives the atmospheric circulation? The answer is rain and infrared radiation from the ocean absorbed by the moist tropical atmosphere. Here's what happens. Sunlight warms the tropical ocean which evaporates water to keep from warming up. The ocean also radiates heat to the atmosphere, but the net radiation term is smaller than the evaporative term. Trade winds carry the heat in the form of water vapour to the tropical convergence zone. There vapour condenses as rain, releasing its latent heat, and heating the atmosphere by as much as 125 W/m^2 averaged over a year.

At first it may seem strange that rain heats the air. After all, we are familiar with summertime thunderstorms cooling the air at ground level. The cool air from thunderstorms is due to downdrafts. Higher in the cumulus cloud, heat released by rain warms the mid-levels of the atmosphere causing air to rise rapidly in the storm. Thunderstorms are large heat engines converting the energy of latent heat into kinetic energy of winds.

The zonal average of the oceanic heat-budget terms (figure 5.7) shows that insolation is greatest in the tropics, that evaporation balances insolation, and that sensible heat flux is small. *Zonal average* is an average along lines of constant latitude. Note that the terms in figure 5.7 don't sum to zero. The areal-weighted integral of the curve for total heat flux is not zero. Because the net heat flux into the ocean averaged over several years must be less than a few watts per square meter, the non-zero value must be due to errors in the various terms in the heat budget.

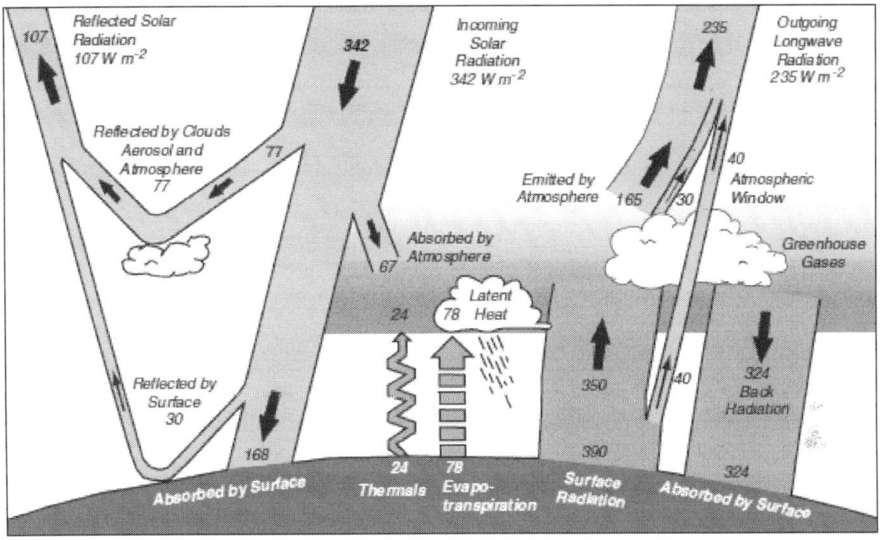

Figure 5.6 The mean annual radiation and heat balance of the earth. After Houghton et al. (1996: 58), which used data from Kiehl and Trenberth (1996).

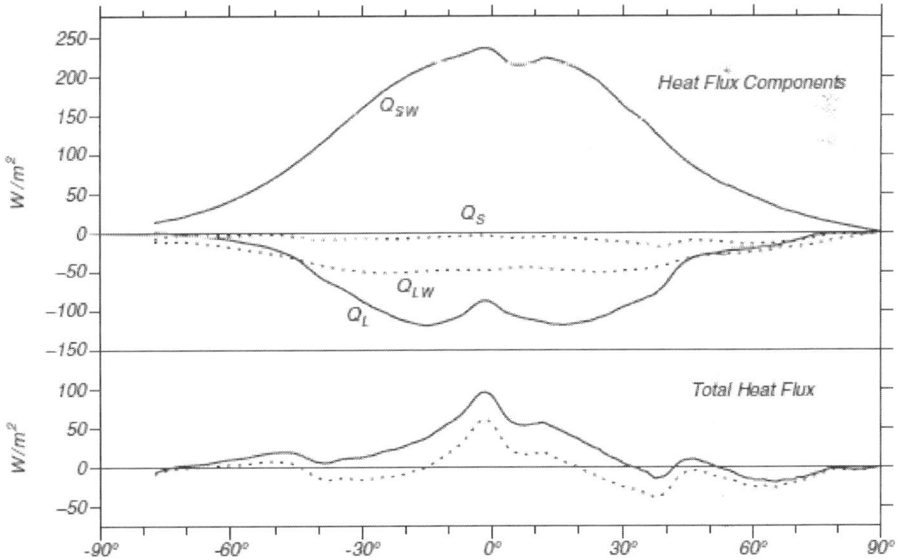

Figure 5.7 Upper: Zonal averages of heat transfer to the ocean by insolation QSW, and loss by infrared radiation QLW, sensible heat flux QS, and latent heat flux QL, calculated by DaSilva, Young, and Levitus (1995) using the ICOADS data set. Lower: Net heat flux through the sea surface calculated from the data above (solid line) and net heat flux constrained to give heat and other transports that match independent calculations of these transports. The area under the lower curves ought to be zero, but it is 16 W/m^2 for the unconstrained case and -3 W/m^2 for the constrained case.

Errors in the heat budget terms can be reduced by using additional information. For example, we know roughly how much heat and other quantities are transported by the ocean and atmosphere, and the known values for these transports can be used to constrain the calculations of net heat fluxes (figure 5.7). The constrained fluxes show that the heat gained by the ocean in the tropics is balanced by heat lost by the ocean at high latitudes.

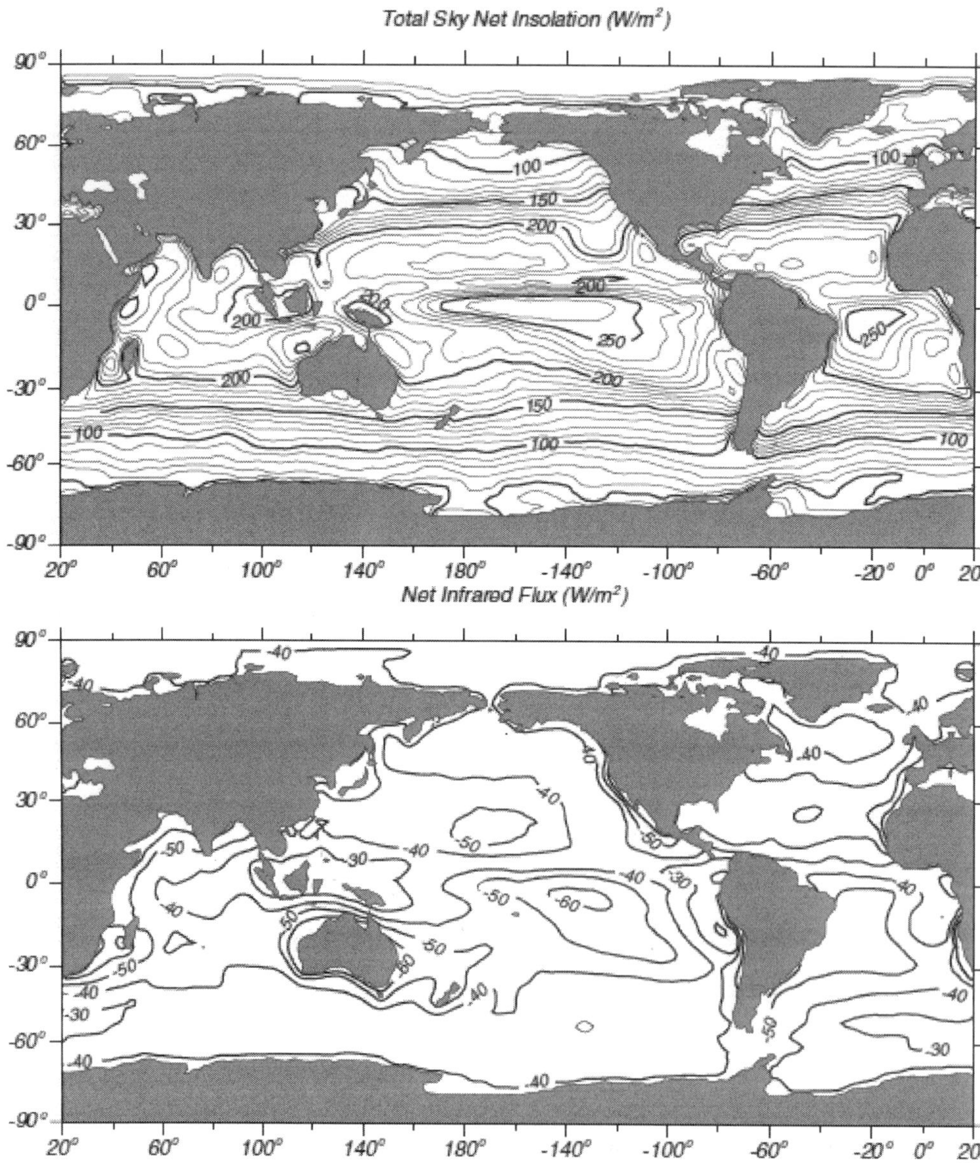

Figure 5.8 Annual-mean insolation *QSW* (top) and infrared radiation *QLW* (bottom) through the sea surface during 1989 calculated by the Satellite Data Analysis Center at the NASA Langley Research Center (Darnell et al., 1992) using data from the International Satellite Cloud Climatology Project. Units are W/m^2, contour interval is $10\ W/m^2$.

Maps of the regional distribution of fluxes give clues to the processes producing the fluxes. Clouds regulate the amount of sunlight reaching the sea surface (figure 5.8 top), and solar heating is everywhere positive. The net infrared heat flux (figure 5.8 bottom) is largest in regions with the least clouds, such as the center of the ocean and the eastern central Pacific. The net infrared flux is everywhere negative. Latent heat fluxes (figure 5.9) are dominated by evaporation in the trade wind regions and the offshore flow of cold air masses behind cold fronts in winter offshore of Japan and North America. Sensible heat fluxes (figure 5.10 top) are dominated by cold air blowing off continents. The net heating gain (figure 5.10 bottom) is largest in equatorial regions and net heat loss is largest downwind on Asia and North America.

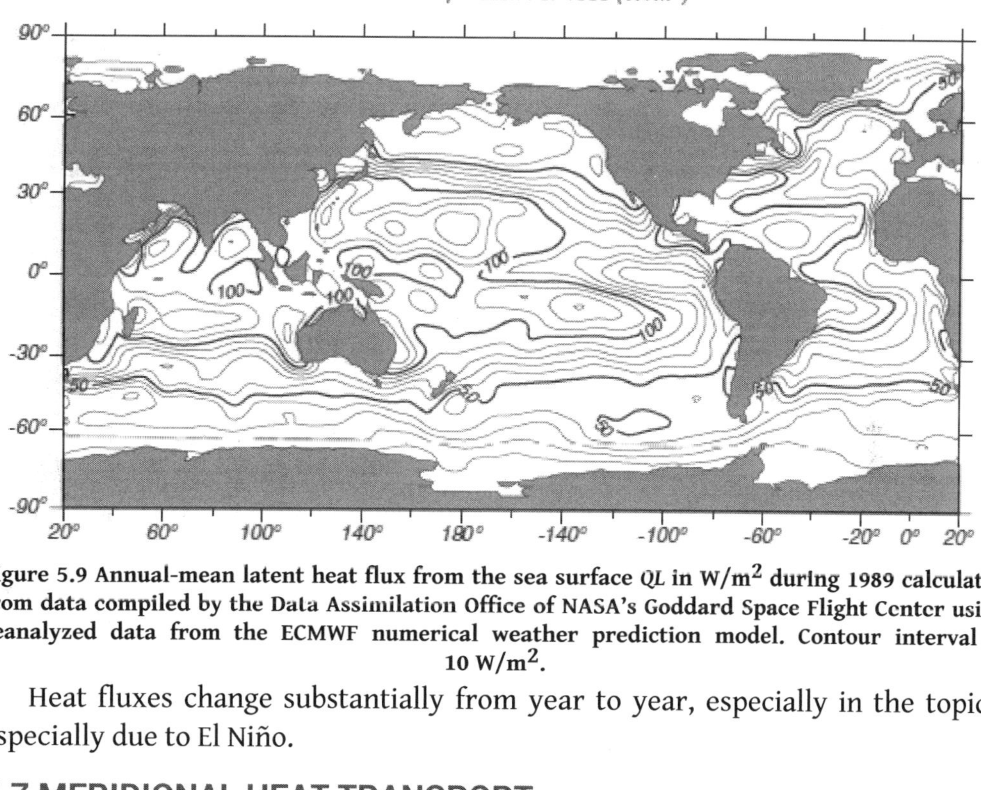

Figure 5.9 Annual-mean latent heat flux from the sea surface QL in W/m^2 during 1989 calculated from data compiled by the Data Assimilation Office of NASA's Goddard Space Flight Center using reanalyzed data from the ECMWF numerical weather prediction model. Contour interval is 10 W/m^2.

Heat fluxes change substantially from year to year, especially in the topics, especially due to El Niño.

5.7 MERIDIONAL HEAT TRANSPORT

Overall, earth gains heat at the top of the tropical atmosphere, and it loses heat at the top of the polar atmosphere. The atmospheric and oceanic circulation together must transport heat from low to high latitudes to balance the gains and losses. This north-south transport is called the *meridional heat transport* .

The sum of the meridional heat transport in the ocean and atmosphere is calculated from the zonal average of the net heat flux through the top of the atmosphere measured by satellites. In making the calculation, we assume that transports averaged over a few years are steady. Thus any long-term, net heat gain or loss through the top of the atmosphere must be balanced by a meridional *Corrected Sensible Heat Flux (W/m 2).*

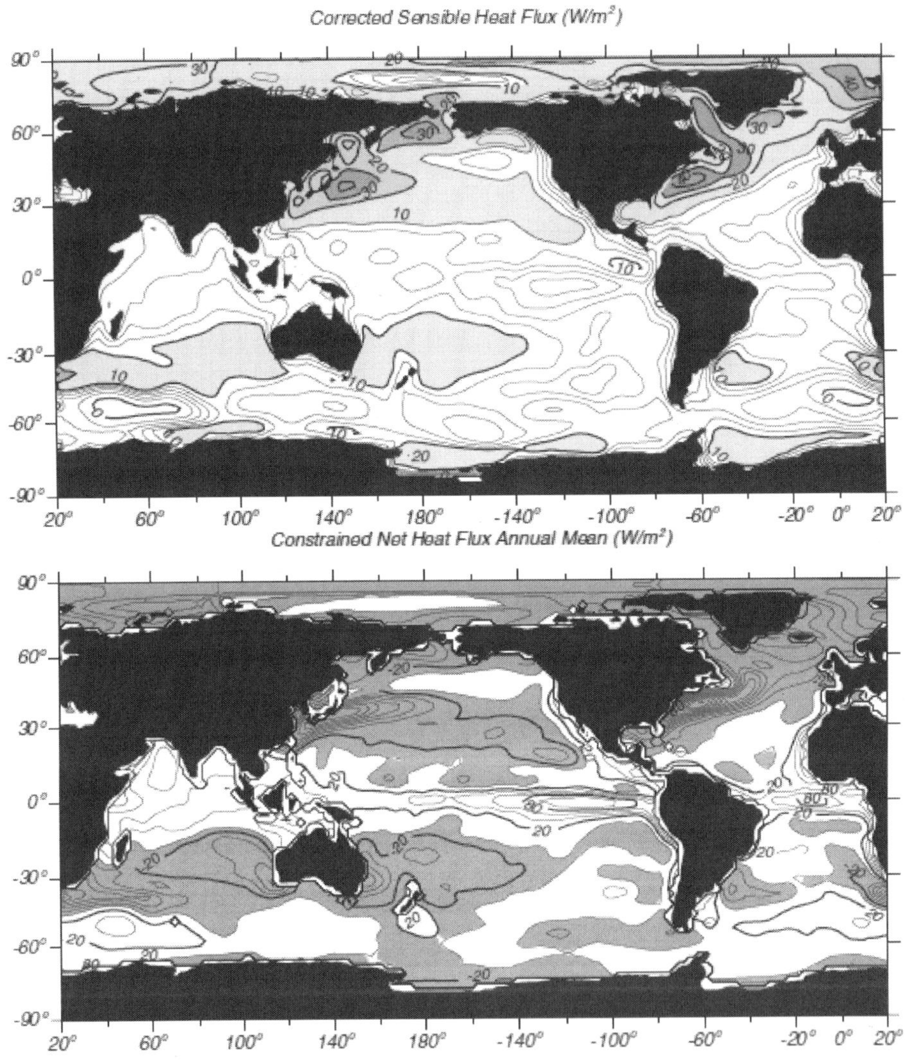

Figure 5.10 Annual-mean upward sensible heat flux *QS* (top) and constrained, net, downward heat flux (bottom) through the sea surface in W/m^2 calculated by DaSilva, Young, and Levitus (1995) using the ICOADS data set from 1945 to 1989. Contour interval is 2 W/m^2 (top) and 20 W/m^2 (bottom).

Net Heat Flux at the Top of the Atmosphere Heat flux through the top of the atmosphere is measured very accurately by radiometers on satellites.

- Insolation is calculated from the solar constant and observations of reflected sunlight made by meteorological satellites and by special satellites of the Earth Radiation Budget Experiment.

- Infrared radiation is measured by infrared radiometers on the satellites.
- The difference between insolation and net infrared radiation is the net heat flux across the top of the atmosphere.

Net Meridional Heat Transport: To calculate the meridional heat transport in the atmosphere and the ocean, we first average the net heat flux through the top of the atmosphere in a zonal band. Because the meridional derivative of the transport is the zonal-mean flux, we calculate the transport from the meridional integral of the zonal-mean flux. The integral must be balanced by the heat transported by the atmosphere and the ocean across each latitude band. Calculations by Trenberth and Caron (2001) show that the total, annual- mean, meridional heat transport by the ocean and atmosphere peaks at 6 PW toward each pole at 35° latitude.

Oceanic Heat Transport: The meridional heat transport in the ocean can be calculated three ways:

- *Surface Flux Method* calculates the heat flux through the sea surface from measurements of wind, insolation, air, and sea temperature, and cloudiness. The fluxes are integrated to obtain the zonal average of the heat flux (figure 5.7). Finally, we calculate the transport from the meridional integral of the zonal-mean flux just as we did at the top of the atmosphere.
- *Direct Method* calculates the heat transport from values of current velocity and temperature measured from top to bottom along a zonal section spanning an ocean basin. The flux is the product of northward velocity and heat content derived from the temperature measurement.
- *Residual Method* first calculates the atmospheric heat transport from atmospheric measurements or the output of numerical weather models. This is the direct method applied to the atmosphere. The atmospheric transport is subtracted from the total meridional transport calculated from the top-of-the-atmosphere heat flux to obtain the oceanic contribution as a residual (figure 5.11).

Various calculations of oceanic heat transports, such as those shown in figure 5.11, tend to be in agreement, and the error bars shown in the figure are realistic. The total meridional transport of heat by the ocean is small compared with the total meridional heat transport by the atmosphere except in the tropics. At 35°,where the total meridional heat transport is greatest, the ocean carries only 22% of the heat in the northern hemisphere, and 8% in the southern (Trenberth and Caron, 2001).

5.8 VARIATIONS IN SOLAR CONSTANT

We have assumed so far that the solar constant, the output of light and heat from the sun, is steady. Recent evidence based on variability of sunspots and faculae (bright spots) shows that the output varied by ±0.2% over centuries(Lean, Beer, and Bradley, 1995), and that this variability is correlated with changes in global mean temperature of earth's surface of ±0.4°C. (figure 5.12).

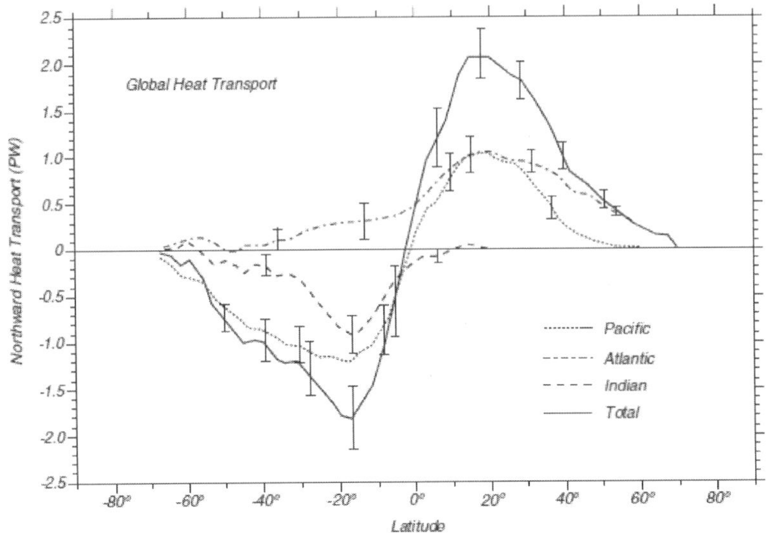

Figure 5.11 Northward heat transport for 1988 in each ocean and the total transport summed over all ocean calculated by the residual method using atmospheric heat transport from ECMWF and top of the atmosphere heat fluxes from the Earth Radiation Budget Experiment satellite. After Houghton et al. (1996: 212), which used data from Trenberth and Solomon (1994). 1 PW = 1 petawatt = 10^{15} W.

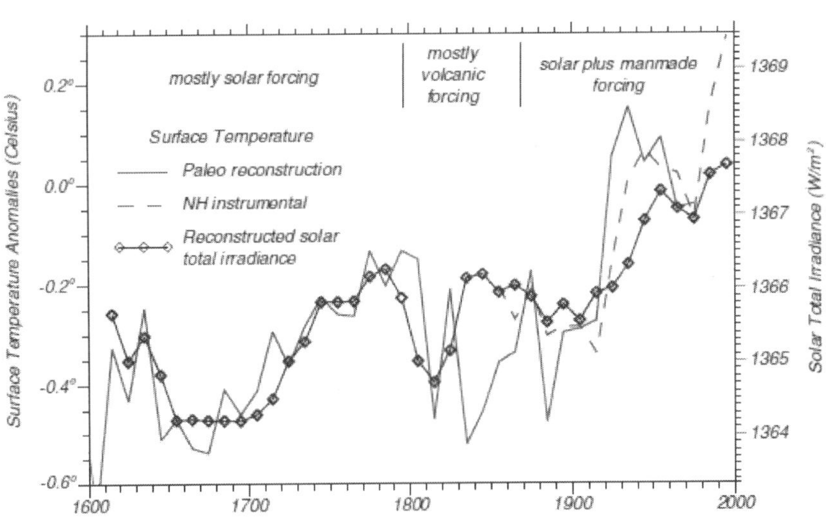

Figure 5.12 Changes in solar constant (total solar irradiance) and global mean temperature of earth's surface over the past 400 years. Except for a period of enhanced volcanic activity in the early 19th century, surface temperature is well correlated with solar variability. After Lean, personal communication.

In addition, White and Cayan (1998) found a small 12 yr, 22 yr, and longer-period variations of sea-surface temperature measured by bathythermographs and ship-board thermometers over the past century. The observed response of earth to solar variability is about that calculated from numerical models of the coupled ocean-atmosphere climate system. Many other changes in climate and weather have been attributed to solar variability. The correlations are somewhat controversial, and much more information can be found in Hoyt and Schatten's (1997) book on the subject.

5.9 CONCISE SUMMARY

- Sunlight is absorbed primarily in the tropical ocean. The amount of sunlight changes with season, latitude, time of day, and cloud cover.
- Most of the heat absorbed by the ocean in the tropics is released as water vapour which heats the atmosphere when water is condenses as rain. Most of the rain falls in the tropical convergence zones, lesser amounts fall in mid-latitudes near the polar front.
- Heat released by rain and absorbed infrared radiation from the ocean are the primary drivers for the atmospheric circulation.
- The net heat flux from the ocean is largest in mid-latitudes and offshore of Japan and New England.
- Heat fluxes can be measured directly using fast response instruments on low-flying aircraft, but this is not useful for measuring heat fluxes over large oceanic regions.
- Heat fluxes through large regions of the sea surface can be calculated from bulk formula. Seasonal, regional, and global maps of fluxes are available based on ship and satellite observations.
- The most widely used data sets for studying heat fluxes are the International Comprehensive Ocean-Atmosphere Data Set and the reanalysis of meteorological data by numerical weather prediction models.
- The atmosphere transports most of the heat needed to warm latitudes higher than 35°. The oceanic meridional transport is comparable to the atmospheric meridional transport only in the tropics.
- Solar output is not constant, and the observed small variations in output of heat and light from the sun seem to produce the changes in global temperature observed over the past 400 years.

06 TEMPERATURE, SALINITY, AND DENSITY

Heat fluxes, evaporation, rain, river inflow, and freezing and melting of sea ice all influence the distribution of temperature and salinity at the ocean's surface. Changes in temperature and salinity can increase or decrease the density of water at the surface, which can lead to convection. If water from the surface sinks into the deeper ocean, it retains a distinctive relationship between temperature and salinity which helps oceanographers track the movement of deep water. In addition, temperature, salinity, and pressure are used to calculate density. The distribution of density inside the ocean is directly related to the distribution of horizontal pressure gradients and ocean currents. For all these reasons, we need to know the distribution of temperature, salinity, and density in the ocean.

Before discussing the distribution of temperature and salinity, let's first define what we mean by the terms, especially salinity.

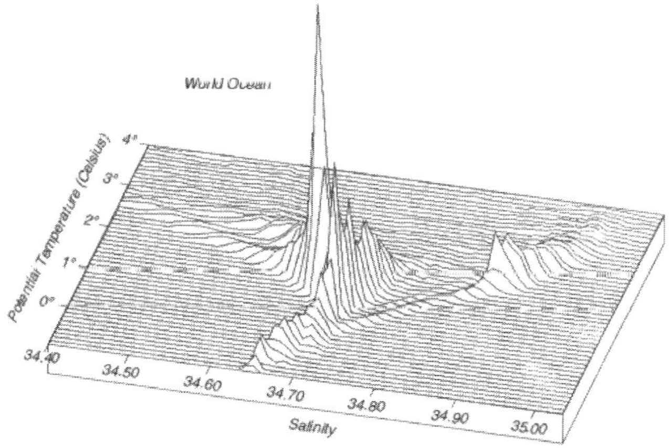

Figure 6.1 Histogram of temperature and salinity of ocean water colder than 4°C. Height is proportional to volume. Height of highest peak corresponds to a volume of 26 million cubic kilometres per bivariate class of 0.1°C and 0.01. After Worthington (1981: 47).

6.1 SALINITY

At the simplest level, salinity is the total amount of dissolved material in grams in one kilogram of sea water. Thus salinity is a dimensionless quantity. It has no units. The variability of dissolved salt is very small, and we must be very careful to define salinity in ways that are accurate and practical. To better understand the need for accuracy, look at figure 6.1. Notice that the range of salinity for most of the ocean's water is from 34.60 to 34.80 parts per thousand, which is 200 parts per million. The variability in the deep North Pacific is even smaller, about 20 parts per million. If we want to classify water with different salinity, we need definitions and instruments accurate to about one part per million. Notice that the range of temperature is much larger, about 1°C, and temperature is easier to measure.

Writing a practical definition of salinity that has useful accuracy is difficult (see Lewis, 1980, for the details), and various definitions have been used.

A Simple Definition : Originally salinity was defined to be the "Total amount of dissolved material in grams in one kilogram of sea water." This is not useful because the dissolved material is almost impossible to measure in practice. For example, how do we measure volatile material like gasses? Nor can we evaporate sea-water to dryness because chlorides are lost in the last stages of drying (Sverdrup, Johnson, and Fleming, 1942: 50).

A More Complete Definition : To avoid these difficulties, the International Council for the Exploration of the Sea set up a commission in 1889 which recommended that salinity be defined as the "Total amount of solid materials in grams dissolved in one kilogram of sea water when all the carbonate has been converted to oxide, the bromine and iodine replaced by chlorine and all organic matter completely oxidized." The definition was published in 1902. *This is useful but difficult to use routinely.*

Salinity Based on Chlorinity: Because the above definition was difficult to implement in practice, because salinity is directly proportional to the amount of chlorine in sea water, and because chlorine can be measured accurately by a simple chemical analysis, salinity S was redefined using chlorinity:

$$S = 0.03 + 1.805\, Cl \qquad (6.1)$$

where *chlorinity* Cl is defined as "the mass of silver required to precipitate completely the halogens in 0.328 523 4 kg of the sea-water sample."

As more and more accurate measurements were made, (6.1) turned out to be too inaccurate. In 1964 UNESCO and other international organizations appointed a Joint Panel on Oceanographic Tables and Standards to produce a more accurate definition. The Joint Panel recommended in 1966 (Wooster, Lee, and Dietrich, 1969) that salinity and chlorinity be related using:

$$S = 1.806\,55\, Cl \qquad (6.2)$$

This is the same as (6.1) for $S = 35$.

Salinity Based on Conductivity: At the same time (6.2) was adopted, oceanographers had began using conductivity meters to measure salinity. The meters were very precise and relatively easy to use compared with the chemical techniques used to measure chlorinity. As a result, the Joint Panel also recommended that salinity be related to conductivity of sea water using:

$$S = -0.089\,96 + 28.297\,29\,R_{15} + 12.808\,32\,R_{15}^2 - 10.678\,69\,R_{15}^3 + 5.986\,24\,R_{15}^4 - 1.323\,11\,R_{15}^5 \quad (6.3a)$$

$$R_{15} = C(S, 15, 0)/C(35, 15, 0) \quad (6.3b)$$

where C(S, 15, 0) is the conductivity of the sea-water sample at 15°C and atmospheric pressure, having a salinity S derived from (6.4), and C(35, 15, 0) is the conductivity of standard "Copenhagen" sea water. Millero (1996) points out that (6.3) is not a new definition of salinity, it merely gives chlorinity as a function of conductivity of seawater relative to standard seawater.

Practical Salinity Scale of 1978: By the early 1970s, accurate conductivity meters could be deployed from ships to measure conductivity at depth. The need to re-evaluate the salinity scale led the Joint Panel to recommend in 1981 (jpots, 1981; Lewis, 1980) that salinity be defined using only conductivity, breaking the link with chlorinity. All water samples with the same conductivity ratio have the same salinity even though the their chlorinity may differ.

The *Practical Salinity Scale of 1978* is now the official definition:

$$S = 0.0080 - 0.1692\,K_{15}^{1/2} + 25.3851\,K_{15} + 14.0941\,K_{15}^{3/2} - 7.0261\,K_{15}^2 + 2.7081\,K_{15}^{5/2} \quad (6.4a)$$

$$K_{15} = C(S, 15, 0)/C(KCl, 15, 0) \quad (6.4b)$$

$$2 \leq S \leq 42$$

where C(S, 15, 0) is the conductivity of the sea-water sample at a temperature of 14.996°C on the International Temperature Scale of 1990 (its-90, see §6.2) and standard atmospheric pressure of 101 325 Pa. *C*(*KCl*, 15, 0) is the conductivity of the standard potassium chloride (KCl) solution at a temperature of 15°C and standard atmospheric pressure. The standard KCl solution contains a mass of32.435 6 grams of KCl in a mass of 1.000 000 kg of solution. Millero (1996: 72) and Lewis (1980) gives equations for calculating salinity at other pressures and temperatures.

Comments: The various definitions of salinity work well because the ratios of the various ions in sea water are nearly independent of salinity and location in the ocean (table 6.1). Only very fresh waters, such as are found in estuaries, have significantly different ratios. The result is based on Dittmar's (1884) chemical analysis of 77 samples of sea water collected by the Challenger Expedition and further studies by Carritt and Carpenter (1959).

The importance of this result cannot be over emphasized, as upon it depends the validity of the chlorinity: salinity: density relationships and, hence, the accuracy of all conclusions based on the distribution of density where the latter is determined by chemical or indirect physical methods such as electrical conductivity. . .— Sverdrup, Johnson, Fleming (1942).

The relationship between conductivity and salinity has an accuracy of around ± 0.003 in salinity. The very small error is caused by variations in constituents such as SiO_2 which cause small changes in density but no change in conductivity.

Table 6.1 Major Constituents of Sea Water

Ion		Atoms	
55.3%	Chlorine	55.3%	Chlorine
30.8%	Sodium	30.8%	Sodium
7.7%	Sulfate	3.7%	Magnesium
3.7%	Magnesium	2.6%	Sulfur
1.2%	Calcium	1.2%	Calcium
1.1%	Potassium	1.1%	Potassium

Reference Seawater and Salinity: The Practical Salinity Scale of 1978 introduced several small problems. It led to confusion about units and to the use of "practical salinity units" that are not part of the definition of Practical Salinity. In addition, absolute salinity differs from salinity by about 0.5%. And, the composition of seawater differs slightly from place to place in the ocean, leading to small errors in measuring salinity.

To avoid these and other problems, Millero et al (2008) defined a new measure of salinity, the Reference Salinity, that accurately represents the Absolute Salinity of an artificial seawater solution. It is based on a Reference Composition of seawater that is much more accurate than the values in Table 6.1 above. The *Reference Composition* of the artificial seawater is defined by a list of solutes and their mole fractions given in Table 4 of their paper. From this, they defined artificial *Reference Seawater* to be seawater having a Reference Composition so- lute dissolved in pure water as the solvent, and adjusted to its thermo-dynamic equilibrium state. Finally, the *Reference Salinity* of Reference Seawater was defined to be exactly 35.16504 g kg^{-1}.

With these definitions, plus many details described in their paper, Millero et al (2008) show Reference Salinity is related to Practical Salinity by:

$$S_R \approx (35.16504/35)\text{g kg}^{-1} \times S \qquad (6.5)$$

The equation is exact at $S = 35$. Reference Salinity is approximately 0.47% larger than Practical Salinity. Reference Salinity *SR* is intended to be used as an SI-based extension of Practical Salinity.

6.2 TEMPERATURE

Many physical processes depend on temperature. A few can be used to define absolute temperature T. The unit of T is the kelvin, which has the symbol K. The fundamental processes used for defining an absolute temperature scale over the range of temperatures found in the ocean include (Soulen and Fogle, 1997):

1) the gas laws relating pressure to temperature of an ideal gas with corrections for the density of the gas; and 2) the voltage noise of a resistance *R*.

The measurement of temperature using an absolute scale is difficult and the measurement is usually made by national standards laboratories. The absolute measurements are used to define a practical temperature scale based on the temperature of a few fixed points and interpolating devices which are calibrated at the fixed points.

For temperatures commonly found in the ocean, the interpolating device is a platinum-resistance thermometer. It consists of a loosely wound, strain-free, pure platinum wire whose resistance is a function of temperature. It is calibrated at fixed points between the triple point of equilibrium hydrogen at 13.8033 K and the freezing point of silver at 961.78 K, including the triple point of water at 0.060°C, the melting point of Gallium at 29.7646°C, and the freezing point of Indium at 156.5985°C (Preston-Thomas, 1990). The triple point of water is the temperature at which ice, water, and water vapour are in equilibrium. The temperature scale in kelvin *T* is related to the temperature scale in degrees Celsius *t*[°C] by:

$$t\ [°\mathrm{C}] = T\ [\mathrm{K}] - 273.15 \tag{6.6}$$

The practical temperature scale was revised in 1887, 1927, 1948, 1968, and 1990 as more accurate determinations of absolute temperature became accepted. The most recent scale is the International Temperature Scale of 1990 (its-90). It differs slightly from the International Practical Temperature Scale of 1968ipts-68. At 0°C they are the same, and above 0°C its-90 is slightly cooler. $t90 - t68 = -0.002$ at 10°C, -0.005 at 20°C, -0.007 at 30°C and -0.010 at 40°C.

Notice that while oceanographers use thermometers calibrated with an accuracy of a milli degree, which is 0.001°C, the temperature scale itself has uncertainties of a few milli degrees.

6.3 GEOGRAPHICAL DISTRIBUTION OF SURFACE TEMPERATURE AND SALINITY

The distribution of temperature at the sea surface tends to be *zonal*, that is, it is independent of longitude (figure 6.2). Warmest water is near the equator, coldest water is near the poles. The deviations from zonal are small. Equator- ward of 40°, cooler waters tend to be on the eastern side of the basin. North of this latitude, cooler waters tend to be on the western side.

The *anomalies* of sea-surface temperature, the deviation from a long term average, are small, less than 1.5°C (Harrison and Larkin, 1998) except in the equatorial Pacific where the deviations can be 3°C (figure 6.3: top).

The annual range of surface temperature is highest at mid-latitudes, especially on the western side of the ocean (figure 6.3: bottom). In the west, cold air blows off the continents in winter and cools the ocean. The cooling dominates the heat budget. In the tropics the temperature range is mostly less than 2°C.

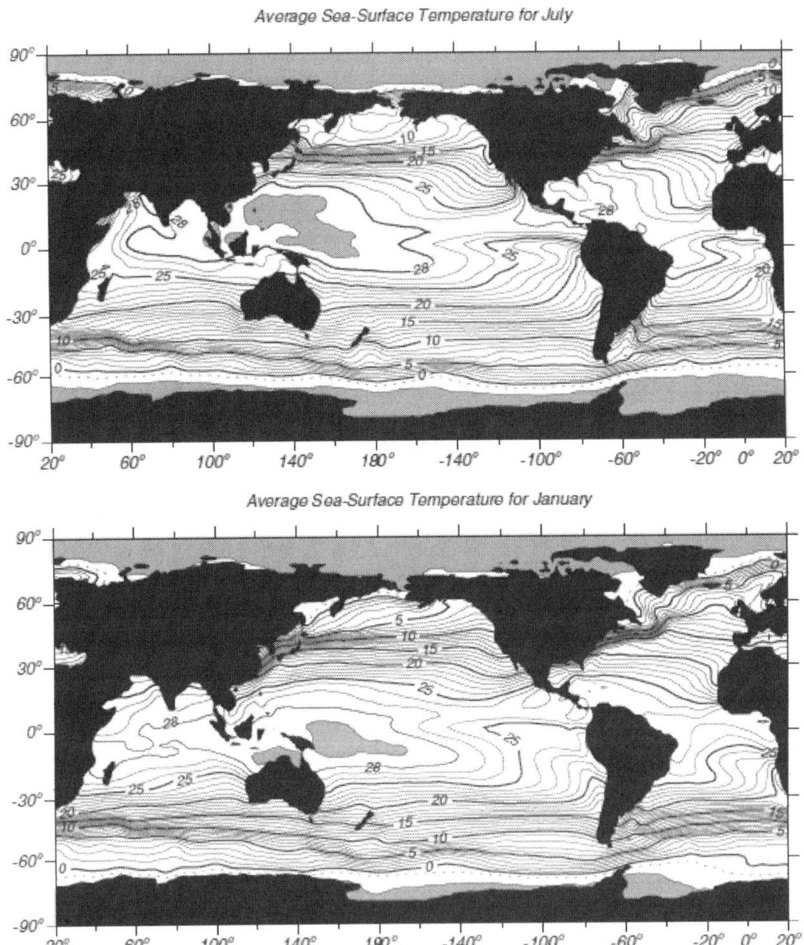

Figure 6.2 Mean sea-surface temperature calculated from the optimal interpolation technique(Reynolds and Smith, 1995) using ship reports and AVHRR measurements of temperature. Contour interval is 1°C with heavy contours every 5°C. Shaded areas exceed 29°C.

The distribution of sea-surface salinity also tends to be zonal. The saltiest waters are at mid-latitudes where evaporation is high. Less salty waters are near the equator where rain freshens the surface, and at high latitudes where melted sea ice freshens the surface (figure 6.4). The zonal (east-west) average of salinity shows a close correlation between salinity and evaporation minus precipitation plus river input (figure 6.5).

Because many large rivers drain into the Atlantic and the Arctic Sea, why is the Atlantic saltier than the Pacific? Broecker (1997) showed that 0.32 Sv of the water evaporated from the Atlantic does not fall as rain on land. Instead, it is carried by winds into the Pacific (figure 6.6). Broecker points out that the quantity is small, equivalent to a little more than the flow in the Amazon River, but "were this flux not compensated by an exchange of more salty Atlantic waters for less salty Pacific waters, the salinity of the entire Atlantic would rise about 1 gram per liter per millennium."

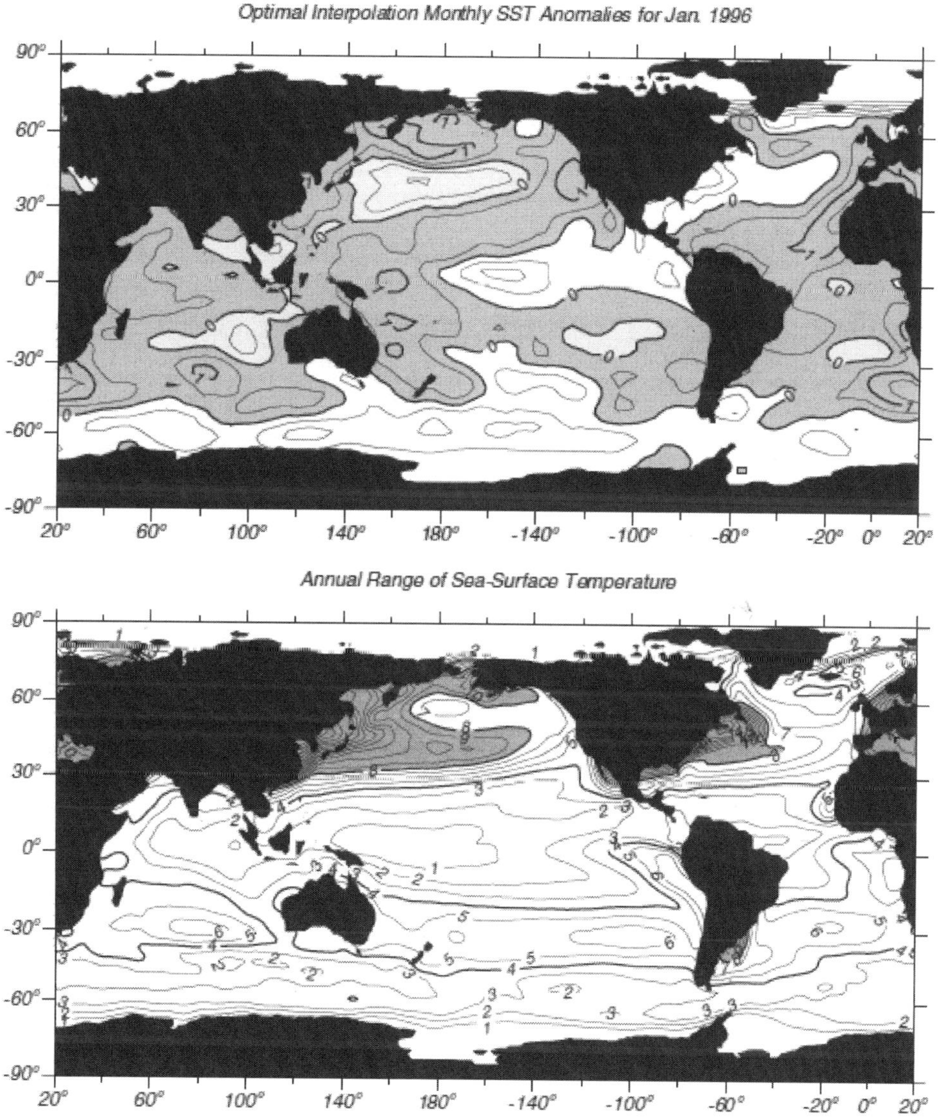

Figure 6.3 Top: Sea-surface temperature anomaly for January 1996 relative to mean temperature shown in figure 6.2 using data published by Reynolds and Smith (1995) in the *Climate Diagnostics Bulletin* for February 1995. Contour interval is 1°C. Shaded areas are positive. Bottom: Annual range of sea-surface temperature in °C calculated from the Reynolds and Smith (1995) mean sea-surface temperature data set. Contour interval is 1°C with heavy contours at 4°C and 8°C. Shaded areas exceed 8°C.

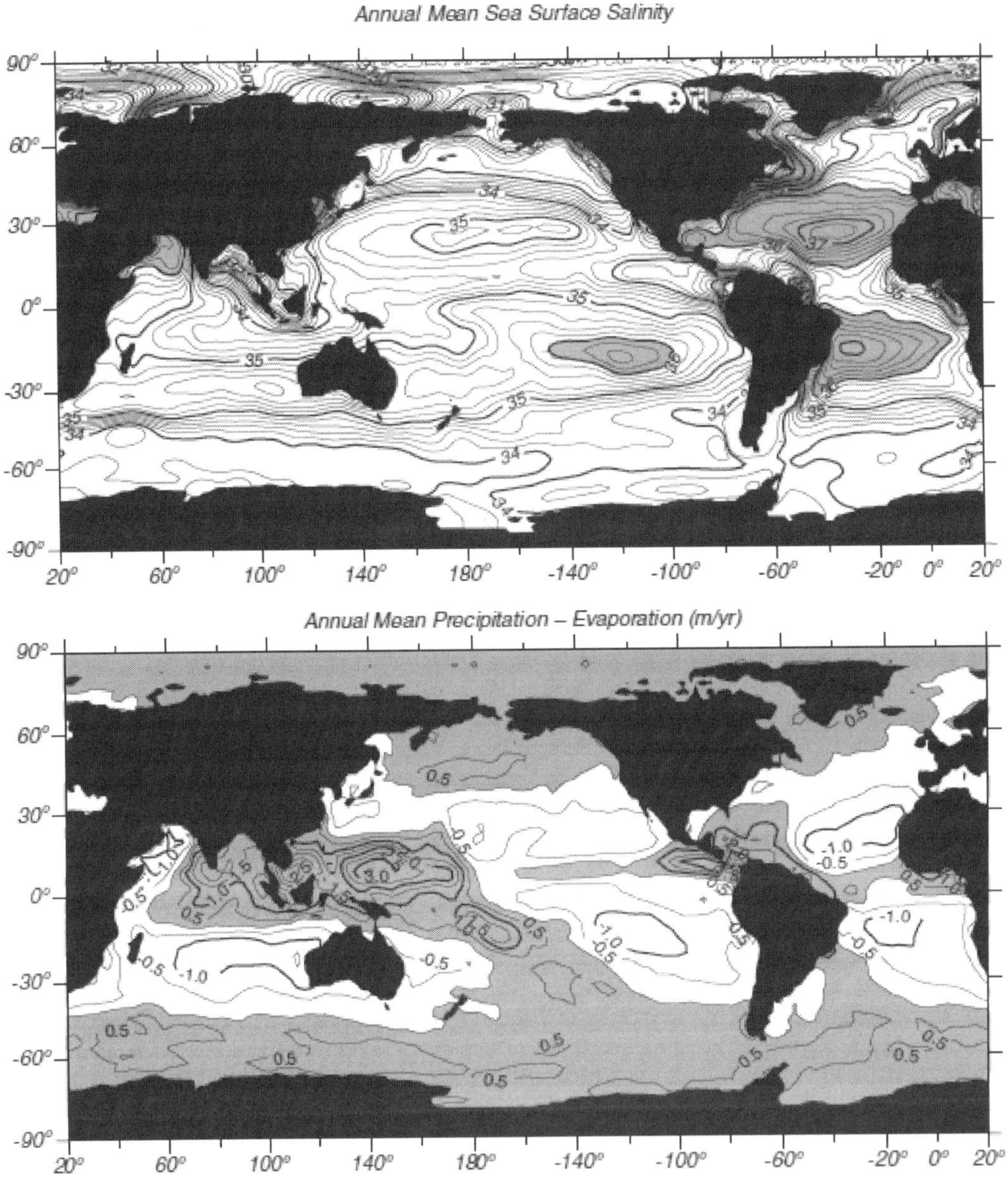

Figure 6.4 Top: Mean sea-surface salinity. Contour interval is 0.25. Shaded areas exceed a salinity of 36. From Levitus (1982). Bottom: Precipitation minus evaporation in meters per year calculated from global rainfall by the Global Precipitation Climatology Project and latent heat flux calculated by the Data Assimilation Office, both at NASA's Goddard Space Flight Center. Precipitation exceeds evaporation in the shaded regions, contour interval is 0.5 m.

Ocean's waters is: t = 3.5°C. The mean salinity is S = 34.7. The distribution about the mean is small: 50% of the water is in the range:

$$1.3°\text{C} < t < 3.8°\text{C}$$

$$34.6 < S < 34.8$$

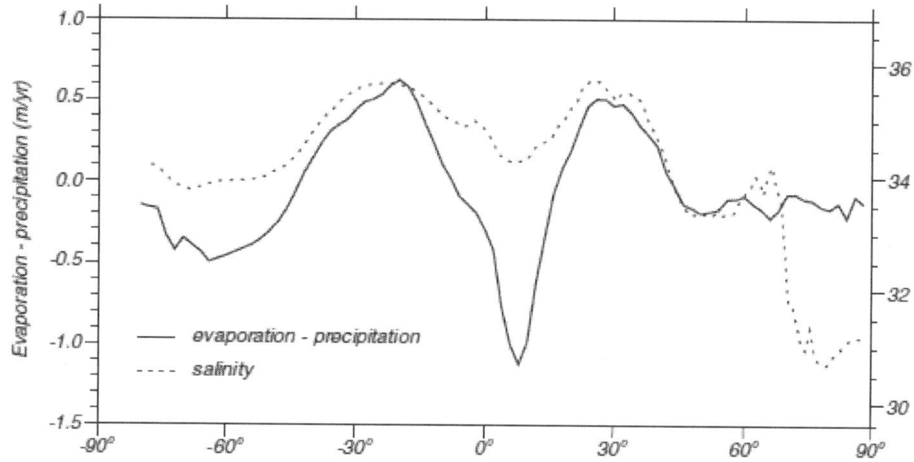

Figure 6.5 Zonal average of sea-surface salinity calculated for all the ocean from Levitus (1982) and the difference between evaporation and precipitation (E P) calculated from data shown in figure 6.4 (bottom).

6.4 THE OCEANIC MIXED LAYER AND THERMOCLINE

Wind blowing on the ocean stirs the upper layers leading to a thin *mixed layer* at the sea surface having constant temperature and salinity from the surface down to a depth where the values differ from those at the surface. The magnitude of the difference is arbitrary, but typically the temperature at the bottom of the layer must be no more than 0.02–0.1° colder than at the surface.

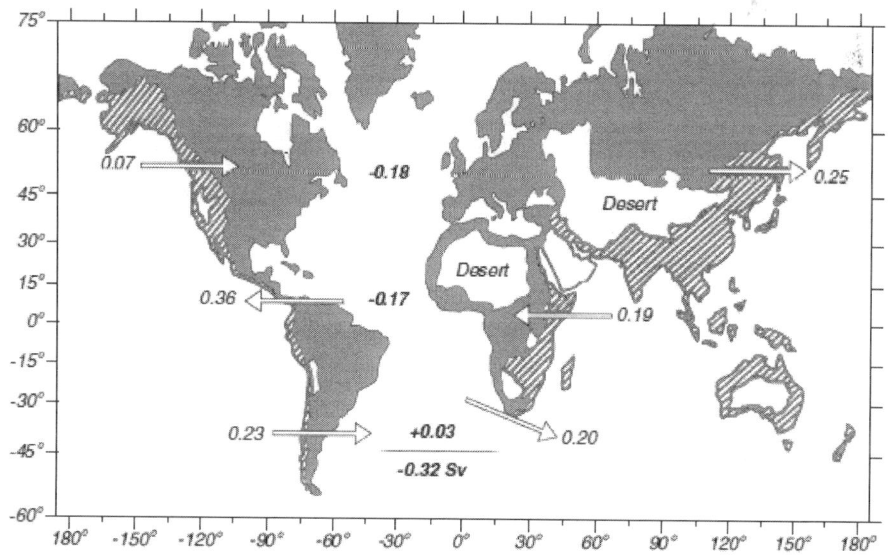

Figure 6.6 Water transported by the atmosphere into and out of the Atlantic. Basins draining into the Atlantic are black, deserts are white, and other drainage basins are shaded. Arrows give direction of water transport by the atmosphere, and values are in Sverdrups. Bold numbers give the net transport for the Atlantic at each latitude band. Overall, the Atlantic loses 0.32 Sv, an amount approximately equal to the flow in the Amazon River. After Broecker (1997).

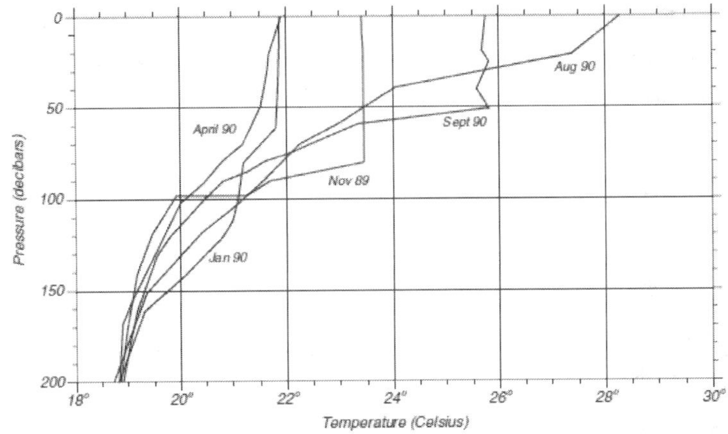

Figure 6.7 Growth and decay of the mixed layer and seasonal thermocline from November 1989 to September 1990 at the Bermuda Atlantic Time-series Station (bats) at 31.8°N 64.1°W. Data were collected by the Bermuda Biological Station for Research, Inc. Note that pressure in decibars is nearly the same as depth in meters (see §6.8 for a definition of decibars).

Note that both temperature and salinity must be constant in the mixed layer. We will see later that mean velocity is not constant. The mixed layer is roughly 10–200 m thick over most of the tropical and mid-latitude belts.

The depth and temperature of the mixed layer varies from day to day and from season to season in response to two processes:

- Heat fluxes through the surface heat and cool the surface waters. Changes in temperature change the density contrast between the mixed layer and deeper waters. The greater the contrast, the more work is needed to mix the layer downward and vice versa.
- Turbulence in the mixed layer mixes heat downward. The turbulence depends on the wind speed and on the intensity of breaking waves. Turbulence mixes water in the layer, and it mixes the water in the layer with water in the thermocline.

The mid-latitude mixed layer is thinnest in late summer when winds are weak, and sunlight warms the surface layer (figure 6.7). At times, the heating is so strong, and the winds so weak, that the layer is only a few meters thick. In fall, the first storms of the season mix the heat down into the ocean thickening the mixed layer, but little heat is lost. In winter, heat is lost, and the mixed layer continues to thicken, becoming thickest in late winter. In spring, winds weaken, sunlight increases, and a new mixed layer forms.

Below the mixed layer, water temperature decreases rapidly with depth except at high latitudes. The range of depths where the rate of change, the gradient of temperature, is large is called the *thermocline*. Because density is closely related to temperature, the thermocline also tends to be the layer where density gradient is greatest, the *pycnocline*.

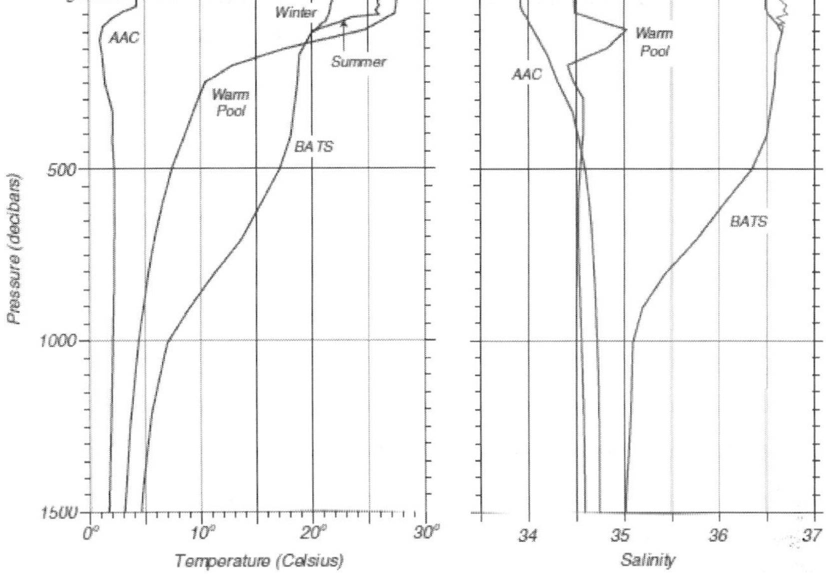

Figure 6.8 Typical temperature and salinity profiles in the open ocean. AAC: At 62.0°S, 170.0° E in the Antarctic Circumpolar Current on 16 January 1969 as measured by the *R/V Hakuho Maru*. Warm Pool: At 9.5° N 176.3° E in the tropical west Pacific warm pool on 12 March 1989 as measured by Bryden and Hall on the *R/V Moana Wave*. bats: At 31.8° N 64.1° W near Bermuda on 17 April and 10 September 1990 as measured by the Bermuda Biological Station for Research, Inc. Data are included with Java Ocean Atlas.

The shape of the thermocline varies slightly with the seasons (figure 6.7). This is the *seasonal thermocline*. The *permanent thermocline* extends from below the seasonal thermocline to depths of 1500–2000 meters (figure 6.8). At high latitudes, such as at the AAC station in the figure, there may be a cooler, fresher layer above the permanent thermocline.

The mixed layer tends to be saltier than the thermocline between 10° and 40° latitude, where evaporation exceeds precipitation. At high latitudes the mixed layer is fresher because rain and melting ice reduce salinity. In some tropical regions, such as the warm pool in the western tropical Pacific, rain also produces a thin fresher mixed layer.

6.5 DENSITY, POTENTIAL TEMPERATURE, AND NEUTRAL DENSITY

During winter, cold water formed at the surface sinks to a depth determined by its density relative to the density of the deeper water. Currents then carry the water to other parts of the ocean. At all times, the water parcel moves to stay below less dense water and above more dense water. The distribution of currents within the ocean depends on the distribution of pressure, which depends on the variations of density inside the ocean as outlined in §10.4. So, if we want to follow water movement within the ocean, we need to know the distribution of density within the ocean.

Density and sigma-t: The calculation of water movement requires measurements of density with an accuracy of a few parts per million. This is not easy.

Absolute Density of water can only be measured in special laboratories, and only with difficulty. The best accuracy is $1 : 2.5 \times 10^5 = 4$ parts per million.

To avoid the difficulty of working with absolute density, oceanographers use density relative to density of pure water. Density $\rho(S, t, p)$ is now defined using Standard Mean Ocean Water of known isotopic composition, assuming saturation of dissolved atmospheric gasses. Here S, t, p refers to salinity, temperature, and pressure.

In practice, density is not measured, it is calculated from *in situ* measurements of pressure, temperature, and conductivity using the equation of state for sea water. This can be done with an accuracy of two parts per million.

Density of water at the sea surface is typically 1027 kg/m^3. For simplification, physical oceanographers often quote only the last 2 digits of the density, a quantity they call *density anomaly* or *Sigma (S,t,p)*:

$$\sigma(S,t,p) = \rho(S,t,p) - 1000 \text{ kg/m}^3 \tag{6.7}$$

The Working Group on Symbols, Units and Nomenclature in Physical Oceanography (sun, 1985) recommends that σ be replaced by γ because σ was originally defined relative to pure water and it was dimensionless. Here, however, I will follow common practice and use σ.

If we are studying surface layers of the ocean, we can ignore compressibility, and we use a new quantity sigma-t (written σ*t*):

$$\sigma_t = \sigma(S,t,0) \tag{6.8}$$

This is the density anomaly of a water sample when the total pressure on it has been reduced to atmospheric pressure (*i.e.* zero water pressure), but the temperature and salinity are *in situ* values.

Potential Temperature: As a water parcel moves within the ocean below the mixed layer, its salt and heat content can change only by mixing with other water. Thus we can use measurements of temperature and salinity to trace the path of the water. This is best done if we remove the effect of compressibility.

As water sinks, pressure increases, the water is compressed, and the compression does work on the water. This increases the internal energy of the water. To understand how compression increases energy, consider a cube containing a fixed mass of water. As the cube sinks, its sides move inward as the cube is compressed. Recalling that work is force times distance, the work is the distance the side moves times the force exerted on the side by pressure. The change in internal energy may or may not result in a change in temperature (McDougall and Feistel, 2003). The internal energy of a fluid is the sum of molecular kinetic energy (temperature) and molecular potential energy. In sea water, the later term dominates, and the change of internal energy produces the temperature change shown in figure 6.9. At a depth of 8 km, the increase in temperature is almost 0.9°C.

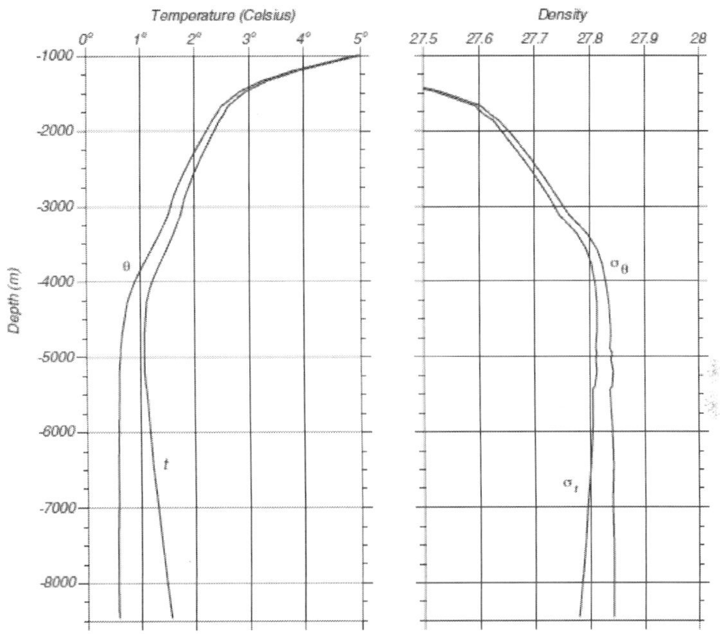

Figure 6.9 Profiles of Left *in situ* t and potential θ temperature and Right sigma-t and sigma-theta in the Kermadec Trench in the Pacific measured by the R/V Eltanin during the Scorpio Expedition on 13 July 1967 at 175.825° E and 28.258° S. Data from Warren (1973).

To remove the influence of compressibility from measurements of temperature, oceanographers (and meteorologists who have the same problem in the atmosphere) use the concept of potential temperature. *Potential temperature* Θ is defined as the temperature of a parcel of water at the sea surface after ithas been raised adiabatically from some depth in the ocean. Raising the parcel *adiabatically* means that it is raised in an insulated container so it does not exchange heat with its surroundings. Of course, the parcel is not actually brought to the surface. Potential temperature is calculated from the temperature in the water at depth, the *in situ* temperature.

Potential Density : If we are studying intermediate layers of the ocean, say at depths near a kilometre, we cannot ignore compressibility. Because changes in pressure primarily influence the temperature of the water, the influence ofpressure can be removed, to a first approximation, by using the *potential density.*

Potential density $\rho\Theta$ is the density a parcel of water would have if it were raised adiabatically to the surface without change in salinity. Written as sigma, $\sigma\Theta = \sigma(S, \Theta, 0)$ (6.9)$\sigma\Theta$ is especially useful because it is a conserved thermodynamic property.

Potential density is not useful for comparing density of water at great depths. If we bring water parcels to the surface and compare their densities, the calculation of potential density ignores the effect of pressure on the coefficients for thermal and salt expansion. As a result, two water samples having the same density but different temperature and salinity at a depth of four kilometres can have noticeably different potential density. In some regions the use of $\rho(\Theta)$ can lead to an apparent decrease of density with depth (figure 6.10) although we know that this is not possible because such a column of water would be unstable.

To compare samples from great depths, it is better to bring both samples to a nearby depth instead of to the surface $p = 0$. For example, we can bring both parcels to a pressure of 4,000 decibars, which is near a depth of 4 km:

$$\sigma_4 = \sigma(S, \Theta, 4000) \qquad (6.10)$$

where $\sigma 4$ is the density of a parcel of water brought adiabatically to a pressure of 4,000 decibars. More generally, oceanographers sometimes use ρr

$$\sigma_r = \sigma(S, \Theta, p, p_r) \qquad (6.11)$$

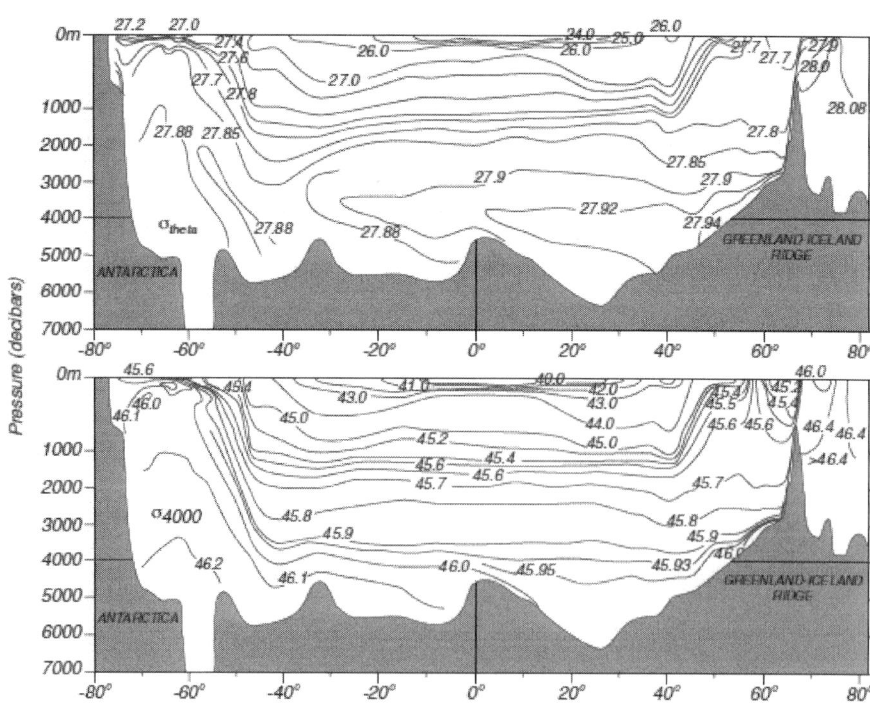

Figure 6.10 Vertical sections of density in the western Atlantic. Note that the depth scale changes at 1000 m depth. Upper: σΘ, showing an apparent density inversion below 3,000 m. Lower: σ4 showing continuous increase in density with depth. After Lynn and Reid (1968).

where p is pressure, and pr is pressure at some reference level. In (6.8) the level is $pr = 0$ decibars, and in (6.9) $pr = 4000$ decibels.

The use of σ_r leads to problems. If we wish to follow parcels of water deep in the ocean, we might use σ_3 in some areas, and σ_4 in others. But what happens when a parcel moves from a depth of 3 km in one area to a depth of 4 km in another? There is a small discontinuity between the density of the parcel expressed as σ_3 compared with density expressed as σ_4. To avoid this difficulty, Jackett and McDougall (1997) proposed a new variable they called neutral density.

Neutral Surfaces and Density: A parcel of water moves locally along a path of constant density so that it is always below less dense water and above more dense water. More precisely, it moves along a path of constant potential density σr referenced to the local depth r. Such a path is called a *neutral path* (Eden and Will brand, 1999). A *neutral surface element* is the surface tangent to the neutral paths through a point in the water. No work is required to move a parcel on this surface because there is no buoyancy force acting on the parcel as it moves (if we ignore friction).

Now let's follow the parcel as it moves away from a local region. At first we might think that because we know the tangents to the surface everywhere, we can define a surface that is the envelope of the tangents. But an exact surface is not mathematically possible in the real ocean, although we can come very close. Jackett and McDougall (1997) developed a practical neutral density variable γ^n and surface that stays within a few tens meters of an ideal surface anywhere in the world. They constructed their variables using data in the Levitus (1982) atlas. The neutral density values were then used to label the data in the Levitus atlas. This pre-labelled data set is used to calculate γ^n at new locations where t, S are measured as a function of depth by interpolation to the four closest points in the Levitus atlas. Through this practice, neutral density γ^n is a function of salinity S, *in situ* temperature t, pressure p, longitude, and latitude.

The neutral surface defined above differs only slightly from an ideal neutral surface. If a parcel moves around a gyre on the neutral surface and returns to its starting location, its depth at the end will differ by around 10 meters from the depth at the start. If potential density surfaces are used, the difference can be hundreds of meters, a far larger error.

Equation of state of sea water: Density of sea water is rarely measured. Density is calculated from measurements of temperature, conductivity, or salinity, and pressure using the equation of state of sea water. The equation of state is an equation relating density to temperature, salinity, and pressure.

The equation is derived by fitting curves through laboratory measurements of density as a function of temperature, pressure, and salinity, chlorinity, or conductivity. The International Equation of State (1980) published by the Joint Panel on Oceanographic Tables and Standards (1981) is now used. See also Millero and Poisson (1981) and Millero et al (1980). The equation has an accuracy of 10 parts per million, which is 0.01 units of $\sigma(\Theta)$.

I have not actually written out the equation of state because it consists of three polynomials with 41 constants (JPOTS, 1991).

Accuracy of Temperature, Salinity, and Density: If we want to distinguish between different water masses in the ocean, and if the total range of temperature and salinity is as small as the range in figure 6.1, then we must measure temperature, salinity, and pressure very carefully. We will need an accuracy of a few parts per million.

Such accuracy can be achieved only if all quantities are carefully defined, if all measurements are made with great care, if all instruments are carefully calibrated, and if all work is done according to internationally accepted standards.

The standards are laid out in *Processing of Oceanographic Station Data* (JPOTS, 1991) published by UNESCO. The book contains internationally accepted definitions of primary variables such as temperature and salinity and methods for the measuring the primary variables. It also describes accepted methods for calculating quantities derived from primary variables, such as potential temperature, density, and stability.

6.6 MEASUREMENT OF TEMPERATURE

Temperature in the ocean is measured many ways. Thermistors and mercury thermometers are commonly used on ships and buoys. These are calibrated in the laboratory before being used, and after use if possible, using mercury or platinum thermometers with accuracy traceable to national standards laboratories. Infrared radiometers on satellites measure the ocean's surface temperature.

Mercury Thermometer: This is the most widely used, non-electronic thermometer. It was widely used in buckets dropped over the side of a ship to measure the temperature of surface waters, on Nansen bottles to measure sub-sea temperatures, and in the laboratory to calibrate other thermometers. Accuracy of the best thermometers is about ±0.001°C with very careful calibration.

One very important mercury thermometer is the reversing thermometer (figure 6.11) carried on Nansen bottles, which are described in the next section. It is a thermometer that has a constriction in the mercury capillary that causes the thread of mercury to break at a precisely determined point when the thermometer is turned upside down. The thermometer is lowered deep into the ocean in the normal position, and it is allowed to come to equilibrium with the water. Mercury expands into the capillary, and the amount of mercury in the capillary is proportional to temperature. The thermometer is then flipped upside down, the thread of mercury breaks trapping the mercury in the capillary, and the thermometer is brought back.

The mercury in the capillary of the reversed thermometer is read on deck along with the temperature of a normal thermometer, which gives the temperature at which the reversed thermometer is read. The two readings give the temperature of the water at the depth where the thermometer was reversed.

The reversing thermometer is carried inside a glass tube which protects the thermometer from the ocean's pressure because high pressure can squeeze additional mercury into the capillary. If the thermometer is unprotected, the apparent temperature read on deck is proportional to temperature and pressure at the depth where the thermometer was flipped. A pair of protected and unprotected thermometers gives temperature and pressure of the water at the depth the thermometer was reversed.

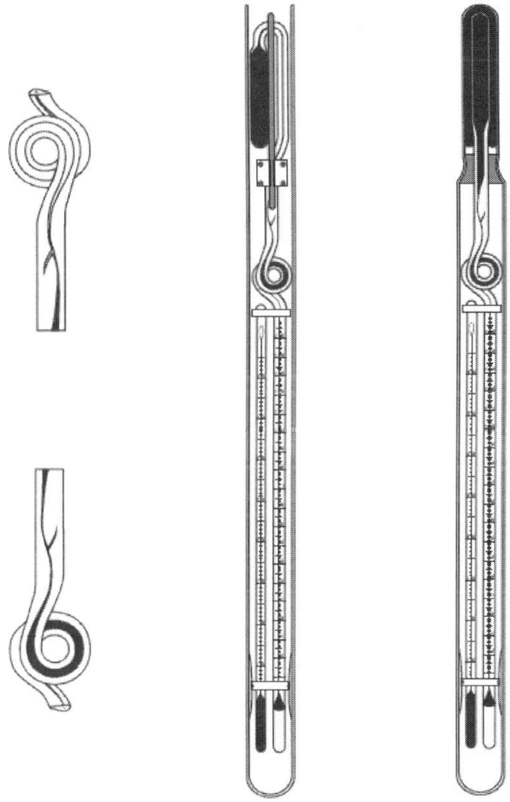

Figure 6.11 Left: Protected and unprotected reversing thermometers is set position, before reversal. Right: The constricted part of the capillary in set and reversed positions. After von Arx (1962: 259).

Pairs of reversing thermometers carried on Nansen bottles were the primary source of sub-sea measurements of temperature as a function of pressure from around 1900 to 1970.

Platinum Resistance Thermometer : This is the standard for temperature. It is used by national standards laboratories to interpolate between defined points on the practical temperature scale. It is used primarily to calibrate other temperature sensors.

Thermistor : A thermistor is a semiconductor having resistance that varies rapidly and predictably with temperature. It has been widely used on moored instruments and on instruments deployed from ships since about 1970. It has high resolution and an accuracy of about $\pm 0.001^{\circ}$C when carefully calibrated.

Bucket temperatures: The temperature of surface waters has been routinely measured at sea by putting a mercury thermometer into a bucket which is lowered into the water, letting it sit at a depth of about a meter for a few minutes until the thermometer comes to equilibrium, then bringing it aboard and reading the temperature before water in the bucket has time to change temperature. The accuracy is around 0.1°C. This is a very common source of direct surface temperature measurements.

Ship Injection Temperature : The temperature of the water drawn into the ship to cool the engines has been recorded routinely for decades. These recorded values of temperature are called injection temperatures. Errors are due to ship's structure warming water before it is recorded. This happens when the temperature recorder is not placed close to the point on the hull where water is brought in. Accuracy is 0.5°–1°C.

Advanced Very High Resolution Radiometer : The most commonly used instrument to measure sea-surface temperature from space is the Advanced Very High Resolution Radiometer AVHRR. The instrument has been carried on all polar-orbiting meteorological satellites operated by NOAA since Tiros-N was launched in 1978. The instrument was originally designed to measure cloud temperatures and hence cloud height. The instrument had, however, sufficient accuracy and precision that it was soon used to measure regional and global temperature patterns at the sea surface.

The instrument is a radiometer that converts infrared radiation into an electrical voltage. It includes a mirror that scans from side to side across the sub-satellite track and reflects radiance from the ground into a telescope, a telescope that focuses the radiance on detectors, detectors sensitive to different wavelengths that convert the radiance at those wavelengths into electrical signals, and electronic circuitry to digitize and store the radiance values. The instruments observes a 2700-km wide swath centered on the sub-satellite track. Each observation along the scan is from a pixel that is roughly one kilometre in diameter near the center of the scan and that increases in size with distance from the sub-satellite track.

The radiometers measures infrared radiation emitted from the surface in five wavelength bands: three infrared bands: 3.55–3.99 μm, 10.3–11.3 μm, and 11.5–12.5

μm; a near-infrared band at 0.725–1.10 μm; and a visible-light band at 0.55–0.90 μm. All infrared bands include radiation emitted from the sea and from water vapour in the air along the path from the satellite to the ground. The 3.7 μm band is least sensitive to water vapour and other errors, but it works only at night because sunlight has radiance in this band. The two longest wavelength bands at 10.8 μm and 12.0 μm are used to observe sea-surface temperature and water vapour along the path in daylight.

Data with 1-km resolution are transmitted directly to ground stations that view the satellite as it passes the station. This is the Local Area Coverage mode. Data are also averaged to produce observations from 4 × 4 km pixels.

These data are stored by the satellite and later transmitted to NOAA receiving stations. This is the Global Area Coverage mode.

The swath width is sufficiently wide that the satellite views the entire earth twice per day, at approximately 09:00 AM and 9:00 PM local time. Areas at high latitudes may be observed as often as eight or more times per day.

The most important errors are due to:

- Unresolved or undetected clouds: Large, thick clouds are obvious in the images of water temperature Thin clouds such as low stratus and high cirrus produce much small errors that are difficult or almost impossible to detect. Clouds smaller in diameter than 1 km, such as trade-wind cumuli, are also difficult to detect. Special techniques have been developed for detecting small clouds (figure 6.12).

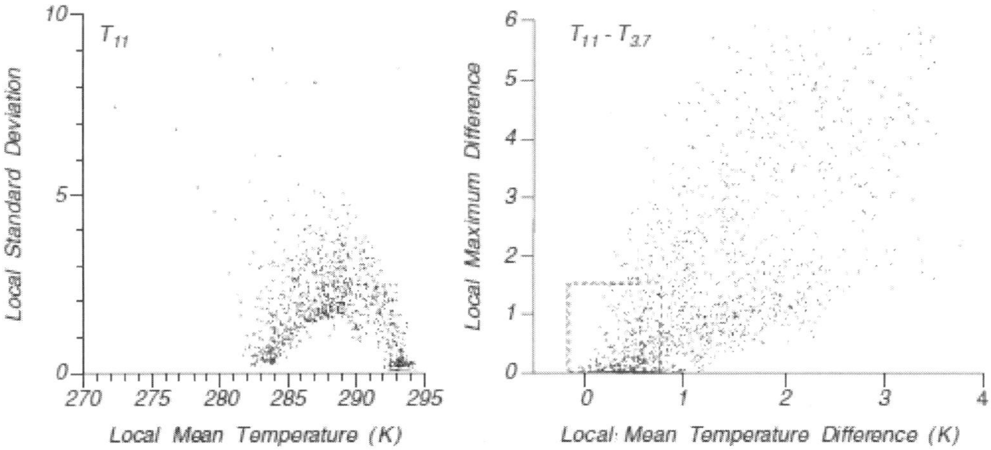

Figure 6.12 The influence of clouds on infrared observations. Left: The standard deviation of the radiance from small, partly cloudy areas each containing 64 pixels. The feet of the arch-like distribution of points are the sea-surface and cloud-top temperatures. After Coakley and Bretherton (1982). Right: The maximum difference between local values of *T*11 - *T*3.7 and the local mean values of the same quantity. Values inside the dashed box indicate cloud-free pixels. *T*11 and *T*3.7 are the apparent temperatures at 11.0 and 3.7 μm (data from K. Kelly). After Stewart (1985: 137).

- Water vapour, which absorbs part of the energy radiated from the sea surface: Water vapour reduces the apparent temperature of the sea surface. The influence is different in the 10.8 μm and 12.0 μm channels, allowing the difference in the two signals to be used to reduce the error.
- Aerosols, which absorb infrared radiation. They radiate at temperatures found high in the atmosphere. Stratospheric aerosols generated by volcanic eruptions can lower the observed temperatures by up to a few degrees Celsius. Dust particles carried over the Atlantic from Saharan dust storms can also cause errors.
- Skin temperature errors. The infrared radiation seen by the instrument comes from a layer at the sea surface that is only a few micrometers thick. The temperature in this layer is not quite the same as temperature a meter below the sea surface. They can differ by several degrees when winds are light (Emery and Schussel, 1989). This error is greatly reduced when AVHRR data are used to interpolate between ship measurements of surface temperature.

Maps of temperature processed from Local Area Coverage of cloud-free regions show variations of temperature with a precision of 0.1°C. These maps are useful for observing local phenomena including patterns produced by local currents. Figure 10.16 shows such patterns off the California coast.

Global maps are made by the U.S. Naval Oceanographic Office, which receives the global AVHRR data directly from NOAA's National Environmental Satellite, Data and Information Service in near-real time each day. The data are carefully processed to remove the influence of clouds, water vapour, aerosols, and other sources of error. Data are then used to produce global maps between±70° with an accuracy of ±0.6°C (May et al 1998). The maps of sea-surface temperature are sent to the U.S. Navy and to NOAA's National Centers for Environmental Prediction. In addition, the office produces daily 100-km global and 14-km regional maps of temperature.

Global Maps of Sea-Surface Temperature: Global, monthly maps of surface temperature are produced by the National Centers for Environmental Prediction using Reynolds et al (2002) optimal-interpolation method. The technique blends ship and buoy measurements of sea-surface temperature with AVHRR data processed by the Naval Oceanographic Office in 1°areas for a month. Essentially, AVHRR data are interpolated between buoy and ship reports using previous information about the temperature field. Overall accuracy ranges from approximately ±0.3° C in the tropics to ±0.5° C near western boundary currents in the northern hemisphere where temperature gradients are large. Maps are available from November 1981. Figures 6.2–6.4 were made by NOAA using Reynolds' technique. Other data sets have been produced by the NOAA/NASA Pathfinder program (Kilpatrick, Podesta, and Evans, 2001).

Maps of mean temperature have also been made from ICOADS data (Smith and Reynolds, 2004). Because the data are poorly distributed in time and space, errors also vary in time and space. Smith and Reynolds (2004) estimated the error in the global mean temperature and found the 95% confidence uncertainty for the near-global average is 0.48° C or more in the nineteenth century, near 0.28° C for the first half of the twentieth century, and 0.18° C or less after1950. Anomalies of sea-surface temperature were calculated using mean sea- surface temperature from the period 1854–1997 using ICOADS supplemented with satellite data since 1981.

6.7 MEASUREMENT OF CONDUCTIVITY OR SALINITY

Conductivity is measured by placing platinum electrodes in seawater and measuring the current that flows when there is a known voltage between the electrodes. The current depends on conductivity, voltage, and volume of sea water in the path between electrodes. If the electrodes are in a tube of non-conducting glass, the volume of water is accurately known, and the current is independent of other objects near the conductivity cell (figure 6.13). The best measurements of salinity from conductivity give salinity with an accuracy of±0.005.

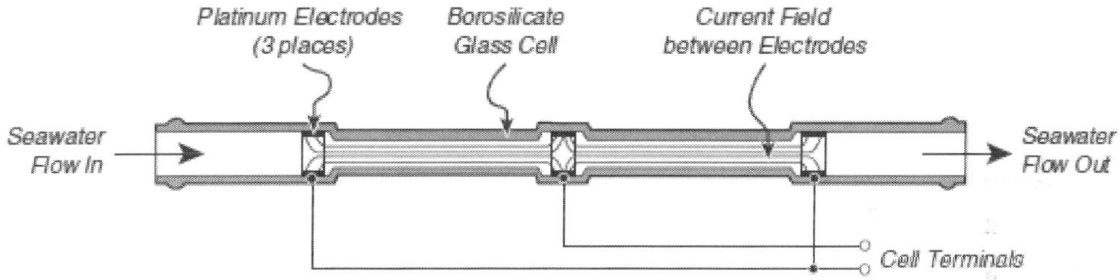

Figure 6.13 A conductivity cell. Current flows through the seawater between platinum electrodes in a cylinder of borosilicate glass 191 mm long with an inside diameter between the electrodes of 4 mm. The electric field lines (solid lines) are confined to the interior of the cell in this design making the measured conductivity (and instrument calibration) independent of objects near the cell. This is the cell used to measure conductivity and salinity shown in figure 6.15. From Sea-Bird Electronics.

Before conductivity measurements were widely used, salinity was measured using chemical titration of the water sample with silver salts. The best measurements of salinity from titration give salinity with an accuracy of ±0.02.

Individual salinity measurements are calibrated using standard seawater. Long-term studies of accuracy use data from measurements of deep water masses of known, stable, salinity. For example, Saunders (1986) noted that temperature is very accurately related to salinity for a large volume of water contained in the deep basin of the northwest Atlantic under the Mediterranean outflow. He used the consistency of measurements of temperature and salinity made at many hydrographic stations in the area to estimate the accuracy of temperature, salinity and oxygen measurements. He concluded that the most careful measurements made since 1970 have an accuracy of 0.005 for salinity and 0.005°C for temperature. The largest source of salinity error was the error in determination of the standard water used for calibrating the salinity measurements.

Gouretski and Jancke (1995) estimated accuracy of salinity measurements as a function of time. Using high quality measurements from 16,000 hydrographic stations in the south Atlantic from 1912 to 1991, they estimated accuracy by plotting salinity as a function of temperature using all data collected below 1500 m in twelve regions for each decade from 1920 to 1990. A plot of accuracy as a function of time since 1920 shows consistent improvement in accuracy since 1950 (figure 6.14). Recent measurements of salinity are the most accurate. The standard deviation of salinity data collected from all areas in the south Atlantic from 1970 to 1993 adjusted as described by Gouretski and Jancke (1995) was 0.0033. Recent instruments such as the Sea-Bird Electronics Model 911 Plus have an accuracy of better than 0.005 without adjustments. A comparison of salinity measured at 43° 10′N, 14° 4.5′W by the 911 Plus with historic data collected by Saunders (1986) gives an accuracy of 0.002 (figure 6.15).

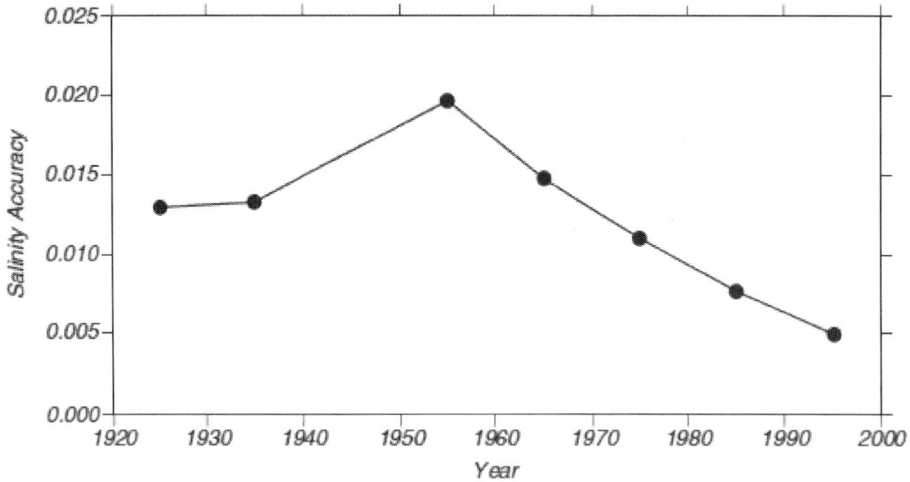

Figure 6.14. Standard deviation of salinity measurements below 1500 m in the south Atlantic. Each point is the average for the decade centered on the point. The value for 1995 is an estimate of the accuracy of recent measurements. From Gouretski and Jancke (1995).

6.8 MEASUREMENT OF PRESSURE

Pressure is routinely measured by many different types of instruments. The SI unit of pressure is the pascal (Pa), but oceanographers normally report pressure in decibars (dbar), where:

$$1 \text{ dbar} = 10^4 \text{ Pa} \tag{6.12}$$

because the pressure in decibars is almost exactly equal to the depth in meters. Thus 1000 dbar is the pressure at a depth of about 1000 m.

$$1 \text{ dbar} = 10^4 \text{ Pa} \tag{6.12}$$

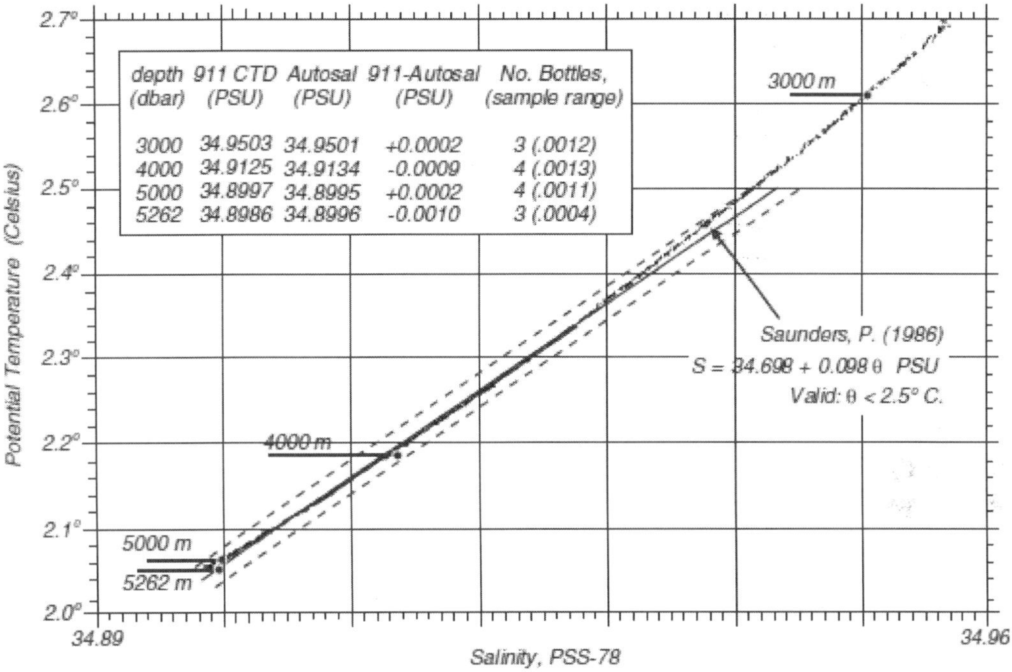

Figure 6.15. Results from a test of the Sea-Bird Electronics 911 Plus CTD in the North Atlantic Deep Water in 1992. Data were collected at 43.17° N and 14.08° W from the R/V Poseidon. From Sea-Bird Electronics (1992).

Strain Gage : This is the simplest and cheapest instrument, and it is widely used. Accuracy is about ±1%.

Vibratron: Much more accurate measurements of pressure can be made by measuring the natural frequency of a vibrating tungsten wire stretched in a magnetic field between diaphragms closing the ends of a cylinder. Pressure distorts the diaphragm, which changes the tension on the wire and its frequency. The frequency can be measured from the changing voltage induced as the wire vibrates in the magnetic field. Accuracy is about ±0.1%, or better when temperature controlled. Precision is 100–1000 times better than accuracy. The instrument is used to detect small changes in pressure at great depths. Snodgrass (1964) obtained a precision equivalent to a change in depth of ±0.8 mmat a depth of 3 km.

Quartz crystal : Very accurate measurements of pressure can also be made by measuring the natural frequency of a quartz crystal cut for minimum temperature dependence. The best accuracy is obtained when the temperature of the crystal is held constant. The accuracy is ±0.015%, and precision is ±0.001% of full-scale values.

Quartz Bourdon Gage: It has accuracy and stability comparable to quartz crystals. It too is used for long-term measurements of pressure in the deep sea.

6.9 MEASUREMENT OF TEMPERATURE AND SALINITY WITH DEPTH

Temperature, salinity, and pressure are measured as a function of depth using various instruments or techniques, and density is calculated from the measurements.

Bathythermograph (BT): was a mechanical device that measured temperature vs depth on a smoked glass slide. The device was widely used to map the thermal structure of the upper ocean, including the depth of the mixed layer before being replaced by the expendable bathythermograph in the 1970s.

Expendable Bathythermograph (XBT): is an electronic device that measures temperature vs depth using a thermistor on a free-falling streamlined weight. The thermistor is connected to an ohm-meter on the ship by a thin copper wire that is spooled out from the sinking weight and from the moving ship. The xbt is now the most widely used instrument for measuring the thermal structure of the upper ocean. Approximately 65,000 are used each year.

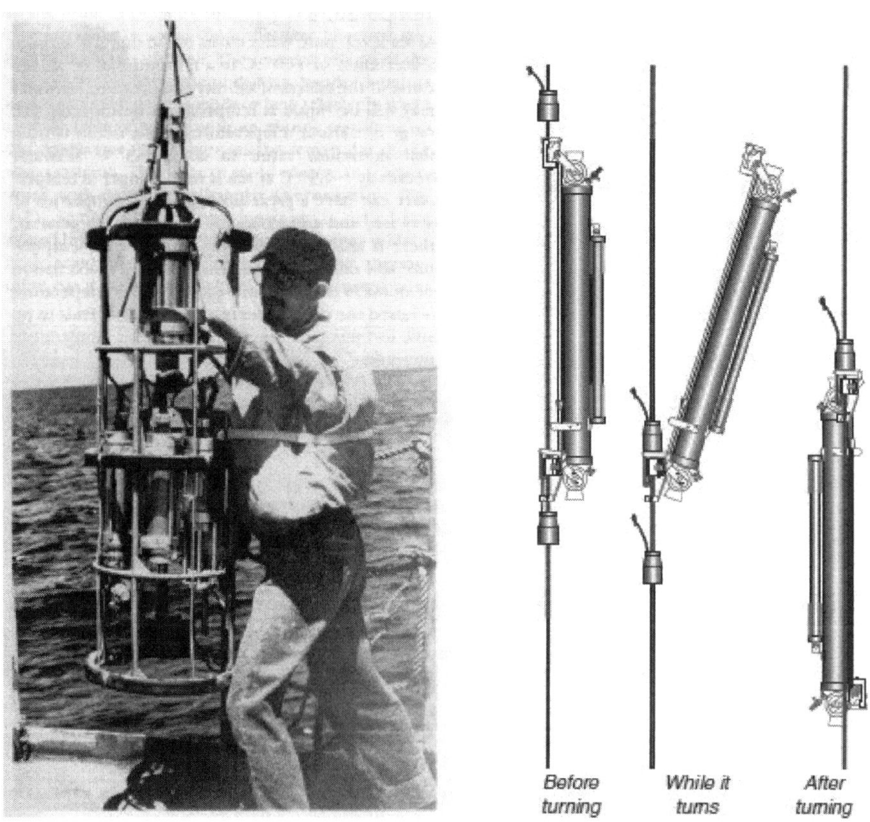

Figure 6.16 Left A CTD ready to be lowered over the side of a ship. From Davis (1987). Right Nansen water bottles before (I), during (II), and after (III) reversing. Both instruments are shown at close to the same scale. After Defant (1961: 33).

The streamlined weight falls through the water at a constant velocity. So depth can be calculated from fall time with an accuracy of ±2%. Temperature accuracy is

±0.1°C. And, vertical resolution is typically 65 cm. Probes reach to depths of 200 m to 1830 m depending on model.

Nansen Bottles : (figure 6.16) were deployed from ships stopped at hydrographic stations. *Hydrographic stations* are places where oceanographers measure water properties from the surface to some depth, or to the bottom, using instruments lowered from a ship. Usually 20 bottles were attached at intervals of a few tens to hundreds of meters to a wire lowered over the side of the ship. The distribution with depth was selected so that most bottles are in the upper layers of the water column where the rate of change of temperature in the vertical is greatest. A protected reversing thermometer for measuring temperature was attached to each bottle along with an unprotected reversing thermometer for measuring depth. The bottle contains a tube with valves on each end to collect sea water at depth. Salinity was determined by laboratory analysis of water sample collected at depth.

After bottles had been attached to the wire and all had been lowered to their selected depths, a lead weight was dropped down the wire. The weight tripped a mechanism on each bottle, and the bottle flipped over, reversing the thermometers, shutting the valves and trapping water in the tube, and releasing another weight. When all bottles had been tripped, the string of bottles was recovered. The deployment and retrieval typically took several hours.

CTD Mechanical instruments on Nansen bottles were replaced beginning in the 1960s by an electronic instrument, called a CTD, that measured conductivity, temperature, and depth (figure 6.16). The measurements are recorded in digital form either within the instrument as it is lowered from a ship or on the ship. Temperature is usually measured by a thermistor. Conductivity is measured by a conductivity cell. Pressure is measured by a quartz crystal. Modern instruments have accuracy summarized in table 6.2.

Table 6.2 Summary of Measurement Accuracy

Variable	Range	Best Accuracy
Temperature Salinity	42 °C	± 0.001 °C
Pressure Density	1	± 0.02 by titration
		± 0.005 by conductivity
	10,00 dbar	± 0.65 dbar
	2 kg/m^3	± 0.005 kg/m^3
Equation of State		± 0.005 kg/m^3

CTD on Drifters : Perhaps the most common source of temperature and salinity as a function of depth in the upper two kilometers of the ocean is the set of profiling argo floats described in §11.8. The floats drift at a depth of 1 km, sinkto 2 km, then rise to the surface. They profile temperature and salinity while changing depth using instruments very similar to those on a CTD. Data are sent to shore via the Argos system on the NOAA polar-orbiting satellites. In 2006,nearly 2500 floats were producing one profile every 10 days throughout most of the ocean. The accuracy of data from the floats is 0.005°C for temperature, 5decibars for pressure, and 0.01 for salinity (Riser et al (2008).

Data Sets : Data are in the Marine Environment and Security For European Area mersea Enact/Ensembles (en3 Quality Controlled in situ Ocean Temparature and Salinity Profiles database. As of 2008 the database contained about one million xbt profiles, 700,000 CTD profiles, 60,000 argos profiles, 1,100,000 Nansen bottle data of high quality in the upper 700 m of the ocean (Domingues et al, 2008).

6.10 LIGHT IN THE OCEAN AND ABSORPTION OF LIGHT

Sunlight in the ocean is important for many reasons: It heats sea water, warming the surface layers; it provides energy required by phytoplankton; it is used for navigation by animals near the surface; and reflected subsurface light is used for mapping chlorophyll concentration from space.

Light in the ocean travels at a velocity equal to the velocity of light in a vacuum divided by the index of refraction (n), which is typically $n = 1.33$.Hence the velocity in water is about 2.25×10^8 m/s. Because light travels slower in water than in air, some light is reflected at the sea surface. For light shining straight down on the sea, the reflectivity is $(n - 1)^2/(n + 1)^2$. For seawater, the reflectivity is 0.02 = 2%. Hence most sunlight reaching the sea surface is transmitted into the sea, little is reflected. This means that sunlight incident on the ocean in the tropics is mostly absorbed below the sea surface.

The rate at which sunlight is attenuated determines the depth which is lighted and heated by the sun. Attenuation is due to absorption by pigments and scattering by molecules and particles. Attenuation depends on wavelength. Blue light is absorbed least, red light is absorbed most strongly. Attenuation per unit distance is proportional to the radiance or the irradiance of light:

$$\frac{dI}{dx} = -c\,I \qquad (6.13)$$

where x is the distance along beam, c is an attenuation coefficient (figure 6.17), and I is irradiance or radiance.

Radiance is the power per unit area per solid angle. It is useful for describing the energy in a beam of light coming from a particular direction. Sometimes we want to know how much light reaches some depth in the ocean regardless of which direction it is going. In this case we use *irradiance*, which is the power per unit area of surface.

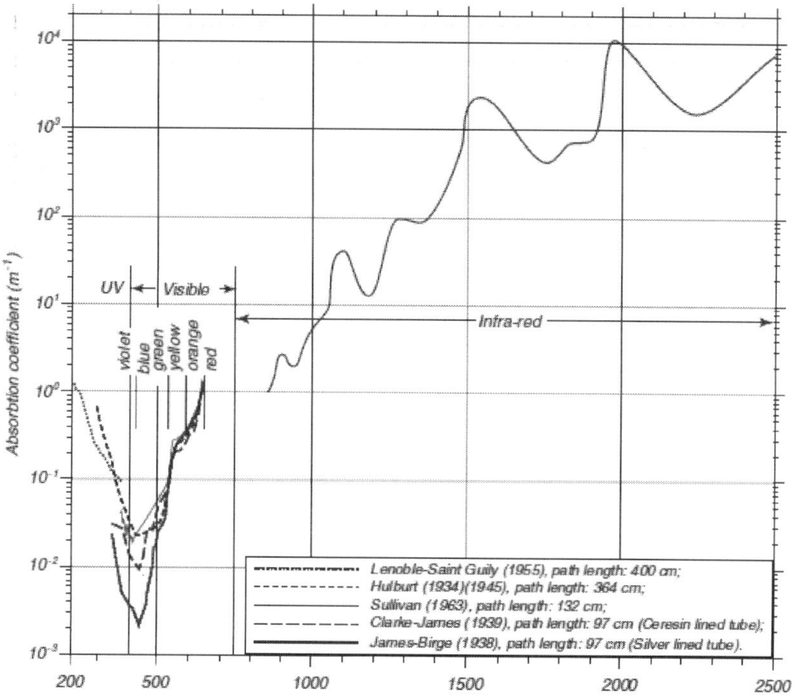

Figure 6.17 Absorption coefficient for pure water as a function of wavelength λ of the radiation. Redrawn from Morel (1974: 18, 19). See Morel (1974) for references.

If the absorption coefficient is constant, the light intensity decreases exponentially with distance.

$$I_2 = I_1 \exp(-cx) \tag{6.14}$$

where I_1 is the original radiance or irradiance of light, and I_2 is the radiance or irradiance of light after absorption.

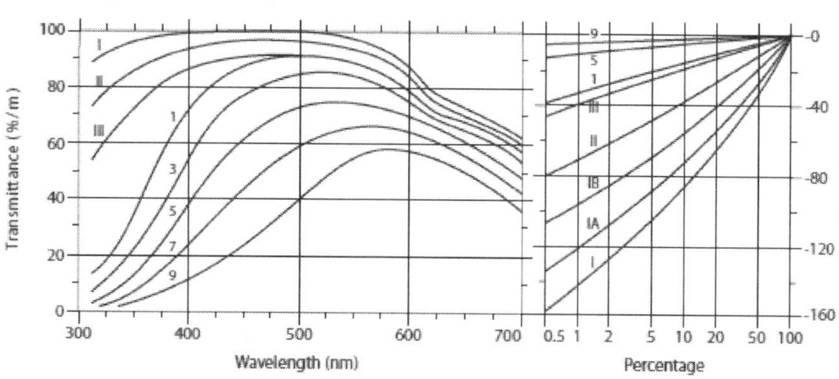

Figure 6.18 Left: Transmittance of daylight in the ocean in % per meter as a function of wavelength. I: extremely pure ocean water; II: turbid tropical-subtropical water; III: mid-latitude water; 1-9: coastal waters of increasing turbidity. Incidence angle is 90° for the first three cases, 45° for the other cases. Right: Percentage of 465 nm light reaching indicated depths for the same types of water. After Jerlov (1976).

Clarity of Ocean Water Sea water in the middle of the ocean is very clear— clearer than distilled water. These waters are a very deep, cobalt, blue—almost black. Thus the strong current which flows northward offshore of Japan carrying very clear water from the central Pacific into higher latitudes is known as the Black Current, or Kuroshio in Japanese. The clearest ocean water is called Type I waters by Jerlov (figure 6.18). The water is so clear that 10% of the light transmitted below the sea surface reaches a depth of 90 m.

In the subtropics and mid-latitudes closer to the coast, sea water contains more phytoplankton than the very clear central-ocean waters. Chlorophyll pigments in phytoplankton absorb light, and the plants themselves scatter light. Together, the processes change the colour of the ocean as seen by observer looking downward into the sea. Very productive waters, those with high concentrations of phytoplankton, appear blue-green or green (figure 6.19). On clear days the colour can be observed from space. This allows ocean-colour scanners, such as those on SeaWiFS, to map the distribution of phytoplankton over large areas.

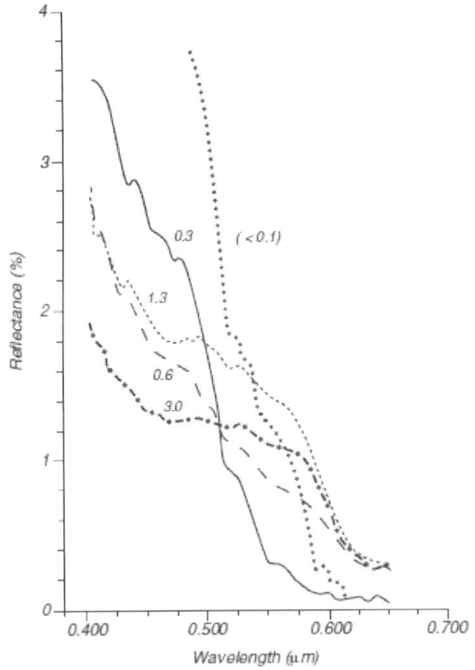

Figure 6.19 Spectral reflectance of sea water observed from an aircraft flying at 305 m over waters of different colours in the Northwest Atlantic. The numerical values are the average chlorophyll concentration in the euphotic (sunlit) zone in units of mg/m^3. The reflectance is for vertically polarized light observed at Brewster's angle of 53°. This angle minimizes reflected skylight and emphasizes the light from below the sea surface. After Clarke, Ewing, and Lorenzen (1970).

As the concentration of phytoplankton increases, the depth where sunlight is absorbed in the ocean decreases. The more turbid tropical and mid-latitude waters are classified as type II and III waters by Jerlov (figure 6.18). Thus the depth where sunlight warms the water depends on the productivity of the waters. This complicates the calculation of solar heating of the mixed layer.

Coastal waters are much less clear than waters offshore. These are the type 1–9 waters shown in figure 6.18. They contain pigments from land, sometimes called gelbstoffe, which just means yellow stuff, muddy water from rivers, and mud stirred up by waves in shallow water. Very little light penetrates more than a few meters into these waters.

Measurement of Chlorophyll from Space : The colour of the ocean, and hence the chlorophyll concentration in the upper layers of the ocean has been measured by the Coastal Zone Colour Scanner carried on the Nimbus-7 satellite launched in 1978, by the Sea-viewing Wide Field-of-view Sensor (Sea WiFS) carried on SeaStar, launched in 1997, and on the Moderate Resolution Imaging Spectrometer (MODIS) carried on the Terra and Aqua satellites launched in 1999 and 2002 respectively. MODIS measures upwelling radiance in 36 wavelength bands between 405 nm and 14,385 nm.

Most of the upwelling radiance seen by the satellite comes from the atmosphere. Only about 10% comes from the sea surface. Both air molecules and aerosols scatter light, and very accurate techniques have been developed to remove the influence of the atmosphere.

The total radiance *Lt* received by an instrument in space is:

$$L_t(\lambda_i) = t(\lambda_i)L_W(\lambda_i) + L_r(\lambda_i) + L_a(\lambda_i) \qquad (6.15)$$

where λi is the wavelength of thc radiation in the band measured by the instrument, *LW* is the radiance leaving the sea surface, *Lr* is radiance scattered by molecules, called the Rayleigh radiance, *La* is radiance scattered from aerosols, and *t* is the transmittance of the atmosphere. *Lr* can be calculated from theory, and *La* can be calculated from the amount of red light received at the instrument because very little red light is reflected from the water. Therefore *LW* can be calculated from the radiance measured at the spacecraft.

Chlorophyll concentration in the water column is calculated from the ratio of *LW* at two frequencies. Using data from the Coastal Zone Colour Scanner, Gordon et al. (1983) proposed

$$C_{13} = 1.1298 \left[\frac{L_W(443)}{L_W(550)}\right]^{-1.71} \qquad (6.16a)$$

$$C_{23} = 3.3266 \left[\frac{L_W(520)}{L_W(550)}\right]^{-2.40} \qquad (6.16b)$$

where *C* is the chlorophyll concentration in the surface layers in mg pigment/m^3,and *LW* (443), *LW* (520), and *LW* (550) is the radiance at wavelengths of 443, 520, and 550 nm. *C13* is used when *C13* ≤ 1.5 mg/m^3, otherwise *C23* is used.

The technique is used to calculate chlorophyll concentration within a factor of 50% over a wide range of concentrations from 0.01 to 10 mg/m^3.

6.11 CONCISE SUMMARY

- Density in the ocean is determined by temperature, salinity, and pressure.
- Density changes in the ocean are very small, and studies of water masses and currents require density with an accuracy of 10 parts per million.
- Density is not measured, it is calculated from measurements of temperature, salinity, and pressure using the equation of state of sea water.
- Accurate calculations of density require accurate definitions of temperature and salinity and an accurate equation of state.
- Salinity is difficult to define and to measure. To avoid the difficulty, oceanographers use conductivity instead of salinity. They measure conductivity and calculate density from temperature, conductivity, and pressure.
- A mixed layer of constant temperature and salinity is usually found in the top 1–100 meters of the ocean. The depth is determined by wind speed and the flux of heat through the sea surface.
- To compare temperature and density of water masses at different depths in the ocean, oceanographers use potential temperature and potential density which remove most of the influence of pressure on density.
- Water parcels below the mixed layer move along neutral surfaces.
- Surface temperature of the ocean was usually measured at sea using bucket or injection temperatures. Global maps of temperature combine these observations with observations of infrared radiance from the sea surface measured by an AVHRR in space.
- Temperature and conductivity are usually measured digitally as a function of pressure using a CTD. Before 1960–1970 the salinity and temperature were measured at roughly 20 depths using Nansen bottles lowered on a line from a ship. The bottles carried reversing thermometers which recorded temperature and depth and they returned a water sample from that depth which was used to determine salinity on board the ship.
- Light is rapidly absorbed in the ocean. 95% of sunlight is absorbed in the upper 100 m of the clearest sea water. Sunlight rarely penetrates deeper than a few meters in turbid coastal waters.
- Phytoplankton change the colour of sea water, and the change in colour can be observed from space. Water colour is used to measure phytoplankton concentration from space.

07 THE EQUATIONS OF MOTION: SOME MATHEMATICS

The Equations of Motion represent a cornerstone in the language of physics, providing a mathematical framework to describe the behaviour of objects in motion. Rooted in classical mechanics, these equations offer a systematic and quantitative approach to understanding how forces influence the motion of particles and bodies. Embarking on a journey through the realm of kinematics and dynamics, the equations of motion delve into the fundamental principles that govern the trajectory, velocity, and acceleration of objects in our physical world.

In this chapter first consider the response of a fluid to internal and external forces. This leads to a derivation of some of the basic equations describing ocean dynamics. In the next chapter, we will consider the influence of viscosity, and in chapter 12 we will consider the consequences of vorticity.

Fluid mechanics used in oceanography is based on Newtonian mechanics modified by our evolving understanding of turbulence. Conservation of mass, momentum, angular momentum, and energy lead to particular equations having names that hide their origins (table 7.1).

Table 7.1 Conservation Laws Leading to Basic Equations of Fluid Motion

Conservation of Mass:	Leads to Continuity Equation.
Conservation of Energy:	Conservation of heat leads to Heat Budgets. Conservation of mechanical energy leads to Wave Equation.
Conservation of Momentum:	Leads to Momentum (Navier-Stokes) Eq. Conservation of Angular Momentum: Leads to Conservation of Vorticity.

7.1 PRIMARY FACTORS INFLUENCING OCEAN DYNAMICS

Only a few forces are important in physical oceanography: gravity, friction, and Coriolis (table 7.2). Remember that forces are vectors. They have magnitude and direction.

- *Gravity* is the dominant force. The weight of the water in the ocean produces pressure. Changes in gravity, due to the motion of sun and moon relative to earth produces tides, tidal currents, and tidal mixing in the interior of the ocean. *Buoyancy* is the upward or downward force due to gravity acting on a parcel of water that is more or less dense than other water at its level. For example, cold air blowing over the sea cools surface waters causing them to be more dense than the water beneath. Gravity acting on the difference in density results in a force that causes the water to sink. *Horizontal pressure gradients* are due to the varying weight of water in different regions of the ocean.

- *Friction* is the force acting on a body as it moves past another body while in contact with that body. The bodies can be parcels of water or air. *Wind stress* is the friction due to wind blowing across the sea surface. It transfers horizontal momentum to the sea, creating currents. Wind blowing over waves on the sea surface leads to an uneven distribution of pressure over the waves. The pressure distribution transfers energy to the waves, causing them to grow into bigger waves.
- *Pseudo-forces* are apparent forces that arise from motion in curvilinear or rotating coordinate systems. For example, Newton's first law states that there is no change in motion of a body unless a resultant force acts on it. Yet a body moving at constant velocity seems to change direction when viewed from a rotating coordinate system. The change in direction is due to a pseudo-force, the Coriolis force. *Coriolis Force* is the dominant pseudo-force influencing motion in a coordinate system fixed to the earth.

Table 7.2 Forces in Geophysical Fluid Dynamics

Dominant Forces	
Gravity	Gives rise to pressure gradients, buoyancy, and tides.
Coriolis	Results from motion in a rotating coordinate system
Friction	Is due to relative motion between two fluid parcels. Wind stress is an important frictional force.
Other Forces	
Atmospheric	Pressure Results in inverted barometer effect.
Seismic	Results in *tsunamis* driven by earthquakes.

Note that the last two forces are much less important than the first three.

7.2 COORDINATE SYSTEM

Coordinate systems allow us to find locations in theory and practice. Various systems are used depending on the size of the features to be described or mapped. I will refer to the simplest systems; descriptions of other systems can be found in geography and geodesy books.

1. *Cartesian Coordinate System* is the one I will use most commonly in the following chapters to keep the discussion as simple as possible. We can describe most processes in Cartesian coordinates without the mathematical complexity of spherical coordinates. The standard convention in geophysical fluid mechanics is x is to the east, y is to the north, and z is up.

f-Plane is a Cartesian coordinate system in which the Coriolis force is assumed constant. It is useful for describing flow in regions small compared with the radius of the earth and larger than a few tens of kilometres. *β-plane* is a Cartesian coordinate system in which the Coriolis force is assumed to vary linearly with latitude. It is useful for describing flow over areas as large as ocean basins.

2. *Spherical coordinates* are used to describe flows that extend over large distances and in numerical calculations of basin and global scale flows.

7.3 TYPES OF FLOW IN THE OCEAN

Many terms are used for describing the ocean circulation. Here are a few of the more commonly used terms for describing currents and waves.

- *General Circulation* is the permanent, time-averaged circulation.
- Abyssal also called the Deep Circulation is the circulation of mass, in the meridional plane, in the deep ocean, driven by mixing.
- *Wind-Driven Circulation* is the circulation in the upper kilometre of the ocean forced by the wind. The circulation can be caused by local winds or by winds in other regions.
- *Gyres* are wind-driven cyclonic or anti cyclonic currents with dimensions nearly that of ocean basins.
- *Boundary Currents* are currents flowing parallel to coasts. Two types of boundary currents are important:
- Western boundary currents on the western edge of the ocean tend to be fast, narrow jets such as the Gulf Stream and Kuroshio.
- Eastern boundary currents are weak, *e.g.* the California Current.
- *Squirts* or *Jets* are long narrow currents, with dimensions of a few hundred kilometres, that are nearly perpendicular to west coasts.
- *Mesoscale Eddies* are turbulent or spinning flows on scales of a few hundred kilometres.

In addition to flow due to currents, there are many types of oscillatory flows due to waves. Normally, when we think of waves in the ocean, we visualize waves breaking on the beach or the surface waves influencing ships at sea. But many other types of waves occur in the ocean.

- *Planetary Waves* depend on the rotation of the earth for a restoring force, and they including Rossby, Kelvin, Equatorial, and Yanai waves.
- *Surface Waves* sometimes called gravity waves, are the waves that eventually break on the beach. The restoring force is due to the large density contrast between air and water at the sea surface.

- *Internal Waves* are sub-sea wave similar in some respects to surface waves. The restoring force is due to change in density with depth.
- *Tsunamis* are surface waves with periods near 15 minutes generated by earthquakes.
- *Tidal Currents* are horizontal currents and currents associated with internal waves driven by the tidal potential.
- *Edge Waves* are surface waves with periods of a few minutes confined to shallow regions near shore. The amplitude of the waves drops off exponentially with distance from shore.

7.4 CONSERVATION OF MASS AND SALT

Conservation of mass and salt can be used to obtain very useful information about flows in the ocean. For example, suppose we wish to know the net loss of fresh water, evaporation minus precipitation, from the Mediterranean Sea. We could carefully calculate the latent heat flux over the surface, but there are probably too few ship reports for an accurate application of the bulk formula. Or we could carefully measure the mass of water flowing in and out of the sea through the Strait of Gibraltar, but the difference is small and perhaps impossible to measure accurately.

We can, however, calculate the net evaporation knowing the salinity of the flow in *Si* and out *So*, together with a rough estimate of the volume of water V_o flowing out, where V_o is a volume flow in units of m^3/s (figure 7.1).

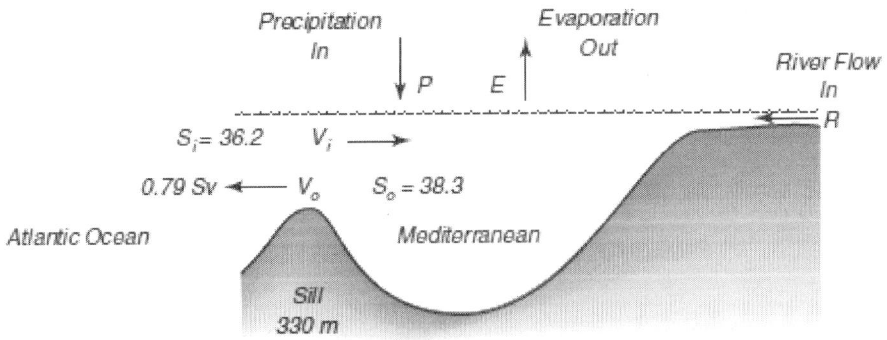

Figure 7.1 Schematic diagram of flow into and out of a basin. Values from Bryden and Kinder (1991).

The mass flowing out is, by definition, $\rho o\ Vo$. If the volume of the sea does not change, conservation of mass requires:

$$\rho_i V_i = \rho_o V_o \tag{7.1}$$

where, ρi, ρo are the densities of the water flowing in and out. We can usually assume, with little error, that $\rho i = \rho o$.

If there is precipitation *P* and evaporation *E* at the surface of the basin and river inflow *R*, conservation of mass becomes:

$$V_i + R + P = V_o + E \qquad (7.2)$$

Solving for (*Vo* - *Vi*):

$$V_o - V_i = (R + P) - E \qquad (7.3)$$

which states that the net flow of water into the basin must balance precipitation plus river inflow minus evaporation when averaged over a sufficiently long time.

Because salt is not deposited or removed from the sea, conservation of salt requires :

$$\rho_i V_i S_i = \rho_o V_o S_o \qquad (7.4)$$

Where ρ_i, S_i are the density and salinity of the incoming water, and ρ_o, *So* are density and salinity of the outflow. With little error, we can again assume that $\rho_i = \rho_o$.

An Example of Conservation of Mass and Salt: Using the values for the flow at the Strait of Gibraltar measured by Bryden and Kinder (1991) and shown in figure 7.1, solving (7.4) for *Vi* assuming that $\rho_i = \rho_o$, and using the estimated value of *Vo*, gives *Vi* = 0.836 Sv = 0.836×10^6 m^3/s, where Sv = Sverdrup = 10^6m^3/s is the unit of volume transport used in oceanography. Using *Vi* and *Vo* in (7.3) gives $(R + P - E) = -4.6 \times 10^4$ m^3/s.

Knowing *Vi* we can also calculate a minimum flushing time for replacing water in the sea by incoming water. The minimum flushing time *Tm* is the volume of the sea divided by the volume of incoming water. The Mediterranean has a volume of around 4×10^6 km^3. Converting 0.836×10^6 m^3/s to km^3/yr we obtain 2.64×10^4 km^3/yr. Then, $Tm = 4 \times 10^6$ km$^3/2.64 \times 10^4$ km^3/yr = 151 yr. The actual time depends on mixing within the sea. If the waters are well mixed, the flushing time is close to the minimum time, if they are not well mixed, the flushing time is longer.

Our example of flow into and out of the Mediterranean Sea is an example of a *box model*. A box model replaces large systems, such as the Mediterranean Sea, with boxes. Fluids or chemicals or organisms can move between boxes, and conservation equations are used to constrain the interactions within systems.

7.5 THE TOTAL DERIVATIVE (D/DT)

If the number of boxes in a system increases to a very large number as the size of each box shrinks, we eventually approach limits used in differential calculus. For example, if we subdivide the flow of water into boxes a few meters on a side, and if we use conservation of mass, momentum, or other properties within each box, we can derive the differential equations governing fluid flow.

Consider the example of acceleration of flow in a small box of fluid. The resulting equation is called the *total derivative.* It relates the acceleration of a particle *Du/Dt* to derivatives of the velocity field at a fixed point in the fluid.

We will use the equation to derive the equations for fluid motion from Newton's Second Law which requires calculating the acceleration of a particles passing a fixed point in the fluid.

We begin by considering the flow of a quantity *q in* into and *qout* out of the small box sketched in figure 7.2. If *q* can change continuously in time and space, the relationship between *qin* and *qout* is:

$$q_{out} = q_{in} + \frac{\partial q}{\partial t}\,\delta t + \frac{\partial q}{\partial x}\,\delta x \tag{7.5}$$

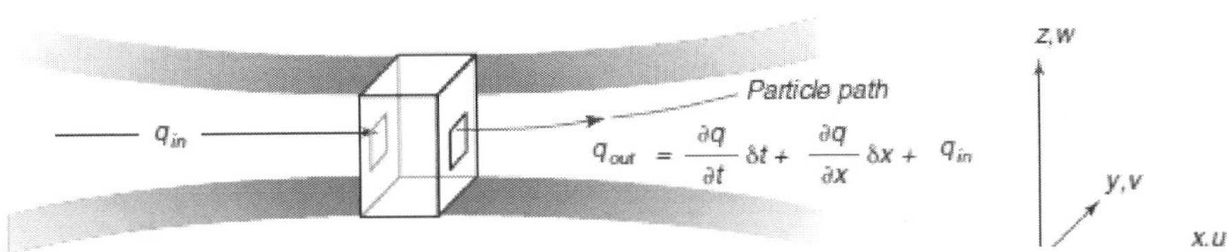

Figure 7.2 Sketch of flow used for deriving the total derivative.

The rate of change of the quantity *q* within the volume is:

$$\frac{Dq}{Dt} = \frac{q_{out} - q_{in}}{\delta t} = \frac{\partial q}{\partial t} + \frac{\partial q}{\partial x}\frac{\delta x}{\delta t} \tag{7.6}$$

But $\delta x/\delta t$ is the velocity *u*, and therefore:

$$\frac{Dq}{Dt} = \frac{\partial q}{\partial t} + u\frac{\partial q}{\partial x}$$

In three dimensions, the total derivative becomes:

$$\frac{D}{Dt} = \frac{\partial}{dt} + u\frac{\partial}{\partial x} + v\frac{\partial}{\partial y} + w\frac{\partial}{\partial z} \tag{7.7a}$$

$$\frac{D}{Dt} = \frac{\partial}{dt} + \mathrm{u}\cdot\nabla() \tag{7.7b}$$

where u is the vector velocity and ∇ is the operator *del* of vector field theory (See Feynman, Leighton, and Sands 1964: 2–6).

This is an amazing result. Transforming coordinates from one following a particle to one fixed in space converts a simple linear derivative into a non- linear partial derivative. Now let's use the equation to calculate the change of momentum of a parcel of fluid.

Momentum Equation; Newton's Second Law relates the change of the momentum of a fluid mass due to an applied force. The change is:

$$\frac{D(m\mathbf{v})}{Dt} = \mathbf{F} \tag{7.8}$$

where **F** is force, *m* is mass, and **v** is velocity. I have emphasized the need to use the total derivative because we are calculating the force on a particle. We can assume that the mass is constant, and (7.8) can be written:

$$\frac{D\mathbf{v}}{Dt} = \frac{\mathbf{F}}{m} = \mathbf{f}_m \tag{7.9}$$

where **f***m* is force per unit mass.

Four forces are important: pressure gradients, Coriolis force, gravity, and friction. Without deriving the form of these forces (the derivations are given in the next section), we can write (7.9) in the following form.

$$\frac{D\mathbf{v}}{Dt} = -\frac{1}{\rho}\nabla p - 2\boldsymbol{\Omega} \times \mathbf{v} + \mathbf{g} + \mathbf{F_r} \tag{7.10}$$

Acceleration equals the negative pressure gradient minus the Coriolis force plus gravity plus other forces. Here g is acceleration of gravity, Fr is friction, and the magnitude Ω of Ω is the *Rotation Rate of earth*, 2π radians per sidereal day or

$$\boxed{\Omega = 7.292 \times 10^{-5} \text{ radians/s}} \tag{7.11}$$

Momentum Equation in Cartesian coordinates: Expanding the derivative in (7.10) and writing the components in a Cartesian coordinate system gives the *Momentum Equation*:

$$\frac{\partial u}{\partial t} + u\frac{\partial u}{\partial x} + v\frac{\partial u}{\partial y} + w\frac{\partial u}{\partial z} = -\frac{1}{\rho}\frac{\partial p}{\partial x} + 2\,\Omega\,v\,\sin\varphi + F_x \tag{7.12a}$$

$$\frac{\partial v}{\partial t} + u\frac{\partial v}{\partial x} + v\frac{\partial v}{\partial y} + w\frac{\partial v}{\partial z} = -\frac{1}{\rho}\frac{\partial p}{\partial y} - 2\,\Omega\,u\,\sin\varphi + F_y \tag{7.12b}$$

$$\frac{\partial w}{\partial t} + u\frac{\partial w}{\partial x} + v\frac{\partial w}{\partial y} + w\frac{\partial w}{\partial z} = -\frac{1}{\rho}\frac{\partial p}{\partial z} + 2\,\Omega\,u\,\cos\varphi - g + F_z \tag{7.12c}$$

where *Fi* are the components of any frictional force per unit mass, and ϕ is latitude. In addition, we have assumed that *w* << *v*, so the 2 Ω *w* cos ϕ has been dropped from equation in (7.12a).

Equation (7.12) appears under various names. Leonhard Euler (1707–1783) first wrote out the general form for fluid flow with external forces, and the equation is sometimes called the *Euler equation* or the *acceleration equation*. Louis Marie Henri

Navier (1785–1836) added the frictional terms, and so the equation is sometimes called the *Navier-Stokes equation*. The term $2\,\Omega\,u\cos\phi$ in (7.12c) is small compared with g, and it can be ignored in ocean dynamics. It cannot be ignored, however, for gravity surveys made with gravimeters on moving ships.

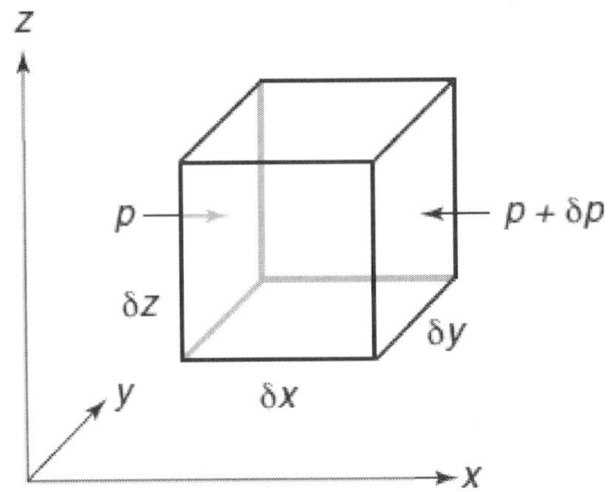

Figure 7.3 Sketch of flow used for deriving the pressure term in the momentum equation.

Derivation of Pressure Term: Consider the forces acting on the sides of a small cube of fluid (figure 7.3). The net force $\delta F x$ in the x direction is

$$\delta F_x = p\,\delta y\,\delta z - (p + \delta p)\,\delta y\,\delta z$$
$$\delta F_x = -\delta p\,\delta y\,\delta z$$

But

$$\delta p = \frac{\partial p}{\partial x}\,\delta x$$

and therefore

$$\delta F_x = -\frac{\partial p}{\partial x}\,\delta x\,\delta y\,\delta z$$
$$\delta F_x = -\frac{\partial p}{\partial x}\,\delta V$$

Dividing by the mass of the fluid δm in the box, the acceleration of the fluid in the x direction is:

$$a_x = \frac{\delta F_x}{\delta m} = -\frac{\partial p}{\partial x}\frac{\delta V}{\delta m}$$

$$\boxed{a_x = -\frac{1}{\rho}\frac{\partial p}{\partial x}} \qquad (7.13)$$

The pressure forces and the acceleration due to the pressure forces in the *y* and *z* directions are derived in the same way.

The Coriolis Term in the Momentum Equation: The Coriolis term exists because we describe currents in a reference frame fixed on earth. The derivation of the Coriolis terms is not easy. Henry Stommel, the noted oceanographer at the Woods Hole Oceanographic Institution even wrote a book on the subject with Dennis Moore (Stommel & Moore, 1989).

Usually, we just state that the force per unit mass, the acceleration of a parcel of fluid in a rotating system, can be written:

$$\mathbf{a}_{fixed} = \left(\frac{D\mathbf{v}}{Dt}\right)_{fixed} = \left(\frac{D\mathbf{v}}{Dt}\right)_{rotating} + (2\boldsymbol{\Omega} \times \mathbf{v}) + \boldsymbol{\Omega} \times (\boldsymbol{\Omega} \times \mathbf{R}) \qquad (7.14)$$

where **R** is the vector distance from the center of earth, **Ω** is the angular velocity vector of earth, and **v** is the velocity of the fluid parcel in coordinates fixed to earth. The term $2\boldsymbol{\Omega} \times \mathbf{v}$ is the Coriolis force, and the term $\boldsymbol{\Omega} \times (\boldsymbol{\Omega} \times \mathbf{R})$ is the centrifugal acceleration. The latter term is included in gravity (figure 7.4).

The Gravity Term in the Momentum Equation: The gravitational attraction of two masses *M*1 and *m* is:

$$\mathbf{F}_g = \frac{G M_1 m}{R^2}$$

where *R* is the distance between the masses, and *G* is the gravitational constant. The vector force $\mathbf{F}_g$ is along the line connecting the two masses.

The force per unit mass due to gravity is:

$$\frac{\mathbf{F}_g}{m} = \mathbf{g}_f = \frac{G M_E}{R^2} \qquad (7.15)$$

where *ME* is the mass of earth. Adding the centrifugal acceleration to (7.15) gives gravity **g** (figure 7.4):

$$\mathbf{g} = \mathbf{g}_f - \boldsymbol{\Omega} \times (\boldsymbol{\Omega} \times \mathbf{R}) \qquad (7.16)$$

Note that gravity does not point toward earth's center of mass. The centrifugal acceleration causes a plumb bob to point at a small angle to the line directed to earth's center of mass. As a result, earth's surface including the ocean's surface is not spherical but it is a oblate ellipsoid. A rotating fluid planet has an equatorial bulge.

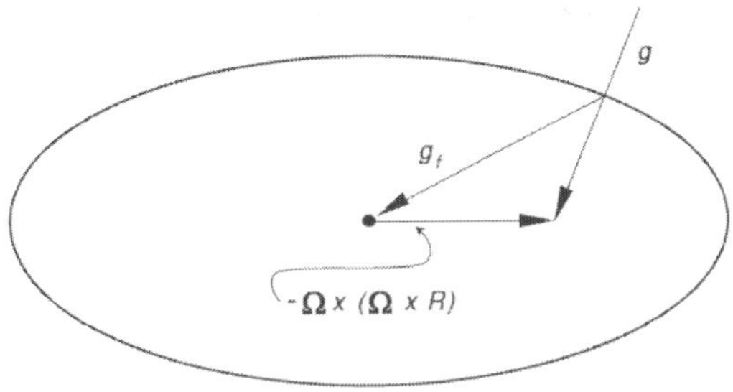

Figure 7.4 Downward acceleration *g* of a body at rest on earth's surface is the sum of gravitational acceleration between the body and earth's mass *gf* and the centrifugal acceleration due to earth's rotation $\Omega \times (\Omega \times R)$. The surface of an ocean at rest must be perpendicular to *g*, and such a surface is close to an ellipsoid of rotation. earth's elasticity is greatly exaggerated here.

7.6 CONSERVATION OF MASS: THE CONTINUITY EQUATION

Now let's derive the equation for the conservation of mass in a fluid. We begin by writing down the flow of mass into and out of a small box (figure 7.5).

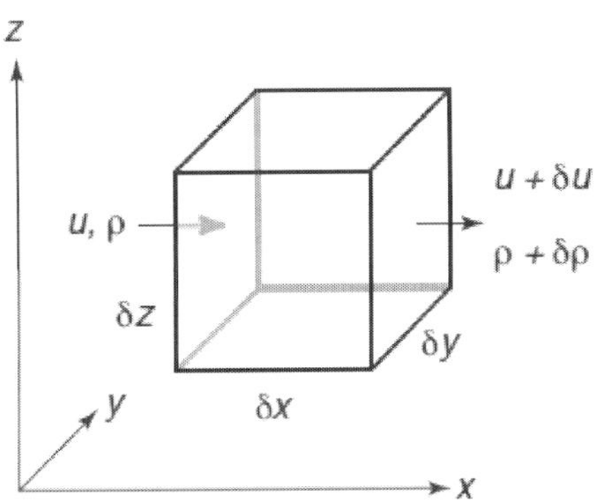

Figure 7.5 Sketch of flow used for deriving the continuity equation.

$$\text{Mass flow in} = \rho\, u\, \delta z\, \delta y$$
$$\text{Mass flow out} = (\rho + \delta\rho)(u + \delta u)\delta z\, \delta y$$

The mass flux into the volume must be (mass flow out) - (mass flow in). There- fore,

$$\text{Mass flux} = (\rho\, \delta u + u\, \delta\rho + \delta\rho\, \delta u)\delta z\, \delta y$$

But

$$\delta u = \frac{\partial u}{\partial x}\delta x\,; \quad \delta\rho = \frac{\partial \rho}{\partial x}\delta x$$

Therefore

$$\text{Mass flux} = \left(\rho\frac{\partial u}{\partial x} + u\frac{\partial \rho}{\partial x} + \frac{\partial \rho}{\partial x}\frac{\partial u}{\partial x}\delta x\right)\delta x\, \delta y\, \delta z$$

The third term inside the parentheses becomes much smaller than the first two terms as $\delta x \rightarrow 0$, and

$$\text{Mass flux} = \frac{\partial(\rho u)}{\partial x}\,\delta x\,\delta y\,\delta z$$

$$\text{Mass flux} = \left(\frac{\partial(\rho u)}{\partial x} + \frac{\partial(\rho v)}{\partial y} + \frac{\partial(\rho w)}{\partial z}\right)\delta x\,\delta y\,\delta z$$

The mass flux must be balanced by a change of mass inside the volume, which is:

$$\frac{\partial \rho}{\partial t}\,\delta x\,\delta y\,\delta z$$

and conservation of mass requires:

$$\frac{\partial \rho}{\partial t} + \frac{\partial(\rho u)}{\partial x} + \frac{d(\rho v)}{\partial y} + \frac{\partial(\rho w)}{\partial z} = 0 \qquad (7.17)$$

This is the *continuity equation* for compressible flow, first derived by Leonhard Euler (1707–1783).

The equation can be put in an alternate form by expanding the derivatives of products and rearranging terms to obtain:

$$\frac{\partial \rho}{\partial t} + u\frac{\partial \rho}{\partial x} + v\frac{\partial \rho}{\partial y} + w\frac{\partial \rho}{\partial z} + \rho\frac{\partial u}{\partial x} + \rho\frac{\partial v}{\partial y} + \rho\frac{\partial w}{\partial z} = 0$$

The first four terms constitute the total derivative of density $D\rho/Dt$ from (7.7), and we can write (7.17) as:

$$\boxed{\frac{1}{\rho}\frac{D\rho}{Dt} + \frac{\partial u}{\partial x} + \frac{\partial v}{\partial y} + \frac{\partial w}{\partial z} = 0} \qquad (7.18)$$

This is the alternate form for the continuity equation for a compressible fluid.

The Boussinesq Approximation: Density is nearly constant in the ocean, and Joseph Boussinesq (1842–1929) noted that we can safely assume density is constant except when it is multiplied by g in calculations of pressure in the ocean. The assumption greatly simplifies the equations of motion.

Boussinesq's assumption requires that:

- Velocities in the ocean must be small compared to the speed of sound c. This ensures that velocity does not change the density. As velocity approaches the speed of sound, the velocity field can produces large changes of density such as shock waves.
- The phase speed of waves or disturbances must be small compared with

- c. Sound speed in incompressible flows is infinite, and we must assume the fluid is compressible when discussing sound in the ocean. Thus the approximation is not true for sound. All other waves in the ocean have speeds small compared to sound.
- The vertical scale of the motion must be small compared with c^2/g, where g is gravity. This ensures that as pressure increases with depth in the ocean, the increase in pressure produces only small changes in density.

The approximations are true for oceanic flows, and they ensure that oceanic flows are incompressible. See Kundu (1990: 79 and 112), Gill (1982: 85), Batchelor (1967: 167), or other texts on fluid dynamics for a more complete description of the approximation.

Compressibility The Boussinesq approximation is equivalent to assuming seawater is incompressible. Now let's see how the assumption simplifies the continuity equation. We define the *coefficient of compressibility*

$$\beta \equiv -\frac{1}{V}\frac{\partial V}{\partial p} = -\frac{1}{V}\frac{dV}{dt} \Big/ \frac{dp}{dt}$$

where V is volume, and p is pressure. For incompressible flows, $\beta = 0$, and:

$$\frac{1}{V}\frac{dV}{dt} = 0$$

because $dp/dt /= 0$. Remembering that density is mass m per unit volume V, and that mass is constant:

$$\frac{1}{V}\frac{dV}{dt} = -V\frac{d}{dt}\left(\frac{1}{V}\right) = -\frac{V}{m}\frac{d}{dt}\left(\frac{m}{V}\right) = -\frac{1}{\rho}\frac{d\rho}{dt} = -\frac{1}{\rho}\frac{D\rho}{Dt} = 0$$

If the flow is incompressible, (7.18) becomes:

$$\boxed{\frac{\partial u}{\partial x} + \frac{\partial v}{\partial y} + \frac{\partial w}{\partial z} = 0} \qquad (7.19)$$

This is the *Continuity Equation for Incompressible Flows.*

7.7 SOLUTIONS TO THE EQUATIONS OF MOTION

Equations (7.12) and (7.19) are four equations, the three components of the momentum equation plus the continuity equation, with four unknowns: u, v, w, p. Note, however, that these are non-linear partial differential equations. Conservation of momentum, when applied to a fluid, converted a simple, first- order, ordinary, differential equation for velocity (Newton's Second Law), which is usually easy to solve, into a non-linear partial differential equation, which is almost impossible to solve.

Boundary Conditions: In fluid mechanics, we generally assume:

- No velocity normal to a boundary, which means there is no flow through the boundary; and
- No flow parallel to a solid boundary, which means no slip at the solid boundary.

Solutions We expect that four equations in four unknowns plus boundary conditions give a system of equations that can be solved in principle. In practice, solutions are difficult to find even for the simplest flows. First, as far as I know, there are no exact solutions for the equations with friction. There are very few exact solutions for the equations without friction. Those who are interested in ocean waves might note that one such exact solution is Gerstner's solution for water waves (Lamb, 1945: 251). Because the equations are almost impossible to solve, we will look for ways to greatly simplify the equations. Later, we will find that even numerical calculations are difficult.

Analytical solutions can be obtained for much simplified forms of the equations of motion. Such solutions are used to study processes in the ocean, including waves. Solutions for oceanic flows with realistic coasts and bathymetric features must be obtained from numerical solutions. In the next few chapters we seek solutions to simplified forms of the equations. In Chapter 15 we will consider numerical solutions.

7.8 CONCISE SUMMARY

- Gravity, buoyancy, and wind are the dominant forces acting on the ocean.
- Earth's rotation produces a pseudo force, the Coriolis force.
- Conservation laws applied to flow in the ocean lead to equations of motion. Conservation of salt, volume and other quantities can lead to deep insights into oceanic flow.
- The transformation from equations of motion applied to fluid parcels to equations applied at a fixed point in space greatly complicates the equations of motion. The linear, first-order, ordinary differential equations describing Newtonian dynamics of a mass accelerated by a force become nonlinear, partial differential equations of fluid mechanics.
- Flow in the ocean can be assumed to be incompressible except when de-scribing sound. Density can be assumed to be constant except when density is multiplied by gravity g. The assumption is called the Boussinesq approximation.
- Conservation of mass leads to the continuity equation, which has an especially simple form for an incompressible fluid.

08 EQUATIONS OF MOTION WITH VISCOSITY

The study of fluid dynamics, particularly in the context of oceanic and atmospheric phenomena, necessitates a comprehensive understanding of the equations governing motion, especially when viscosity is a significant factor. Viscosity, a property of fluids that determines their resistance to deformation or flow, plays a crucial role in shaping the dynamics of fluid motion. In this exploration of the equations of motion with viscosity, we delve into the mathematical framework that captures the intricate interplay between forces, velocities, and the dissipative effects of viscosity, offering valuable insights into the complex behaviour of fluids in motion.

Throughout most of the interior of the ocean and atmosphere friction is relatively small, and we can safely assume that the flow is frictionless. At the boundaries, friction, in the form of viscosity, becomes important. This thin, viscous layer is called a *boundary layer*. Within the layer, the velocity slows from values typical of the interior to zero at a solid boundary. If the boundary is not solid, then the boundary layer is a thin layer of rapidly changing velocity whereby velocity on onc side of the boundary changes to match the velocity on the other side of the boundary. For example, there is a boundary layer at the bottom of the atmosphere, the planetary boundary layer I described in Chapter 3. In the planetary boundary layer, velocity goes from many meters per second in the free atmosphere to tens of centimetre's per second at the sea surface. Below the sea surface, another boundary layer, the Ekman layer described in Chapter 9, matches the flow at the sea surface to the deeper flow.

In this chapter first consider the role of friction in fluid flows, and the stability of the flows to small changes in velocity or density.

8.1 THE INFLUENCE OF VISCOSITY

Viscosity is the tendency of a fluid to resist shear. In the last chapter I wrote the *x*–component of the momentum equation for a fluid in the form (7:12a):

$$\frac{\partial u}{\partial t} + u\frac{\partial u}{\partial x} + v\frac{\partial u}{\partial y} + w\frac{\partial u}{\partial z} = -\frac{1}{\rho}\frac{\partial p}{\partial x} + 2\Omega v \sin\vartheta + F_x \quad (8.1)$$

where *Fx* was a frictional force per unit mass. Now we can consider the form of this term if it is due to viscosity.

Molecules in a fluid close to a solid boundary sometime strike the boundary and transfer momentum to it (figure 8.1). Molecules further from the boundary collide with molecules that have struck the boundary, further transferring the change in momentum into the interior of the fluid. This transfer of momentum is *molecular viscosity* . Molecules, however, travel only micrometers between collisions, and the process is very inefficient for transferring momentum even a few centimetres. Molecular viscosity is important only within a few millimetres of a boundary.

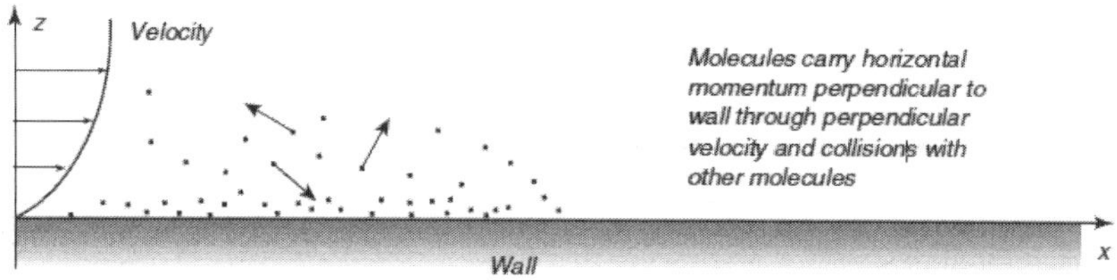

Figure 8.1 Molecules colliding with the wall and with each other transfer momentum from the fluid to the wall, slowing the fluid velocity.

Molecular viscosity ρν is the ratio of the stress *T* tangential to the boundary of a fluid and the velocity shear at the boundary. So the stress has the form:

$$T_{xz} = \rho\nu \frac{\partial u}{\partial z} \tag{8.2}$$

for flow in the (*x*, *z*) plane within a few millimetres of the surface, where ν is the kinematic molecular viscosity. Typically $\nu = 10^{-6}$ m^2/s for water at 20°C.

Generalizing (8.2) to three dimensions leads to a stress tensor giving the nine components of stress at a point in the fluid, including pressure, which is a normal stress, and shear stresses. A derivation of the stress tensor is beyond the scope of this book, but you can find the details in Lamb (1945: §328) or Kundu (1990: p. 93). For an incompressible fluid, the frictional force per unit mass in (8.1) takes the form:

$$F_x = \frac{\partial}{\partial x}\left[\nu\frac{\partial u}{\partial x}\right] + \frac{\partial}{\partial y}\left[\nu\frac{\partial u}{\partial y}\right] + \frac{\partial}{\partial z}\left[\nu\frac{\partial u}{\partial z}\right] = \frac{1}{\rho}\left[\frac{\partial T_{xx}}{\partial x} + \frac{\partial T_{xy}}{\partial y} + \frac{\partial T_{xz}}{\partial z}\right] \tag{8.3}$$

8.2 TURBULENCE

If molecular viscosity is important only over distances of a few millimetres, and if it is not important for most oceanic flows, unless of course you are a zooplankter trying to swim in the ocean, how then is the influence of a boundary transferred into the interior of the flow? The answer is: through turbulence.

$$Re = \frac{\text{Non-linear Terms}}{\text{Viscous Terms}} = \frac{\left(u\dfrac{\partial u}{\partial x}\right)}{\left(\nu\dfrac{\partial^2 u}{\partial x^2}\right)} \approx \frac{U\dfrac{U}{L}}{\nu\dfrac{U}{L^2}} = \frac{UL}{\nu} \tag{8.4}$$

Turbulence arises from the non-linear terms in the momentum equation (u $\partial u/\partial x$, *etc.*). The importance of these terms is given by a non-dimensional number, the Reynolds Number *Re*, which is the ratio of the non-linear terms to the viscous terms:

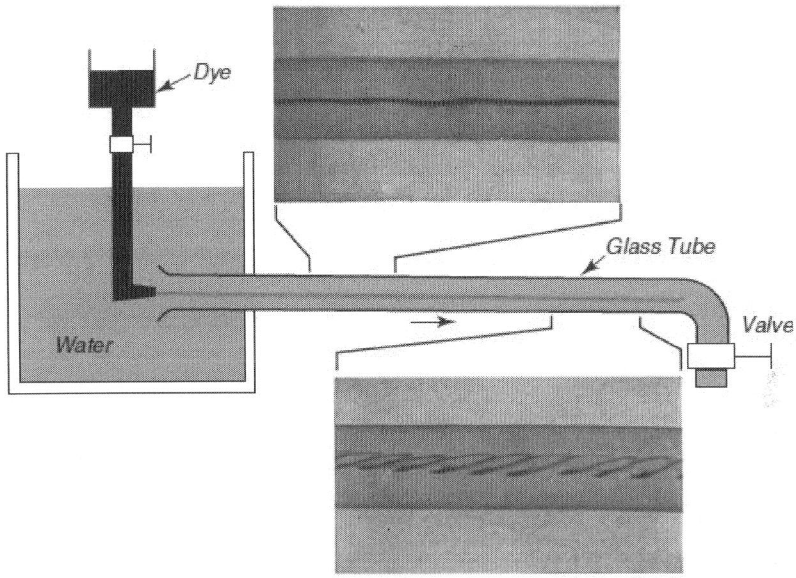

Figure 8.2 Reynolds apparatus for investigating the transition to turbulence in pipe flow, with photographs of near-laminar flow (left) and turbulent flow (right) in a clear pipe much like the one used by Reynolds. After Binder (1949: 88-89).

where, U is a typical velocity of the flow and L is a typical length describing the flow. You are free to pick whatever U, L might be typical of the flow. For example L can be either a typical cross-stream distance, or an along-stream distance. Typical values in the open ocean are $U = 0.1$ m/s and $L = 1$ megameter, so $Re = 10^{11}$. Because non-linear terms are important if Re > 10 – 1000, they are certainly important in the ocean. The ocean is turbulent.

The Reynolds number is named after Osborne Reynolds (1842–1912) who conducted experiments in the late 19th century to understand turbulence. In one famous experiment (Reynolds 1883), he injected dye into water flowing at various speeds through a tube (figure 8.2). If the speed was small, the flow was smooth. This is called *laminar flow*. At higher speeds, the flow became irregular and turbulent. The transition occurred at $Re = V D/\nu \approx 2000$, where V is the average speed in the pipe, and D is the diameter of the pipe.

As Reynolds number increases above some critical value, the flow becomes more and more turbulent. Note that flow pattern is a function of Reynold's number. All flows with the same geometry and the same Reynolds number have the same flow pattern. Thus flow around all circular cylinders, whether 1 mm or 1 m in diameter, look the same as the flow at the top of figure 8.3 if the Reynolds number is 20. Furthermore, the boundary layer is confined to a very thin layer close to the cylinder, in a layer too thin to show in the figure.

Turbulent Stresses: The Reynolds Stress: Prandtl, Karman and others who studied fluid mechanics in the early 20th century, hypothesized that parcels of fluid in a turbulent flow played the same role in transferring momentum within the flow that molecules played in laminar flow. The work led to the idea of turbulent stresses.

To see how these stresses might arise, consider the momentum equation for a flow with mean (U, V, W) and turbulent (u^l, v^l, w^l) components:

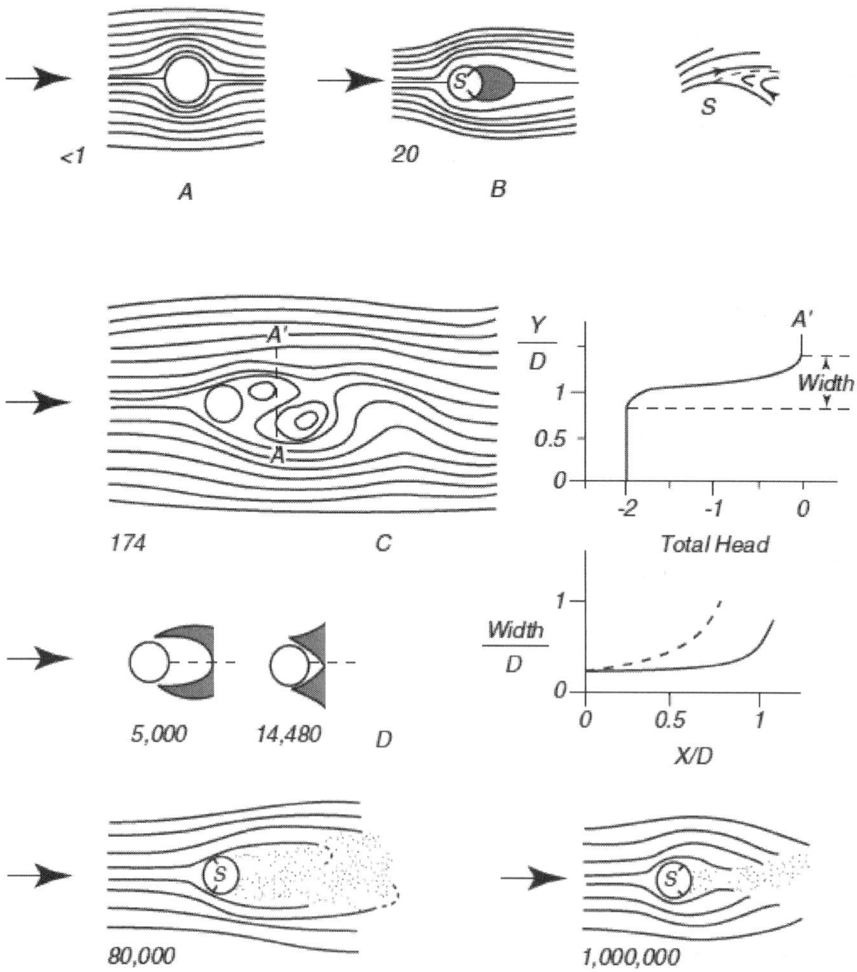

Figure 8.3 Flow past a circular cylinder as a function of Reynolds number between one and a million. From Richardson (1961). The appropriate flows are: A—a toothpick moving at 1 mm/s; B—finger moving at 2 cm/s; F—hand out a car window at 60 mph. All flow at the same Reynolds number has the same streamlines.

Flow past a 10 cm diameter cylinder at 1 cm/s looks the same as 10 cm/s flow past a cylinder 1 cm in diameter because in both cases Re = 1000.

$$u = U + u'; \quad v = V + v'; \quad w = W + w'; \quad p = P + p' \tag{8.5}$$

where the mean value *U* is calculated from a time or space average:

$$U = \langle u \rangle = \frac{1}{T}\int_0^T u(t)\, dt \quad \text{or} \quad U = \langle u \rangle = \frac{1}{X}\int_0^X u(x)\, dx \tag{8.6}$$

The non-linear terms in the momentum equation can be written:

$$\left\langle (U+u')\frac{\partial (U+u')}{\partial x} \right\rangle = \left\langle U\frac{\partial U}{\partial x} \right\rangle + \left\langle U\frac{\partial u'}{\partial x} \right\rangle + \left\langle u'\frac{\partial U}{\partial x} \right\rangle + \left\langle u'\frac{\partial u'}{\partial x} \right\rangle$$

$$\left\langle (U+u')\frac{\partial (U+u')}{\partial x} \right\rangle = \left\langle U\frac{\partial U}{\partial x} \right\rangle + \left\langle u'\frac{\partial u'}{\partial x} \right\rangle \tag{8.7}$$

The second equation follows from the first since $(U\, \partial u'/\partial x) = 0$ and $(u'\, \partial U/\partial x) = 0$, which follow from the definition of U : $(U\partial u'/\partial x) = U\partial(u')/\partial x = 0$.

$$\frac{\partial U}{\partial x} + \frac{\partial V}{\partial y} + \frac{\partial W}{\partial z} + \frac{\partial u'}{\partial x} + \frac{\partial v'}{\partial y} + \frac{\partial w'}{\partial z} = 0 \tag{8.8}$$

Subtracting the mean of (8.8) from (8.8) splits the continuity equation into two equations:

$$\frac{\partial U}{\partial x} + \frac{\partial V}{\partial y} + \frac{\partial W}{\partial z} = 0 \tag{8.9a}$$

$$\frac{\partial u'}{\partial x} + \frac{\partial v'}{\partial y} + \frac{\partial w'}{\partial z} = 0 \tag{8.9b}$$

Using (8.5) in (8.1) taking the mean value of the resulting equation, then simplifying using (8.7), the x component of the momentum equation for the mean flow becomes:

$$\frac{DU}{Dt} = -\frac{1}{\rho}\frac{\partial P}{\partial x} + 2\Omega V \sin\varphi$$
$$+ \frac{\partial}{\partial x}\left[\nu\frac{\partial U}{\partial x} - \langle u'u' \rangle\right] + \frac{\partial}{\partial y}\left[\nu\frac{\partial U}{\partial y} - \langle u'v' \rangle\right] + \frac{\partial}{\partial z}\left[\nu\frac{\partial U}{\partial z} - \langle u'w' \rangle\right] \tag{8.10}$$

The derivation is not as simple as it seems. See Hinze (1975: 22) for details. Thus the additional force per unit mass due to the turbulence is:

The terms ρ(*u'w'*), ρ(*u'w'*), and ρ(*u'w'*) transfer eastward momentum ($\rho u'$) in the *x*, *y*, and *z* directions. For example, the term ρ(*u'w'*) gives the downward transport of eastward momentum across a horizontal plane. Because they transfer momentum, and because they were first derived by Osborne Reynolds, they are called *Reynolds Stresses.*

8.3 CALCULATION OF REYNOLDS STRESS

The Reynolds stresses such as *∂(u'w')/∂z* are called virtual stresses (cf. Goldstein, 1965: §69 & §81) because we assume that they play the same role as the viscous terms in the equation of motion. To proceed further, we need values or functional form for the Reynolds stress. Several approaches are used.

From Experiments: We can calculate Reynolds stresses from direct measurements of *(u' v' w')* made in the laboratory or ocean. This is accurate, but hard to generalize to other flows. So we seek more general approaches.

By Analogy with Molecular Viscosity: Let's return to the example in figure 8.1, which shows flow above a surface in the *x, y* plane. Prandtl, in a revolutionary paper published in 1904, stated that turbulent viscous effects are only important in a very thin layer close to the surface, the boundary layer. Prandtl's invention of the boundary layer allows us to describe very accurately turbulent flow of wind above the sea surface, or flow at the bottom boundary layer in the ocean, or flow in the mixed layer at the sea surface.

To calculate flow in a boundary layer, we assume that flow is constant in the *x, y* direction, that the statistical properties of the flow vary only in the *z* direction, and that the mean flow is steady. Therefore *∂/∂t = ∂/∂x = ∂/∂y* = 0, and (8.10) can be written:

$$2\Omega V \sin\varphi + \frac{\partial}{\partial z}\left[\nu\frac{\partial U}{\partial z} - \langle u'w'\rangle\right] = 0 \tag{8.12}$$

We now assume, in analogy with (8.2)

$$-\rho\langle u'w'\rangle = T_{xz} = \rho A_z \frac{\partial U}{\partial z} \tag{8.13}$$

where *Az* is an *eddy viscosity* or *eddy diffusivity* which replaces the molecular viscosity ν in (8.2). With this assumption,

$$\frac{\partial T_{xz}}{\partial z} = \frac{\partial}{\partial z}\left(A_z\frac{\partial U}{\partial z}\right) \approx A_z\frac{\partial^2 U}{\partial z^2} \tag{8.14}$$

where I have assumed that *Az* is either constant or that it varies more slowly in the *z* direction than *∂U/∂z*. Later, I will assume that *Az* ≈ *z*.

Because eddies can mix heat, salt, or other properties as well as momentum, I will use the term eddy diffusivity. It is more general than eddy viscosity, which is the mixing of momentum.

The x and y momentum equations for a homogeneous, steady-state, turbulent boundary layer above or below a horizontal surface are:

$$\rho f V + \frac{\partial T_{xz}}{\partial z} = 0 \tag{8.15a}$$

$$\rho f U - \frac{\partial T_{yz}}{\partial z} = 0 \tag{8.15b}$$

where $f = 2\omega \sin \phi$ is the Coriolis parameter, and I have dropped the molecular viscosity term because it is much smaller than the turbulent eddy viscosity. Note, (8.15b) follows from a similar derivation from the y-component of the momentum equation. We will need (8.15) when we describe flow near the surface. The assumption that an eddy viscosity Az can be used to relate the Reynolds stress to the mean flow works well in turbulent boundary layers. However Az cannot be obtained from theory. It must be calculated from data collected in wind tunnels or measured in the surface boundary layer at sea. See Hinze (1975, §5–2 and §7–5) and Goldstein (1965: §80) for more on the theory of turbulence flow near a flat plate.

8.4 THE TURBULENT BOUNDARY LAYER OVER A FLAT PLATE

The revolutionary concept of a boundary layer was invented by Prandtl in 1904 (Anderson, 2005). Later, the concept was applied to flow over a flat plate by G.I. Taylor (1886–1975), L. Prandtl (1875–1953), and T. von Karman (1881–1963) who worked independently on the theory from 1915 to 1935. Their empirical theory, sometimes called the *mixing-length theory* predicts well the mean velocity profile close to the boundary. Of interest to us, it predicts the mean flow of air above the sea. Here's a simplified version of the theory applied to a smooth surface.

We begin by assuming that the mean flow in the boundary layer is steady and that it varies only in the z direction. Within a few millimetres of the boundary, friction is important and (8.2) has the solution

$$U = \frac{T_x}{\rho \nu} z \tag{8.16}$$

and the mean velocity varies linearly with distance above the boundary. Usually (8.16) is written in dimensionless form:

$$\frac{U}{u^*} = \frac{u^* z}{\nu} \tag{8.17}$$

where $u^{*2} \equiv Tx/\rho$ is the *friction velocity*.

Further from the boundary, the flow is turbulent, and molecular friction is not important. In this regime, we can use (8.13), and

Prandtl and Taylor assumed that large eddies are more effective in mixing momentum than small eddies, and therefore Az ought to vary with distance from the wall. Karman assumed that it had the particular functional form $Az = \kappa z u^*$, where κ is a dimensionless constant. With this assumption, the equation for the mean velocity profile becomes

$$\kappa z u^* \frac{\partial U}{\partial z} = u^{*2} \qquad (8.19)$$

$$U = \frac{u^*}{\kappa} \ln\left(\frac{z}{z_0}\right)$$ nction only of z, we can write $dU = u^*/(\kappa z)\, dz$, which has the solution (8.20)

where z_0 is distance from the boundary at which velocity goes to zero.

For airflow over the sea, κ = 0.4 and *zo* is given by Charnock's (1955) relationz0 = 0.0156 u^{*2}/g. The mean velocity in the boundary layer just above the sea surface described in §4.3 fits well the logarithmic profile of (8.20), as does the mean velocity in the upper few meters of the sea just below the sea surface. Furthermore, using (4.2) in the definition of the friction velocity, then using (8.20) gives Charnock's form of the drag coefficient as a function of wind speed.

Prandtl's theory based on assumption (8.13) works well only where friction is much larger than the Coriolis force. This is true for air flow within tens of meters of the sea surface and for water flow within a few meters of the surface. The application of the technique to other flows in the ocean is less clear. For example, the flow in the mixed layer at depths below about ten meters is less well described by the classical turbulent theory. Tennekes and Lumley (1990:57)write:

Mixing-length and eddy viscosity models should be used only to generate analytical expressions for the Reynolds stress and mean-velocity profile if those are desired for curve fitting purposes in turbulent flows characterized by a single length scale and a single velocity scale. The use of mixing-length theory in turbulent flows whose scaling laws are not known beforehand should be avoided.

Problems with the eddy-viscosity approach:

- Except in boundary layers a few meters thick, geophysical flows may be influenced by several characteristic scales. For example, in the atmospheric boundary layer above the sea, at least three scales may be important: i) the height above the sea z, ii) the Monin-Obukhov scale *L* discussed in§4.3, and iii) the typical velocity *U* divided by the Coriolis parameter *U/f*.
- The velocities u', w' are a property of the *fluid*, while *Az* is a property of the *flow*;
- Eddy viscosity terms are not symmetric:

$$\langle u'v' \rangle = \langle v'u' \rangle; \quad \text{but}$$
$$A_x \frac{\partial V}{\partial x} \neq A_y \frac{\partial U}{\partial y}$$

From a Statistical Theory of Turbulence: The Reynolds stresses can be calculated from various theories which relate $(u'u')$ to higher order correlations of the form $(u'u'u')$. The problem then becomes: How to calculate the higher order terms? This is the *closure problem* in turbulence. There is no general solution, but the approach leads to useful understanding of some forms of turbulence such as isotropic turbulence downstream of a grid in a wind tunnel (Batchelor 1967). *Isotropic turbulence* is turbulence with statistical properties that are independent of direction.

The approach can be modified somewhat for flow in the ocean. In the idealized case of high Reynolds flow, we can calculate the statistical properties of a flow in thermodynamic equilibrium. Because the actual flow in the ocean is far from equilibrium, we assume it will evolve towards equilibrium. Holloway (1986) provides a good review of this approach, showing how it can be used to derive the influence of turbulence on mixing and heat transports. One interesting result of the work is that zonal mixing ought to be larger than meridional mixing.

Summary: The turbulent eddy viscosities *Ax*, *Ay*, and *Az* cannot be calculated accurately for most oceanic flows.

- They can be estimated from measurements of turbulent flows. Measurements in the ocean, however, are difficult. Measurements in the lab, although accurate, cannot reach Reynolds numbers of 10^{11} typical of the ocean.
- The statistical theory of turbulence gives useful insight into the role of turbulence in the ocean, and this is an area of active research.

Table 8.1 Some Values for Viscosity

$\nu_{water} =$	$10^{-6}\ m^2/s$
$\nu_{\text{lurul } 15°C} =$	$10^{6}\ m^2/s$
$\nu_{glacier\ ice} =$	$10^{10}\ m^2/s$
$A_{yocean} =$	$10^{4}\ m^2/s$
$A_{zocean} =$	$(10^{-5} - 10^{-3})\ m^2/s$

Mixing in the Ocean: Turbulence in the ocean leads to mixing. Because the ocean has stable stratification, vertical displacement must work against the buoyancy force. Vertical mixing requires more energy than horizontal mixing. As a result, horizontal mixing along surfaces of constant density is much larger than vertical mixing across surfaces of constant density. The latter, however, usually called *diapycnal mixing*, is very important because it changes the vertical structure of the ocean, and it controls to a large extent the rate at which deep water eventually reaches the surface in mid and low latitudes.

The equations describing mixing depend on many processes. See Garrett (2006) for a good overview of the subject. Here I consider some simple flows. A simple equation for vertical mixing by eddies of a tracer Θ such as salt or temperature is:

$$\frac{\partial \Theta}{\partial t} + W\frac{\partial \Theta}{\partial z} = \frac{\partial}{\partial z}\left(A_z \frac{\partial \Theta}{\partial z}\right) + S \qquad (8.21)$$

where *Az* is the vertical eddy diffusivity, *W* is a mean vertical velocity, and *S*is a source term.

Average Vertical Mixing : Walter Munk (1966) used a very simple observation to calculate vertical mixing in the ocean. He observed that the ocean has a thermo cline almost everywhere, and the deeper part of the thermo cline does not change even over decades (figure 8.4). This was a remarkable observation because we expect downward mixing would continuously deepen the thermo- cline. But it doesn't. Therefore, a steady-state thermo cline requires that the downward mixing of heat by turbulence must be balanced by an upward trans- port of heat by a mean vertical current *W*. This follows from (8.21) for steady state with no sources or sinks of heat:

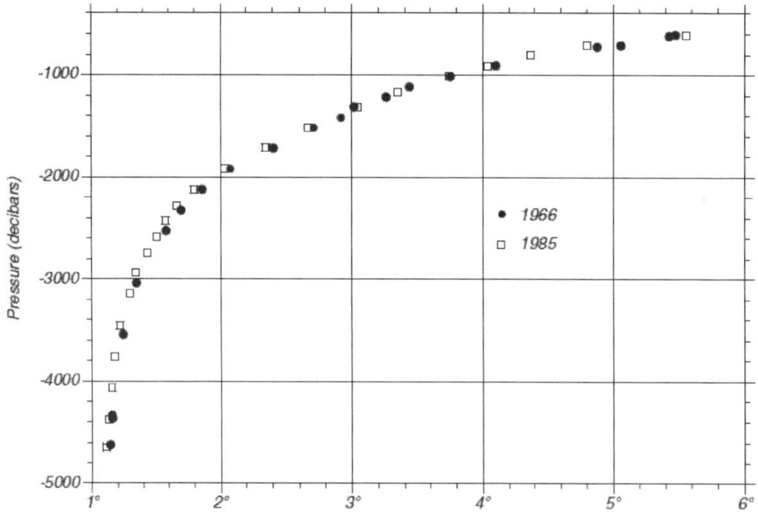

Figure 8.4 Potential temperature measured as a function of depth (pressure) near 24.7°N, 161.4°W in the central North Pacific by the *Yaquina* in 1966 (•), and by the *Thompson* in1985 (D). Data from *Atlas of Ocean Sections* produced by Swift, Rhines, and Schlitzer.

$$W\frac{\partial T}{\partial z} = A_z \frac{\partial^2 T}{\partial z^2} \qquad (8.22)$$

where *T* is temperature as a function of depth in the thermo cline.

The equation has the solution:

$$T \approx T_0 \exp(z/H) \qquad (8.23)$$

where $H = Az/W$ is the scale depth of the thermo cline, and $T0$ is the temperature near the top of the thermo cline. Observations of the shape of the deep thermo cline are indeed very close to a exponential function. Munk used an exponential function fit through the observations of $T(z)$ to get H.

Munk calculated W from the observed vertical distribution of ^{14}C, a radioactive isotope of carbon, to obtain a vertical time scale. In this case, $S = -1.24 \times 10^{-4}$ years^{-1}. The length and time scales gave W = 1.2 cm/day and$\langle Az \rangle = 1.3 \times 10^{-4}$ m^2/s Average Vertical Eddy Diffusivity (8.24) where the brackets denote average eddy diffusivity in the thermo cline.

Munk also used W to calculate the average vertical flux of water through the thermo cline in the Pacific, and the flux agreed well with the rate of formation of bottom water assuming that bottom water upwells almost everywhere at a constant rate in the Pacific. Globally, his theory requires upward mixing of 25 to 30 Sverdrups of water, where one Sverdrup is 10^6 cubic meters per second.

Measured Vertical Mixing: Direct observations of vertical mixing required the development of techniques for measuring: i) the fine structure of turbulence, including probes able to measure temperature and salinity with a spatial resolution of a few centimetres (Gregg 1991), and ii) the distribution of tracers such as sulphur hexafluoride (SF_6) which can be easily detected at concentrations as small as one gram in a cubic kilometre of seawater.

Direct measurements of open-ocean turbulence and the diffusion of SF_6 yield an eddy diffusivity:

$$Az \approx 1 \times 10^{-5} \text{ m}^2/\text{s} \quad \text{Open-Ocean Vertical Eddy Diffusivity (8.25)}$$

For example, Ledwell, Watson, and Law (1998) injected 139 kg of SF_6 in the Atlantic near 26° N, 29° W 1200 km west of the Canary Islands at a depth of310 m. They then measured the concentration for five months as it mixed over hundreds of kilometres to obtain a diapycnal eddy diffusivity of $Az = 1.2 \pm 0.2 \times 10^{-5}$ m^2/s.

The large discrepancy between Munk's calculation of the average eddy diffusivity for vertical mixing and the small values observed in the open ocean has been resolved by recent studies that show:

$$Az \approx 10^{-3} \rightarrow 10^{-1} \text{ m}^2/\text{s} \quad \text{Local Vertical Eddy Diffusivity (8.26)}$$

Polzin et al. (1997) measured the vertical structure of temperature in the Brazil Basin in the south Atlantic. They found $Az > 10^{-3}$ m^2/s close to the bottom when the water flowed over the western flank of the mid-Atlantic ridge at the eastern edge of the basin. Kunze and Toole (1997) calculated enhanced eddy diffusivity as large as $A = 10^{-3}$ m^2/s above Fieberling Guyot in the Northwest Pacific and smaller diffusivity along the flank of the seamount. Garbato et al (2004) calculated even stronger mixing in the Scotia Sea where the Antarctic Circumpolar Current flows between Antarctica and South America.

The results of these and other experiments show that mixing occurs mostly by breaking internal waves and by shear at oceanic boundaries: along continental slopes, above seamounts and mid-ocean ridges, at fronts, and in the mixed layer at the sea surface. To a large extent, the mixing is driven by deep-ocean tidal currents, which become turbulent when they flow past obstacles on the sea floor, including seamounts and mid-ocean ridges (Jayne et al, 2004).

Because water is mixed along boundaries or in other regions (Gnadadesikan, 1999), we must take care in interpreting temperature profiles such as that in figure 8.4. For example, water at 1200 m in the central north Atlantic could move horizontally to the Gulf Stream, where it mixes with water from 1000 m. The mixed water may then move horizontally back into the central north Atlantic at a depth of 1100 m. Thus parcels of water at 1200 m and at 1100 m at some location may reach their position along entirely different paths.

Measured Horizontal Mixing: Eddies mix fluid in the horizontal, and large eddies mix more fluid than small eddies. Eddies range in size from a few meters due to turbulence in the thermo cline up to several hundred kilometres for geostrophic eddies discussed in Chapter 10.

In general, mixing depends on Reynolds number R (Tennekes 1990: p. 11)

$$\frac{A}{\gamma} \approx \frac{A}{\nu} \sim \frac{UL}{\nu} = R \tag{8.27}$$

where γ is the molecular diffusivity of heat, U is a typical velocity in an eddy, and L is the typical size of an eddy. Furthermore, horizontal eddy diffusivity are ten thousand to ten million times larger than the average vertical eddy diffusivity.

Equation (8.27) implies $Ax \sim UL$. This functional form agrees well with Joseph and Sender's (1958) analysis, as reported in (Bowden 1962) of spreading of radioactive tracers, optical turbidity, and Mediterranean Sea water in the north Atlantic. They report

$$A_x = PL \tag{8.28}$$
$$10 \text{ km} < L < 1500 \text{ km}$$
$$P = 0.01 \pm 0.005 \text{ m/s}$$

where L is the distance from the source, and P is a constant.

The horizontal eddy diffusivity (8.28) also agrees well with more recent re- ports of horizontal diffusivity. Work by Holloway (1986) who used satellite altimeter observations of geostrophic currents, Freeland who tracked sofar underwater floats, and Ledwell Watson, and Law (1998) who used observations of currents and tracers to find $Ax \approx 8 \times 10^2$ m^2/s Geostrophic Horizontal Eddy Diffusivity (8.29)

Using (8.28) and the measured Ax implies eddies with typical scales of 80 km, a value near the size of geostrophic eddies responsible for the mixing.

Ledwell, Watson, and Law (1998) also measured a horizontal eddy diffusivity.

They found over scales of meters due to turbulence in the thermo cline probably driven by breaking internal waves. This value, when used in (8.28) implies typical lengths of 100 m for the small eddies responsible for mixing in this experiment.

$$Ax \approx 1 - 3 \text{ m}^2/\text{s} \quad \text{Open-Ocean Horizontal Eddy Diffusivity} \qquad (8.30)$$

8.5 COMMENTS ON HORIZONTAL MIXING:

1. Horizontal eddy diffusivity is 10^5 - 10^8 times larger than vertical eddy diffusivity.
2. Water in the interior of the ocean moves along sloping surfaces of constant density with little local mixing until it reaches a boundary where it is mixed vertically. The mixed water then moves back into the open ocean again along surfaces of constant density (Gregg 1987).
3. One particular case is particularly noteworthy. When water, mixing downward through the base of the mixed layer, flows out into the thermo cline along surfaces of constant density, the mixing leads to the *ventilated thermo cline* model of oceanic density distributions.
4. Observations of mixing in the ocean imply that numerical models of the oceanic circulation should use mixing schemes that have different eddy diffusivity parallel and perpendicular to surfaces of constant density, not parallel and perpendicular to level surfaces of constant z as I used above. Horizontal mixing along surfaces of constant z leads to mixing across layers of constant density because layers of constant density are inclined to the horizontal by about 10^{-3} radians (see §10.7 and figure 10.13).
5. Studies by Danabasoglu, McWilliams, and Gent (1994) show that numerical models using isopycnal and diapycnal mixing leads to much more realistic simulations of the oceanic circulation.
6. Mixing is horizontal and two dimensional for horizontal scales greater than $NH/(2f)$ where H is the water depth, N is the stability frequency (8.36), and f is the Coriolis parameter (Dritschel, Juarez, and Ambaum (1999).

8.6 STABILITY

We saw in section 8.2 that fluid flow with a sufficiently large Reynolds number is turbulent. This is one form of instability. Many other types of instability occur in the in the ocean. Here, let's consider three of the more important ones: i) *static stability* associated with change of density with depth, ii) *dynamic stability* associated with velocity shear, and iii) *double-diffusion* associated with salinity and temperature gradients in the ocean.

Static Stability and the Stability Frequency: Consider first static stability. If more dense water lies above less dense water, the fluid is unstable. The more dense water will sink beneath the less dense. Conversely, if less dense water lies above more dense water, the interface between the two is stable. But how stable? We might guess that the larger the density contrast across the interface, the more stable the interface. This is an example of static stability. Static stability is important in any *stratified* flow where density increases with depth, and we need some criterion for determining the importance of the stability.

Consider a parcel of water that is displaced vertically and adiabatically in a stratified fluid (figure 8.5).The buoyancy force F acting on the displaced parcelis the difference between its weight $V g\rho'$ and the weight of the surrounding water $V g\rho_2$, where V is the volume of the parcel:

$$F = V\,g\,(\rho_2 - \rho')$$

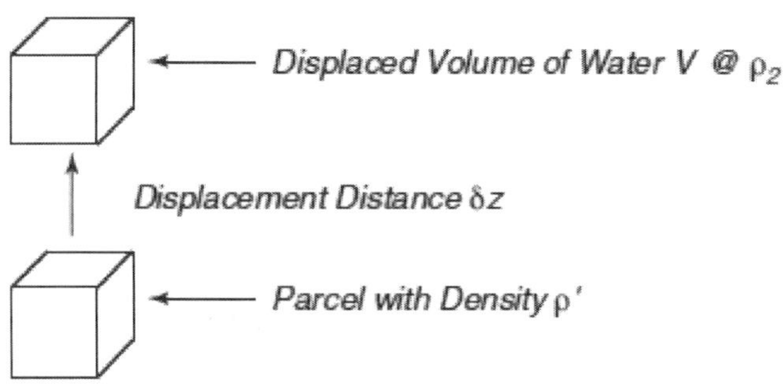

Figure 8.5 Sketch for calculating static stability and stability frequency.

The acceleration of the displaced parcel is:

$$a = \frac{F}{m} = \frac{g\,(\rho_2 - \rho')}{\rho'} \tag{8.31}$$

but

$$\rho_2 = \rho + \left(\frac{d\rho}{dz}\right)_{water} \delta z \tag{8.32}$$

$$\rho' = \rho + \left(\frac{d\rho}{dz}\right)_{parcel} \delta z \tag{8.33}$$

Using (8.32) and (8.33) in (8.31), ignoring terms proportional to δz^2, we obtain:

$$E = -\frac{1}{\rho}\left[\left(\frac{d\rho}{dz}\right)_{water} - \left(\frac{d\rho}{dz}\right)_{parcel}\right] \tag{8.34}$$

where $E \equiv -a/(g\delta z)$ is the *stability* of the water column (McDougall, 1987; Sverdrup, Johnson, and Fleming, 1942: 416; or Gill, 1982: 50).

In the upper kilometre of the ocean stability is large, and the first term in (8.34) is much larger than the second. The first term is proportional to the rate of change of density of the water column. The second term is proportional to the compressibility of sea water, which is very small. Neglecting the second term, we can write the *stability equation*:

$$\boxed{E \approx -\frac{1}{\rho}\frac{d\rho}{dz}} \qquad (8.35)$$

The approximation used to derive (8.35) is valid for $E > 50 \times 10^{-8}$/m.

Below about a kilometre in the ocean, the change in density with depth is so small that we must consider the small change in density of the parcel due to changes in pressure as it is moved vertically.

Stability is defined such that

$E > 0$ Stable

$E = 0$ Neutral Stability

$E < 0$ Unstable

In the upper kilometre of the ocean, $z < 1,000$ m, $E = (50—1000) \times 10^{-8}$/m, and in deep trenches where $z > 7,000$ m, $E = 1 \times 10^{-8}$/m.

The influence of stability is usually expressed by a *stability frequency N*:

$$N^2 \equiv -gE \qquad (8.36)$$

The stability frequency is often called the *buoyancy frequency* or the *Brunt-Vaisala frequency*. The frequency quantifies the importance of stability, and it is a fundamental variable in the dynamics of stratified flow. In simplest terms, the frequency can be interpreted as the vertical frequency excited by a vertical displacement of a fluid parcel. Thus, it is the maximum frequency of internal waves in the ocean. Typical values of N are a few cycles per hour (figure 8.6).

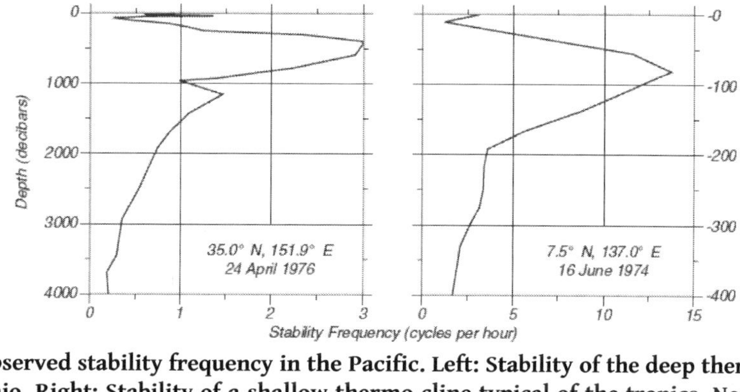

Figure 8.6. Observed stability frequency in the Pacific. Left: Stability of the deep thermo cline east of the Kuroshio. Right: Stability of a shallow thermo cline typical of the tropics. Note the change of scales.

Dynamic Stability and Richardson's Number: If velocity changes with depth in a stable, stratified flow, then the flow may become unstable if the change in velocity with depth, the *current shear* , is large enough. The simplest example is wind blowing over the ocean. In this case, stability is very large across the sea surface. We might say it is infinite because there is a step discontinuity in ρ, and (8.36) is infinite. Yet, wind blowing on the ocean creates waves, and if the wind is strong enough, the surface becomes unstable and the waves break.

This is an example of *dynamic instability* in which a stable fluid is made unstable by velocity shear. Another example of dynamic instability, the Kelvin-Helmholtz instability, occurs when the density contrast in a sheared flow is much less than at the sea surface, such as in the thermo cline or at the top of a stable, atmospheric boundary layer (figure 8.7).

The relative importance of static stability and dynamic instability is ex- pressed by the *Richardson Number*:

$$R_i \equiv \frac{g\,E}{(\partial U/\partial z)^2} \tag{8.37}$$

where the numerator is the strength of the static stability, and the denominator is the strength of the velocity shear.

$R_i > 0.25$ Stable

$R_i < 0.25$ Velocity Shear Enhances Turbulence

Note that a small Richardson number is not the only criterion for instability. The Reynolds number must be large and the Richardson number must be less than 0.25 for turbulence. These criteria are met in some oceanic flows. The turbulence mixes fluid in the vertical, leading to a vertical eddy viscosity and eddy diffusivity. Because the ocean tends to be strongly stratified and currents tend to be weak, turbulent mixing is intermittent and rare. Measurements of density as a function of depth rarely show more dense fluid over less dense fluid as seen in the breaking waves in figure 8.7 (Moum and Caldwell 1985).

Figure 8.7 Billow clouds showing a Kelvin-Helmholtz instability at the top of a stable atmospheric layer. Some billows can become large enough that more dense air overlies less dense air, and then the billows collapse into turbulence. Photography copyright Brooks Martner, NOAA Environmental Technology Laboratory.

Double Diffusion and Salt Fingers: In some regions of the ocean, less dense water overlies more dense water, yet the water column is unstable even if there are no currents. The instability occurs because the molecular diffusion of heat is about 100 times faster than the molecular diffusion of salt. The instability was first discovered by Melvin Stern in 1960 who quickly realized its importance in oceanography.

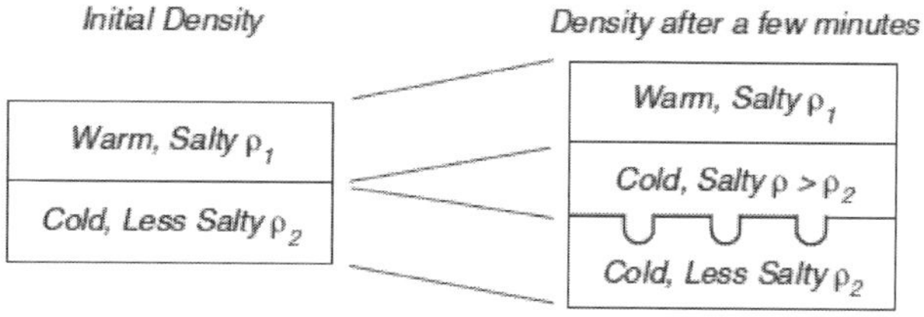

Figure 8.8 Left: Initial distribution of density in the vertical. Right: After some time, the diffusion of heat leads to a thin unstable layer between the two initially stable layers. The thin unstable layer sinks into the lower layer as salty fingers. The vertical scale in the figures is a few centimetres.

Consider two thin layers a few meters thick separated by a sharp interface (figure 8.8). If the upper layer is warm and salty, and if the lower is colder and less salty than the upper layer, the interface becomes unstable even if the upper layer is less dense than the lower.

Here's what happens. Heat diffuses across the interface faster than salt, leading to a thin, cold, salty layer between the two initial layers. The cold salty layer is more dense than the cold, less-salty layer below, and the water in the layer sinks. Because the layer is thin, the fluid sinks in fingers 1–5 cm in diameter and 10s of centimetres long, not much different in size and shape from our fingers. This is *salt fingering* . Because two constituents diffuse across the interface, the process is called *double diffusion.*

There are four variations on this theme. Two variables taken two at a time leads to four possible combinations:

- ***Warm salty over colder less salty.*** This process is called *salt fingering.* It occurs in the thermo cline below the surface waters of sub-tropical gyres and the western tropical north Atlantic, and in the North-east Atlantic beneath the outflow from the Mediterranean Sea. Salt fingering eventually leads to density increasing with depth in a series of steps. Layers of constant-density are separated by thin layers with large changes in density, and the profile of density as a function of depth looks like stair steps. Schmitt et al (1987) observed 5–30 m thick steps in the western, tropical north Atlantic that were coherent over 200–400 km and that lasted for at least eight months. Kerr (2002) reports a recent experiment by Raymond Schmitt, James Leswell, John Toole, and Kurt Polzin showed salt fingering off Barbados mixed water 10 times faster than turbulence.

- ***Colder less salty over warm salty.*** This process is called *diffusive convection.* It is much less common than salt fingering, and it us mostly found at high latitudes. Diffusive convection also leads to a stair step of density as a function of depth. Here's what happens in this case. Double diffusion leads to a thin, warm, less-salty layer at the base of the upper, colder, less-salty layer. The thin layer of water rises and mixes with water in the upper layer. A similar processes occurs in the lower layer where a colder, salty layer forms at the interface. As a result of the convection in the upper and lower layers, the interface is sharpened. Any small gradients of density in either layer are reduced. Neal et al (1969) observed 2–10 m thick layers in the sea beneath the Arctic ice.
- ***Cold salty over warmer less salty.*** Always statically unstable.
- ***Warmer less salty over cold salty.*** Always stable and double diffusion diffuses the interface between the two layers.

Double diffusion mixes ocean water, and it cannot be ignored. Merryfield et al (1999), using a numerical model of the ocean circulation that included double diffusion, found that double-diffusive mixing changed the regional distributions of temperature and salinity although it had little influence on large-scale circulation of the ocean.

8.7 CONCISE SUMMARY

- Friction in the ocean is important only over distances of a few millimetres. For most flows, friction can be ignored.
- The ocean is turbulent for all flows whose typical dimension exceeds a few centimetres, yet the theory for turbulent flow in the ocean is poorly understood.
- The influence of turbulence is a function of the Reynolds number of the flow. Flows with the same geometry and Reynolds number have the same streamlines.
- Oceanographers assume that turbulence influences flows over distances greater than a few centimetres in the same way that molecular viscosity influences flow over much smaller distances.
- The influence of turbulence leads to Reynolds stress terms in the momentum equation.
- The influence of static stability in the ocean is expressed as a frequency, the stability frequency. The larger the frequency, the more stable the water column.

- The influence of shear stability is expressed through the Richardson number. The greater the velocity shear, and the weaker the static stability, the more likely the flow will become turbulent.
- Molecular diffusion of heat is much faster than the diffusion of salt. This leads to a double-diffusion instability which modifies the density distribution in the water column in many regions of the ocean.
- Instability in the ocean leads to mixing. Mixing across surfaces of constant density is much smaller than mixing along such surfaces.
- Horizontal eddy diffusivity in the ocean is much greater than vertical eddy diffusivity.
- Measurements of eddy diffusivity indicate water is mixed vertically near oceanic boundaries such as above seamounts and mid-ocean ridges.

09 THE RESPONSE AND BEHAVIOUR OF THE UPPER OCEAN TO THE INFLUENCE OF WIND

If you have had a chance to travel around the United States, you may have noticed that the climate of the east coast differs considerably from that on the west coast. Why? Why is the climate of Charleston, South Carolina so different from that of San Diego, although both are near 32°N, and both are on or near the ocean? Charleston has 125–150 cm of rain a year, San Diego has 25–50 cm, Charleston has hot summers, San Diego has cool summers. Or why is the climate of San Francisco so different from that of Norfolk, Virginia?

If we look closely at the characteristics of the atmosphere along the two coasts near 32°N, we find more differences that may explain the climate. For example, when the wind blows onshore toward San Diego, it brings a cool, moist, marine, boundary layer a few hundred meters thick capped by much warmer, dry air. On the east coast, when the wind blows onshore, it brings a warm, moist, marine, boundary layer that is much thicker. Convection, which produces rain, is much easier on the east coast than on the west coast. Why then is the atmospheric boundary layer over the water so different on the two coasts? The answer can be found in part by studying the ocean's response to local winds, the subject of this chapter.

9.1 INERTIAL MOTION

To begin our study of currents near the sea surface, let's consider first a very simple solution to the equations of motion, the response of the ocean to an impulse that sets the water in motion. For example, the impulse can be a strong wind blowing for a few hours. The water then moves only under the influence of Coriolis force. No other force acts on the water.

Such motion is said to be inertial. The mass of water continues to move due to its inertia. If the water were in space, it would move in a straight line according to Newton's second law. But on a rotating earth, the motion is much different.

From (7.12) the equations of motion for a parcel of water moving in the ocean without friction are:

$$\frac{du}{dt} = -\frac{1}{\rho}\frac{\partial p}{\partial x} + 2\Omega v \sin\varphi \quad (9.1a)$$
$$\frac{dv}{dt} = -\frac{1}{\rho}\frac{\partial p}{\partial y} - 2\Omega u \sin\varphi \quad (9.1b)$$
$$\frac{dw}{dt} = -\frac{1}{\rho}\frac{\partial p}{\partial z} + 2\Omega u \cos\varphi - g \quad (9.1c)$$

where p is pressure, $\Omega = 2\,\pi/(\text{sidereal day}) = 7.292 \times 10^{-5}$ rad/s is the rotation of the earth in fixed coordinates, and ϕ is latitude.

Let's now look for simple solutions to these equations. To do this we must simplify the momentum equations. First, if only the Coriolis force acts on the water, there must be no horizontal pressure gradient:

$$\frac{\partial p}{\partial x} = \frac{\partial p}{\partial y} = 0$$

Furthermore, we can assume that the flow is horizontal, and (9.1) becomes:

$$\frac{du}{dt} = 2\Omega\, v \sin\varphi = fv \quad (9.2a)$$
$$\frac{dv}{dt} = -2\Omega\, u \sin\varphi = -fu \quad (9.2b)$$

where:

$$\boxed{f = 2\Omega\,\sin\varphi} \quad (9.3)$$

is the *Coriolis Parameter* and $\Omega = 7.292 \times 10^{-5}$/s is the rotation rate of earth. Equations (9.2) are two coupled, first-order, linear, differential equations which can be solved with standard techniques. If we solve the second equation for u, and insert it into the first equation we obtain:

$$\frac{du}{dt} = -\frac{1}{f}\frac{d^2 v}{dt^2} = fv$$

Rearranging the equation puts it into a standard form we should recognize, the equation for the harmonic oscillator:

$$\frac{d^2 v}{dt^2} + f^2 v = 0 \quad (9.4)$$

which has the solution (9.5). This current is called an *inertial current* or *inertial oscillation*:

$$\boxed{\begin{aligned} u &= V\,\sin ft \\ v &= V\,\cos ft \\ V^2 &= u^2 + v^2 \end{aligned}} \quad (9.5)$$

Note that (9.5) are the parametric equations for a circle with diameter $D_i = 2V/f$ and period $T_i = (2\pi)/f = T_{sd}/(2\sin\phi)$ where T_{sd} is a sidereal day.

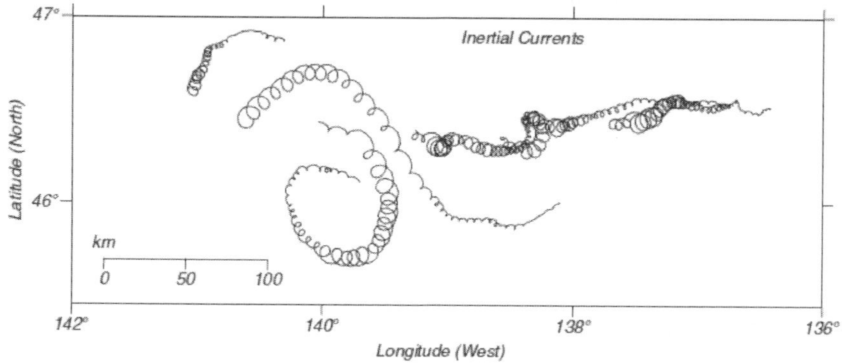

Figure 9.1 Inertial currents in the North Pacific in October 1987 (days 275–300) measured by holey-sock drifting buoys drogued at a depth of 15 meters. Positions were observed 10–12 times per day by the Argos system on NOAA polar orbiting weather satellites and interpolated to positions every three hours. The largest currents were generated by a storm on day277. Note these are not individual eddies. The entire surface is rotating. A drogue placed anywhere in the region would have the same circular motion. After van Meurs (1998).

T_i is the *inertial period*. It is one half the time required for the rotation of a local plane on earth's surface (Table 9.1). The rotation is *anti-cyclonic*: clockwise in the northern hemisphere, counter clockwise in the southern. Inertial currents are the free motion of parcels of water on a rotating plane.

Table 9.1 Inertial Oscillations

Latitude (ϕ)	*Ti* (hr)	D (km)
	for V = 20 cm/s	
90°	11.97	2.7
35°	20.87	4.8
10°	68.93	15.8

Inertial currents are the most common currents in the ocean (figure 9.1). Webster (1968) reviewed many published reports of inertial currents and found that currents have been observed at all depths in the ocean and at all latitudes. The motions are transient and decay in a few days. Oscillations at different depths or at different nearby sites are usually incoherent.

Inertial currents are caused by rapid changes of wind at the sea surface, with rapid changes of strong winds producing the largest oscillations. Although we have derived the equations for the oscillation assuming frictionless flow, friction cannot be completely neglected. With time, the oscillations decay into other surface currents. (See, for example, Apel, 1987: §6.3 for more information.)

9.2 EKMAN LAYER AT THE SEA SURFACE

Steady winds blowing on (the sea surface produce a thin, horizontal boundary layer, the *Ekman layer*. By thin, I mean a layer that is at most a few-hundred meters thick, which is thin compared with the depth of the water in the deep ocean. A similar boundary layer exists at the bottom of the ocean, the *bottom Ekman layer*, and at the bottom of the atmosphere just above the sea surface, the planetary boundary layer or frictional layer described in §4.3. The Ekman layer is named after Professor Walfrid Ekman, who worked out its dynamics for his doctoral thesis.

Table 9.2 Contributions to the Theory of the Wind-Driven Circulation

Fridtjof Nansen	(1898)	Qualitative theory, currents transport water at an angle to the wind.
Vagn Walfrid Ekman	(1902)	Quantitative theory for wind-driven transport at the sea surface.
Harald Sverdrup	(1947)	Theory for wind-driven circulation in the eastern Pacific.
Henry Stommel	(1948)	Theory for westward intensification of wind-driven circulation (western boundary currents).
Walter Munk	(1950)	Quantitative theory for main features of the wind-driven circulation.
Kirk Bryan	(1963)	Numerical models of the oceanic circulation.
Bert Semtner and Robert Chervin	(1988)	Global, eddy-resolving, realistic model of the ocean's circulation.

Ekman's work was the first of a remarkable series of studies conducted during the first half of the twentieth century that led to an understanding of how winds drive the ocean's circulation (Table 9.1). In this chapter we consider Nansen and Ekman's work. The rest of the story is given in chapters 11 and 13.

Nansen's Qualitative Arguments: Fridtjof Nansen noticed that wind tended to blow ice at an angle of 20°–40° to the right of the wind in the Arctic, by which he meant that the track of the iceberg was to the right of the wind looking downwind (See figure 9.2). He later worked out the balance of forces that must exist when wind tried to push icebergs downwind on a rotating earth.

Nansen argued that three forces must be important:

1. Wind Stress, **W**;
2. Friction **F** (otherwise the iceberg would move as fast as the wind);
3. Coriolis Force, **C**.

Nansen argued further that the forces must have the following attributes:

1. Drag must be opposite the direction of the ice's velocity;
2. Coriolis force must be perpendicular to the velocity;
3. The forces must balance for steady flow.

$$\mathbf{W} + \mathbf{F} + \mathbf{C} = 0$$

Ekman's Solution: Nansen asked Vilhelm Bjerknes to let one of Bjerknes' students make a theoretical study of the influence of earth's rotation on wind-driven currents. Walfrid Ekman was chosen, and he presented the results in his thesis at Uppsala (Kullenberg, 1954). Ekman later expanded the study to include the influence of continents and differences of density of water (Ekman, 1905). The following follows Ekman's line of reasoning in that paper.

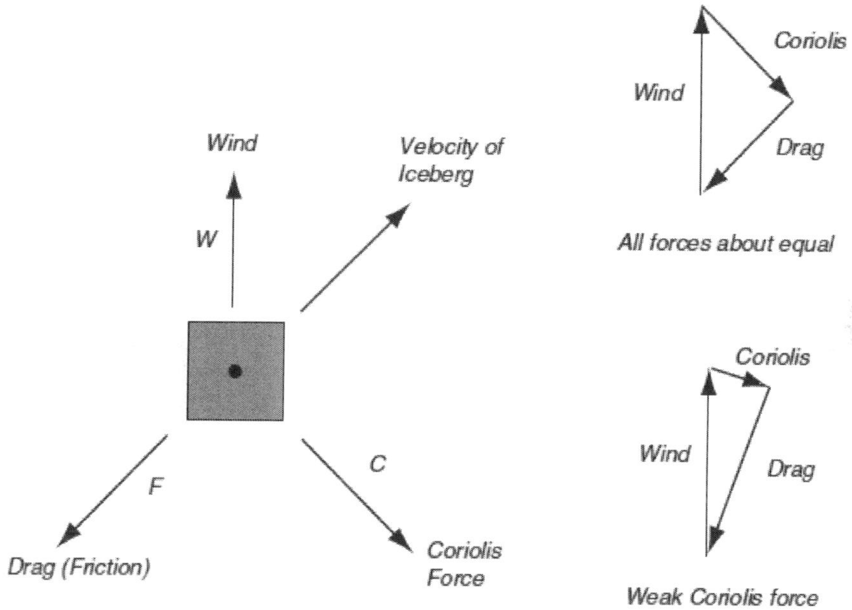

Figure 9.2 The balance of forces acting on an iceberg in a wind on a rotating earth.

Ekman assumed a steady, homogeneous, horizontal flow with friction on a rotating earth. Thus horizontal and temporal derivatives are zero:

$$\frac{\partial}{\partial t} = \frac{\partial}{\partial x} = \frac{\partial}{\partial y} = 0 \tag{9.6}$$

For flow on a rotating earth, this leaves a balance between frictional and Coriolis forces (8.15). Ekman further assumed a constant vertical eddy viscosity of the form (8:13):

$$T_{xz} = \rho A_z \frac{\partial u}{\partial z}, \qquad T_{yz} = \rho A_z \frac{\partial v}{\partial z} \tag{9.7}$$

where Txz , Tyz are the components of the wind stress in the x, y directions, and ρ is the density of sea water.

Using (9.7) in (8.15), the x and y momentum equations are:

$$fv + A_z \frac{\partial^2 u}{\partial z^2} = 0 \tag{9.8a}$$

$$-fu + A_z \frac{\partial^2 v}{\partial z^2} = 0 \tag{9.8b}$$

where f is the Coriolis parameter.

It is easy to verify that the equations (9.9) have solutions:

$$u = V_0 \exp(az) \cos(\pi/4 + az) \quad (9.9a)$$

$$v = V_0 \exp(az) \sin(\pi/4 + az) \quad (9.9b)$$

when the wind is blowing to the north ($T = Tyz$). The constants are and $V0$ is the velocity of the current at the sea surface.

$$V_0 = \frac{T}{\sqrt{\rho_w^2\, f\, A_z}} \quad \text{and} \quad a = \sqrt{\frac{f}{2A_z}} \quad (9.10)$$

Now let's look at the form of the solutions. At the sea surface $z = 0$, $\exp(z = 0) = 1$, and

$$u(0) = V0 \cos(\pi/4) \quad (9.11a)$$

$$v(0) = V0 \sin(\pi/4) \quad (9.11b)$$

The current has a speed of $V0$ to the northeast. In general, the surface current is 45° to the right of the wind when looking downwind in the northern hemisphere. The current is 45° to the left of the wind in the southern hemisphere. Below the surface, the velocity decays exponentially with depth (figure 9.3):

$$\left[u^2(z) + v^2(z)\right]^{1/2} = V_0 \exp(az) \quad (9.12)$$

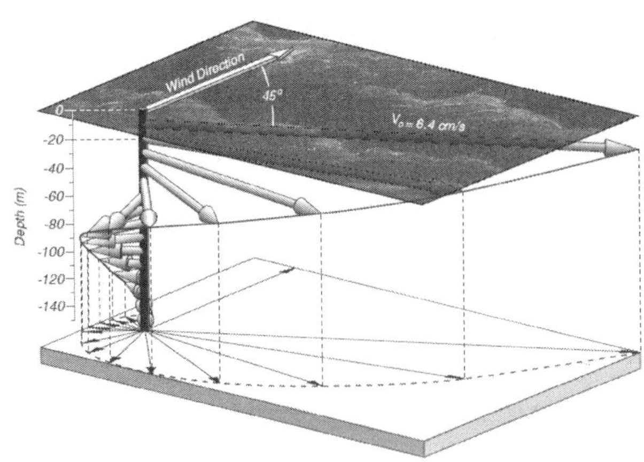

Figure 9.3. Ekman current generated by a 10 m/s wind at 35° N.

Values for Ekman's Constants: To proceed further, we need values for any two of the free parameters: the velocity at the surface, $V0$; the coefficient of eddy viscosity, Az; or the wind stress T.

The wind stress is well known, and Ekman used the bulk formula (4.2):

$$T_{yz} = T = \rho_{air}\, C_D\, U_{10}^2 \quad (9.13)$$

where ρair is the density of air, CD is the drag coefficient, and U10 is the wind speed at 10 m above the sea. Ekman turned to the literature to obtain values for V0 as a function of wind speed. He found:

$$V_0 = \frac{0.0127}{\sqrt{\sin|\varphi|}} U_{10}, \qquad |\varphi| \geq 10 \quad (9.14)$$

With this information, he could then calculate the velocity as a function of depth knowing the wind speed U_{10} and wind direction.

Ekman Layer Depth: The thickness of the Ekman layer is arbitrary because the Ekman currents decrease exponentially with depth. Ekman proposed that the thickness be the depth DE at which the current velocity is opposite the velocity at the surface, which occurs at a depth DE = π/a, and the *Ekman layer depth* is:

$$D_E = \sqrt{\frac{2\pi^2 A_z}{f}} \quad (9.15)$$

Using (9.13) in (9.10), dividing by U10, and using (9.14) and (9.15) gives:

$$D_E = \frac{7.6}{\sqrt{\sin|\varphi|}} U_{10} \quad (9.16)$$

in SI units. Wind in meters per second gives depth in meters. The constant in (9.16) is based onρ =1027 kg/m³, ρair = 1.25 kg/m³, and Ekman's value of CD = 2.6 × 10^{-3} for the drag coefficient.

Using (9.16) with typical winds, the depth of the Ekman layer varies from about 45 to 300 meters (Table 9.3), and the velocity of the surface current varies from 2.5% to 1.1% of the wind speed depending on latitude.

Table 9.3 Typical Ekman Depths

U_{10} [m/s]	Latitude	
	15°	45°
5	75 m	45 m
10	150 m	90 m
20	300 m	180 m

The Ekman Number: Coriolis and Frictional Forces: The depth of the Ekman layer is closely related to the depth at which frictional force is equal to the Coriolis force in the momentum equation (9.9). The Coriolis force is *fu*, and the frictional force is Az ∂²U/∂z². The ratio of the forces, which is non dimensional, is called the *Ekman Number Ez*:

$$E_z = \frac{\text{Friction Force}}{\text{Coriolis Force}} = \frac{A_z \frac{\partial^2 u}{\partial z^2}}{fu} = \frac{A_z \frac{u}{d^2}}{fu}$$

$$E_z = \frac{A_z}{f\,d^2} \quad (9.17)$$

where we have approximated the terms using typical velocities u, and typical depths d. The subscript z is needed because the ocean is stratified and mixing in the vertical is much less than mixing in the horizontal. Note that as depth increases, friction becomes small, and eventually, only the Coriolis force remains.

Solving (9.17) for d gives which agrees with the functional form (9.15) proposed by Ekman. Equating (9.18) and (9.15) requires $Ez = 1/(2\pi^2) = 0.05$ at the Ekman depth. Thus Ekman chose a depth at which frictional forces are much smaller than the Coriolis force.

$$d = \sqrt{\frac{A_z}{f E_z}} \quad (9.18)$$

Bottom Ekman Layer: The Ekman layer at the bottom of the ocean and the atmosphere differs from the layer at the ocean surface. The solution for a bottom layer below a fluid with velocity U in the x-direction is:

$$u = U\,[1 - \exp(-az) \cos az] \quad (9.19a)$$

$$v = U \exp(-az) \sin az \quad (9.19b)$$

The velocity goes to zero at the boundary, $u = v = 0$ at $z = 0$. The direction of the flow close to the boundary is 45° to the left of the flow U outside the boundary layer in the northern hemisphere, and the direction of the flow rotates with distance above the boundary (figure 9.4). The direction of rotation is anti- cyclonic with distance above the bottom.

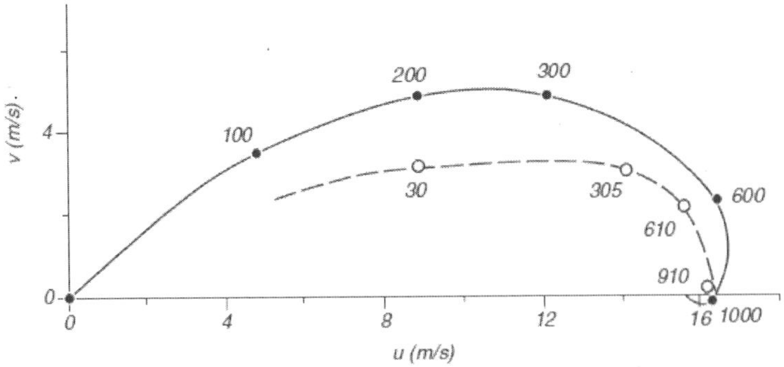

Figure 9.4 Ekman layer in the lowest kilometre of the atmosphere (solid line), with wind velocity measured by Dobson (1914) - - - . The numbers give height above the surface in meters. The Ekman layer at the sea floor has a similar shape. After Houghton (1977: 107).

Winds above the planetary boundary layer are perpendicular to the pressure gradient in the atmosphere and parallel to lines of constant surface pressure. Winds at the surface are 45° to the left of the winds aloft, and surface currents are

45° to the right of the wind at the surface. Therefore we expect current sat the sea surface to be nearly in the direction of winds above the planetary boundary layer and parallel to lines of constant pressure. Observations of surface drifters in the Pacific tend to confirm the hypothesis (figure 9.5).

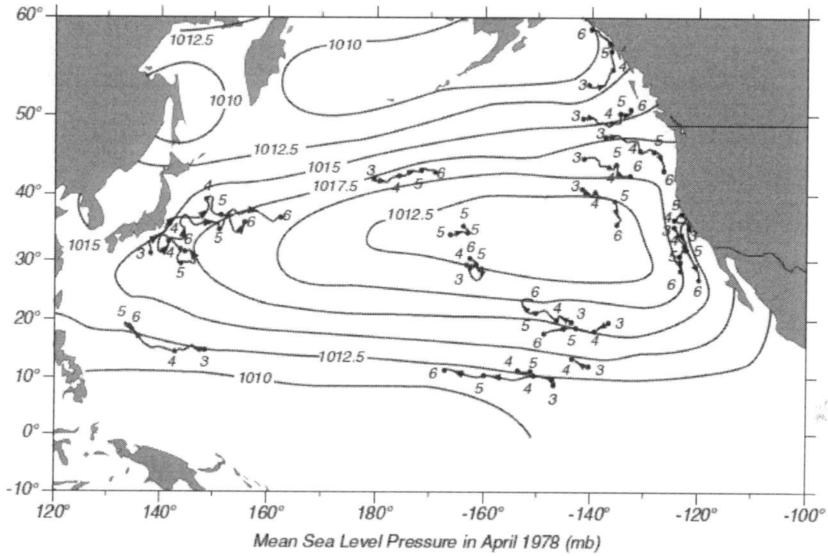

Figure 9.5 Trajectories of surface drifters in April 1978 together with surface pressure in the atmosphere averaged for the month. Note that drifters tend to follow lines of constant pressure except in the Kuroshio where ocean currents are fast compared with velocities in the Ekman layer in the ocean. After McNally et al. (1983).

Examining Ekman's Assumptions: Before considering the validity of Ekman's theory for describing flow in the surface boundary layer of the ocean, let's first examine the validity of Ekman's assumptions. He assumed:

1. No boundaries. This is valid away from coasts.
2. Deep water. This is valid if depth $\gg$ 200 m.
3. f-plane. This is valid.
4. Steady state. This is valid if wind blows for longer than a pendulum day. Note however that Ekman also calculated a time-dependent solution, as did Hasselmann (1970).
5. Az is a function of U_{10}^2 only. It is assumed to be independent of depth. This is not a good assumption. The mixed layer may be thinner than the Ekman depth, and Az will change rapidly at the bottom of the mixed layer because mixing is a function of stability. Mixing across a stable layer is much less than mixing through a layer of a neutral stability. More realistic profiles for the coefficient of eddy viscosity as a function of depth change the shape of the calculated velocity profile. I reconsider this problem below.
6. Homogeneous density. This is probably good, except as it effects stability.

Observations of Flow near the Sea Surface: Does the flow close to the sea surface agree with Ekman's theory? Measurements of currents made during several, very careful experiments indicate that Ekman's theory is remarkably good. The theory accurately describes the flow averaged over many days.

Weller and Plueddmann (1996) measured currents from 2 m to 132 m using 14 vector-measuring current meters deployed from the Floating Instrument Platform flip in February and March 1990 500 km west of point Conception, California. This was the last of a remarkable series of experiments coordinated by Weller using instruments on flip.

Davis, DeSzoeke, and Niiler (1981) measured currents from 2 m to 175 m using 19 vector-measuring current meters deployed from a mooring for 19 days in August and September 1977 at 50°N, 145°W in the northeast Pacific.

Ralph and Niiler (2000) tracked 1503 drifters drogue to 15 m depth in the Pacific from March 1987 to December 1994. Wind velocity was obtained every 6 hours from the European Centre for Medium-Range Weather Forecasts ECMWF. The results of the experiments indicate that:

1. Inertial currents are the largest component of the flow.
2. The flow is nearly independent of depth within the mixed layer for periods near the inertial period. Thus the mixed layer moves like a slab at the inertial period. Current shear is concentrated at the top of the thermo cline.
3. The flow averaged over many inertial periods is almost exactly that calculated from Ekman's theory. The shear of the Ekman currents extends through the averaged mixed layer and into the thermo cline. Ralph and Niiler found (using SI units, *U* in m/s):

$$D_E = \frac{7.12}{\sqrt{\sin|\varphi|}} U_{10} \quad (9.20)$$

$$V_0 = \frac{0.0068}{\sqrt{\sin|\varphi|}} U_{10} \quad (9.21)$$

The Ekman-layer depth *DE* is almost exactly that proposed by Ekman (9.16), but the surface current *V0* is half his value (9.14).

- The angle between the wind and the flow at the surface depends on latitude, and it is near 45° at mid latitudes (figure 9.6).
- The transport is 90° to the right of the wind in the northern hemisphere. The transport direction agrees well with Ekman's theory.

Influence of Stability in the Ekman Layer: Ralph and Niiler (2000) point out that Ekman's choice of an equation for surface currents (9.14), which leads to (9.16), is consistent with theories that include the influence of stability in the upper ocean. Currents with periods near the inertial period produce shear in the thermo cline. The shear mixes the surface layers when the Richardson number falls below the critical value (Pollard et al. 1973). This idea, when included in mixed-layer theories, leads to a surface current *V0* that is proportional to *N/f*

$$V_0 \sim U_{10}\sqrt{N/f} \quad (9.22)$$

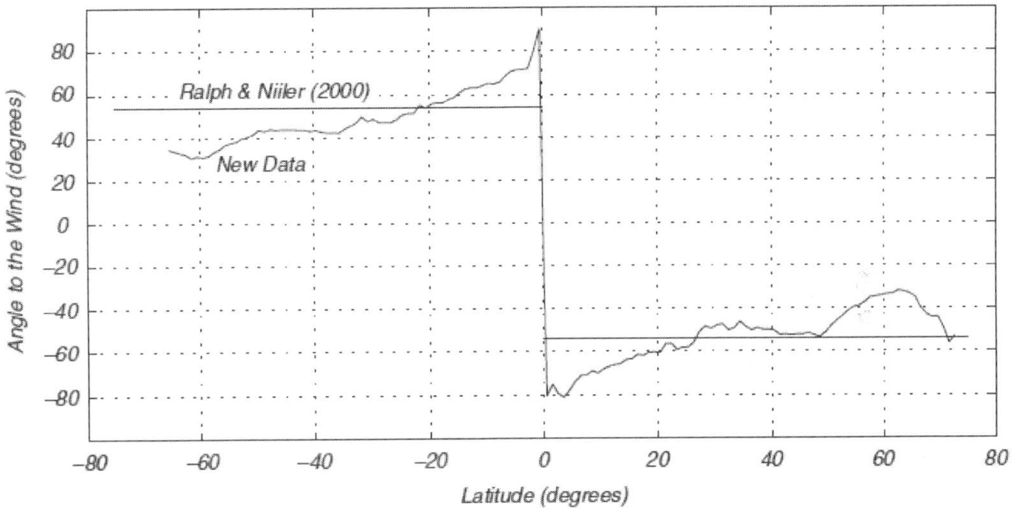

Figure 9.6 Angle between the wind and flow at the surface calculated by Maximenko and Niiler using positions from drifters drogued at 15 m with satellite-altimeter, gravity, and grace data and winds from the NCAR/NCEP reanalysis.

where *N* is the stability frequency defined by (8.36). Furthermore

$$A_z \sim U_{10}^2/N \qquad \text{and} \qquad D_E \sim U_{10}/\sqrt{Nf} \quad (9.23)$$

Notice that (9.22) and (9.23) are now dimensionally correct. The equations used earlier, (9.14), (9.16), (9.20), and (9.21) all required a dimensional coefficient.

9.3 EKMAN MASS TRANSPORT

Flow in the Ekman layer at the sea surface carries mass. For many reasons we may want to know the total mass transported in the layer. The *Ekman mass transport ME* is defined as the integral of the Ekman velocity *UE, VE* from the surface to a depth *d* below the Ekman layer. The two components of

the transport are *MEx, MEy:*

$$M_{Ex} = \int_{-d}^{0} \rho U_E \, dz, \qquad M_{Ey} = \int_{-d}^{0} \rho V_E \, dz \quad (9.24)$$

The transport has units kg/(m·s). It is the mass of water passing through a vertical plane one meter wide that is perpendicular to the transport and extending from the surface to depth *-d* (figure 9.7).

We calculate the Ekman mass transports by integrating (8.15) in (9.24):

$$f \int_{-d}^{0} \rho V_E \, dz = f\, M_{Ey} = -\int_{-d}^{0} dT_{xz}$$
$$f\, M_{Ey} = -T_{xz}\big|_{z=0} + T_{xz}\big|_{z=-d} \quad (9.25)$$

A few hundred meters below the surface the Ekman velocities approach zero, and the last term of (9.25) is zero. Thus mass transport is due only to wind

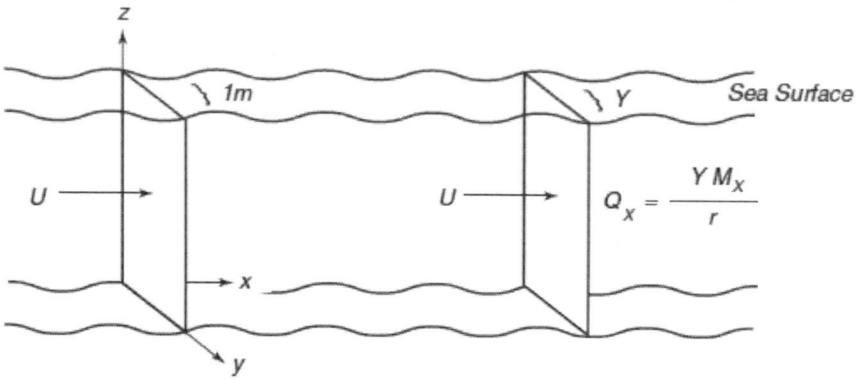

Figure 9.7 Sketch for defining Left: mass transports, and Right: volume transports.

stress at the sea surface ($z = 0$). In a similar way, we can calculate the transport in the x direction to obtain the two components of the *Ekman transport*:

$$f M_{Ey} = -Txz\,(0) \quad (9.26a)$$
$$f M_{Ex} = Tyz\,(0) \quad (9.26b)$$

where $Txz\,(0)$, $Tyz\,(0)$ are the two components of the stress at the sea surface. Notice that the transport is perpendicular to the wind stress, and to the right of the wind in the northern hemisphere. If the wind is to the north in the positive y direction (a south wind), then $Txz\,(0) = 0$, $MEy = 0$, and $MEx = Tyz\,(0)/f$. In the northern hemisphere, f is positive, and the mass transport is in the x direction, to the east.

It may seem strange that the drag of the wind on the water leads to a current at right angles to the drag. The result follows from the assumption that friction is confined to a thin surface boundary layer, that it is zero in the interior of the ocean, and that the current is in equilibrium with the wind so that it is no longer accelerating.

Volume transport Q is the mass transport divided by the density of water and multiplied by the width perpendicular to the transport.

$$Q_x = \frac{Y M_x}{\rho}, \qquad Q_y = \frac{X M_y}{\rho} \quad (9.27)$$

where Y is the north-south distance across which the eastward transport Qx is calculated, and X in the east-west distance across which the northward trans- port Qy is calculated. Volume transport has dimensions of cubic meters per second. A convenient unit for volume transport in the ocean is a million cubic meters per second. This unit is called a *Sverdrup*, and it is abbreviated Sv.

Recent observations of Ekman transport in the ocean agree with the theoretical values (9.26). Chereskin and Roemmich (1991) measured the Ekman volume transport across 11°N in the Atlantic using an acoustic Doppler current profiler described in Chapter 10. They calculated a transport of $Q_y = 12.0 \pm 5.5$ Sv (northward) from direct measurements of current, $Q_y = 8.8 \pm 1.9$ Sv from measured winds using (9.26) and (9.27), and $Q_y = 13.5 \pm 0.3$ Sv from mean winds averaged over many years at 11°N.

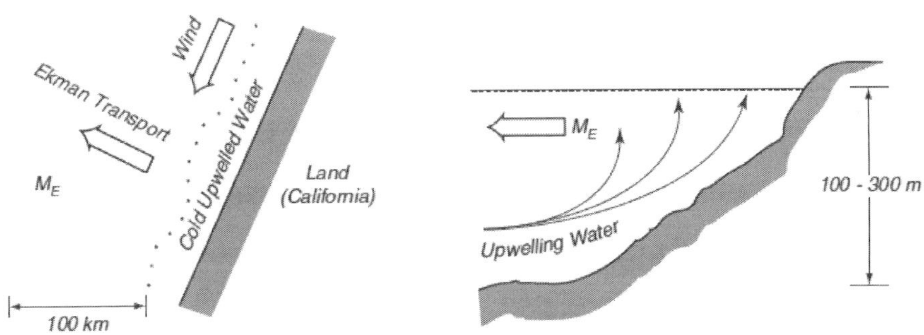

Figure 9.8 Sketch of Ekman transport along a coast leading to upwelling of cold water along the coast. Left: Plan view. North winds along a west coast in the northern hemisphere cause Ekman transports away from the shore. Right: Cross section. The water transported offshore must be replaced by water upwelling from below the mixed layer.

Use of Transports: Mass transports are widely used for two important reasons. First, the calculation is much more robust than calculations of velocities in the Ekman layer. By robust, I mean that the calculation is based on fewer assumptions, and that the results are more likely to be correct. Thus the calculated mass transport does not depend on knowing the distribution of velocity in the Ekman layer or the eddy viscosity.

Second, the variability of transport in space has important consequences.

Let's look at a few applications.

9.4 UTILIZING THE EKMAN THEORY IN PRACTICAL SCENARIOS

Because steady winds blowing on the sea surface produce an Ekman layer that transports water at right angles to the wind direction, any spatial variability of the wind, or winds blowing along some coasts, can lead to upwelling. And upwelling is important:

- Upwelling enhances biological productivity, which feeds fisheries.

- Cold upwelled water alters local weather. Weather onshore of regions of upwelling tend to have fog, low stratus clouds, a stable stratified atmosphere, little convection, and little rain.
- Spatial variability of transports in the open ocean leads to upwelling and downwelling, which leads to redistribution of mass in the ocean, which leads to wind-driven geostrophic currents via Ekman pumping.

Coastal Upwelling: To see how winds lead to upwelling, consider north winds blowing parallel to the California Coast (figure 9.8 left). The winds produce a mass transport away from the shore everywhere along the shore. The water pushed offshore can be replaced only by water from below the Ekman layer.

This is *upwelling* (figure 9.8 right). Because the upwelled water is cold, the upwelling leads to a region of cold water at the surface along the coast. Figure10.16 shows the distribution of cold water off the coast of California.

Upwelled water is colder than water normally found on the surface, and it is richer in nutrients. The nutrients fertilize phytoplankton in the mixed layer, which are eaten by zooplankton, which are eaten by small fish, which are eaten by larger fish and so on to infinity. As a result, upwelling regions are productive waters supporting the world's major fisheries. The important regions are offshore of Peru, California, Somalia, Morocco, and Namibia.

Now I can answer the question I asked at the beginning of the chapter: Why is the climate of San Francisco so different from that of Norfolk, Virginia? Figures 4.2 or 9.8 show that wind along the California and Oregon coasts has a strong southward component. The wind causes upwelling along the coast, which leads to cold water close to shore. The shoreward component of the wind brings warmer air from far offshore over the colder water, which cools the incoming air close to the sea, leading to a thin, cool atmospheric boundary layer. As the air cools, fog forms along the coast. Finally, the cool layer of air is blown over San Francisco, cooling the city. The warmer air above the boundary layer, due to downward velocity of the Hadley circulation in the atmosphere (see figure 4.3), inhibits vertical convection, and rain is rare. Rain forms only when winter storms coming ashore bring strong convection higher up in the atmosphere.

In addition to upwelling, other processes influence weather in California and Virginia.

- The oceanic mixed layer tends to be thin on the eastern side of ocean, and upwelling can easily bring up cold water.
- Currents along the eastern side of the ocean at mid-latitudes tend to bring colder water from higher latitudes.

All these processes are reversed offshore of east coasts, leading to warm water close to shore, thick atmospheric boundary layers, and frequent convective rain. Thus Norfolk is much different that San Francisco due to upwelling and the direction of the coastal currents.

Ekman Pumping: The horizontal variability of the wind blowing on the sea surface leads to horizontal variability of the Ekman transports. Because mass must be conserved, the spatial variability of the transports must lead to vertical velocities at the top of the Ekman layer. To calculate this velocity, we first integrate the continuity equation (7.19) in the vertical:

$$\rho \int_{-d}^{0} \left(\frac{\partial u}{\partial x} + \frac{\partial v}{\partial y} + \frac{\partial w}{\partial z} \right) dz = 0$$

$$\frac{\partial}{\partial x} \int_{-d}^{0} \rho\, u\, dz + \frac{\partial}{\partial y} \int_{-d}^{0} \rho\, v\, dz = -\rho \int_{-d}^{0} \frac{\partial w}{\partial z}\, dz$$

$$\frac{\partial M_{Ex}}{\partial x} + \frac{\partial M_{Ey}}{\partial y} = -\rho\, [w(0) - w(-d)]$$

By definition, the Ekman velocities approach zero at the base of the Ekman layer, and the vertical velocity at the base of the layer *wE* (*-d*) due to divergence of the Ekman flow must be zero. Therefore:

$$\frac{\partial M_{Ex}}{\partial x} + \frac{\partial M_{Ey}}{\partial y} = -\rho\, w_E(0) \qquad (9.28a)$$

$$\nabla_H \cdot \mathbf{M}_E = -\rho\, w_E(0) \quad (9.28b)$$

Where **M***E* is the vector mass transport due to Ekman flow in the upper boundary layer of the ocean, and ∇*H* is the horizontal divergence operator. (9.28)states that the horizontal divergence of the Ekman transports leads to a vertical velocity in the upper boundary layer of the ocean, a process called *Ekman Pumping.*

If we use the Ekman mass transports (9.26) in (9.28) we can relate Ekman pumping to the wind stress.

$$w_E(0) = -\frac{1}{\rho} \left[\frac{\partial}{\partial x} \left(\frac{T_{yz}(0)}{f} \right) - \frac{\partial}{\partial y} \left(\frac{T_{xz}(0)}{f} \right) \right] \quad (9.29a)$$

$$w_E(0) = -\text{curl}_z \left(\frac{\mathbf{T}}{\rho f} \right) \quad (9.29b)$$

where **T** is the vector wind stress and the subscript *z* indicates the vertical component of the curl.

The vertical velocity at the sea surface *w*(0) must be zero because the surface cannot rise into the air, so *wE* (0) must be balanced by another vertical velocity. We will see in Chapter 12 that it is balanced by a geostrophic velocity *wG*(0) at the top of the interior flow in the ocean.

Note that the derivation above follows Pedlosky (1996: 13), and it differs from the traditional approach that leads to a vertical velocity at the base of the Ekman layer. Pedlosky points out that if the Ekman layer is very thin compared with the depth of the ocean, it makes no difference whether the velocity is calculated at the top or bottom of the Ekman layer, but this is usually not true for the ocean. Hence, we must compute vertical velocity at the top of the layer.

9.5 LANGMUIR CIRCULATION

Measurements of surface currents show that winds generate more than Ek- man and inertial currents at the sea surface. They also generate a Langmuir circulation (Langmuir, 1938), a current that spiral around an axis parallel to the wind direction. Weller et al. (1985) observed such a flow during an experiment to measure the wind-driven circulation in the upper 50 meters of the sea. They found that during a period when the wind speed was 14 m/s, surface currents were organized into Langmuir cells spaced 20 m apart, the cells were aligned atan angle of 15° to the right of the wind, and vertical velocity at 23 m depth was concentrated in narrow jets under the areas of surface convergence (figure 9.9). Maximum vertical velocity was -0.18 m/s. The seasonal thermo cline was at 50m, and no downward velocity was observed in or below the thermo cline.

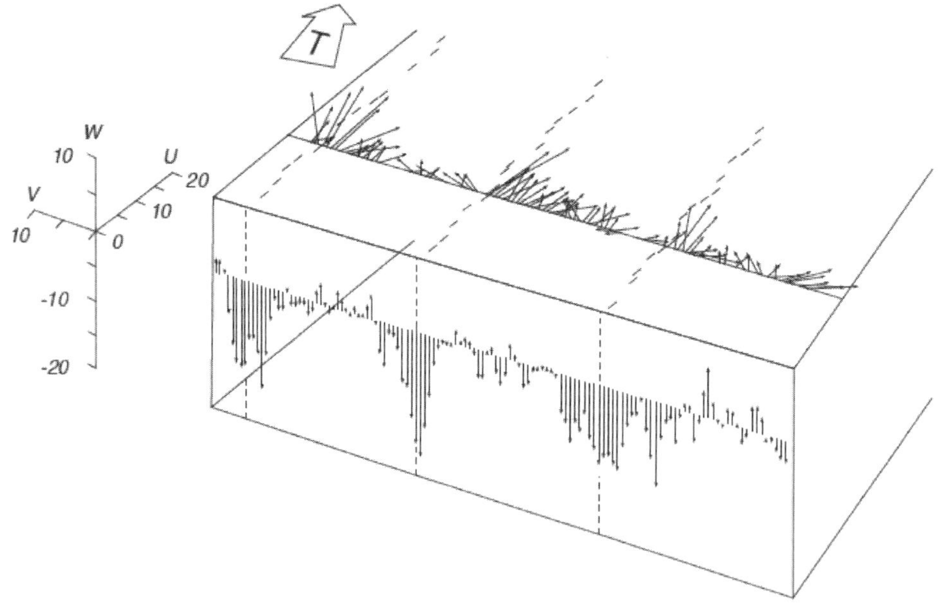

Figure 9.9 A three-dimensional view of the Langmuir circulation at the surface of the Pacific observed from the Floating Instrument Platform flip. The heavy dashed line on the sea surface indicate lines of convergence marked by cards on the surface. Vertical arrows are individual values of vertical velocity measured every 14 seconds at 23 m depth as the platform drifted through the Langmuir currents. Horizontal arrows, which are drawn on the surface for clarity, are values of horizontal velocity at 23 m. The broad arrow gives the direction of the wind. After Weller et al. (1985).

9.6 CONCISE SUMMARY

1. Changes in wind stress produce transient oscillations in the ocean called inertial currents
   - Inertial currents are very common in the ocean.
   - The period of the current is $(2\pi)/f$.
2. Steady winds produce a thin boundary layer, the Ekman layer, at the top of the ocean. Ekman boundary layers also exist at the bottom of the ocean and the atmosphere. The Ekman layer in the atmosphere above the sea surface is called the planetary boundary layer.
3. The Ekman layer at the sea surface has the following characteristics:
   - *Direction*: 45° to the right of the wind looking downwind in the Northern Hemisphere.
   - *Surface Speed* : 1–2.5% of wind speed depending on latitude.
   - *Depth*: approximately 40–300 m depending on latitude and wind velocity.
4. Careful measurements of currents near the sea surface show that:
   - Inertial oscillations are the largest component of the current in the mixed layer.
   - The flow is nearly independent of depth within the mixed layer for periods near the inertial period. Thus the mixed layer moves like a slab at the inertial period.
   - An Ekman layer exists in the atmosphere just above the sea (and land) surface.
   - Surface drifters tend to drift parallel to lines of constant atmospheric pressure at the sea surface. This is consistent with Ekman's theory.
   - The flow averaged over many inertial periods is almost exactly that calculated from Ekman's theory.
5. Transport is 90° to the right of the wind in the northern hemisphere.
6. Spatial variability of Ekman transport, due to spatial variability of winds over distances of hundreds of kilometres and days, leads to convergence and divergence of the transport.
   - Winds blowing toward the equator along west coasts of continents produces upwelling along the coast. This leads to cold, productive waters within about 100 km of the shore.
   - Upwelled water along west coasts of continents modifies the weather along the west coasts.
7. Ekman pumping, which is driven by spatial variability of winds, drives a vertical current, which drives the interior geostrophic circulation of the ocean.

10 MAINTAINING A GEOSTROPHIC BALANCE AND THE RESULTANT GEOSTROPHIC CURRENTS

Within the ocean's interior away from the top and bottom Ekman layers, for horizontal distances exceeding a few tens of kilometres, and for times exceeding a few days, horizontal pressure gradients in the ocean almost exactly balance the Coriolis force resulting from horizontal currents. This balance is known as the *geostrophic balance.*

The dominant forces acting in the vertical are the vertical pressure gradient and the weight of the water. The two balance within a few parts per million. Thus pressure at any point in the water column is due almost entirely to the weight of the water in the column above the point. The dominant forces in the horizontal are the pressure gradient and the Coriolis force. They balance within a few parts per thousand over large distances and times (See Box).

Both balances require that viscosity and nonlinear terms in the equations of motion be negligible. Is this reasonable? Consider viscosity. We know that a rowboat weighing a hundred kilograms will coast for maybe ten meters after the rower stops. A super tanker moving at the speed of a rowboat may coast for kilometres. It seems reasonable, therefore that a cubic kilometre of water weighing 10^{15} kg would coast for perhaps a day before slowing to a stop. And oceanic mesoscale eddies contain perhaps 1000 cubic kilometres of water. Hence, our intuition may lead us to conclude that neglect of viscosity is reasonable. Of course, intuition can be wrong, and we need to refer back to scaling arguments.

10.1 HYDROSTATIC EQUILIBRIUM

Before describing the geostrophic balance, let's first consider the simplest solution of the momentum equation, the solution for an ocean at rest. It gives the hydrostatic pressure within the ocean. To obtain the solution, we assume the fluid is stationary:

$$u = v = w = 0; \quad (10.1)$$

the fluid remains stationary:

$$\frac{du}{dt} = \frac{dv}{dt} = \frac{dw}{dt} = 0; \quad (10.2)$$

Scaling the Equations: The Geostrophic Approximation: We wish to simplify the equations of motion to obtain solutions that describe the deep-sea conditions well away from coasts and below the Ekman boundary layer at the surface. To begin, let's examine the typical size of each term in the equations in the expectation that some will be so small that they can be dropped without changing the dominant characteristics of the solutions. For interior, deep-sea conditions, typical values for distance *L*, horizontal velocity *U*, depth *H*, Coriolis parameter *f*, gravity *g*, and density ρ are:

$$L \approx 10^6 \text{ m} \qquad H_1 \approx 10^3 \text{ m} \qquad f \approx 10^{-4} \text{ s}^{-1} \qquad \rho \approx 10^3 \text{ kg/m}^3$$
$$U \approx 10^{-1} \text{ m/s} \qquad H_2 \approx 1 \text{ m} \qquad \rho \approx 10^3 \text{ kg/m}^3 \qquad g \approx 10 \text{ m/s}^2$$

where H_1 and H_2 are typical depths for pressure in the vertical and horizontal. From these variables we can calculate typical values for vertical velocity W, pressure P, and time T :

$$\frac{\partial W}{\partial z} = -\left(\frac{\partial U}{\partial x} + \frac{\partial v}{\partial y}\right)$$
$$\frac{W}{H_1} = \frac{U}{L}; \quad W = \frac{UH_1}{L} = \frac{10^{-1}\,10^3}{10^6}\text{ m/s} = 10^{-4}\text{m/s}$$
$$P = \rho g H_1 = 10^3\,10^1\,10^3 = 10^7 \text{ Pa}; \quad \partial p/\partial x = \rho g H_2/L = 10^{-2}\text{Pa/m}$$
$$T = L/U = 10^7 \text{ s}$$

The momentum equation for vertical velocity is therefore:

$$\frac{\partial w}{\partial t} + u\frac{\partial w}{\partial x} + v\frac{\partial w}{\partial y} + w\frac{\partial w}{\partial z} = -\frac{1}{\rho}\frac{\partial p}{\partial z} + 2\Omega\, u\cos\varphi - g$$
$$\frac{W}{T} + \frac{UW}{L} + \frac{UW}{L} + \frac{W^2}{H} = \frac{P}{\rho H_1} + fU - g$$
$$10^{-11} + 10^{-11} + 10^{-11} + 10^{-11} = 10 \quad + 10^{-5} - 10$$

and the only important balance in the vertical is hydrostatic:

$$\frac{\partial p}{\partial z} = -\rho g \quad \text{Correct to} \quad 1:10^6.$$

The momentum equation for horizontal velocity in the *x* direction is:

$$\frac{\partial u}{\partial t} + u\frac{\partial u}{\partial x} + v\frac{\partial u}{\partial y} + w\frac{\partial u}{\partial z} = -\frac{1}{\rho}\frac{\partial p}{\partial x} + fv$$
$$10^{-8} + 10^{-8} + 10^{-8} + 10^{-8} = 10^{-5} + 10^{-5}$$

Thus the Coriolis force balances the pressure gradient within one part per thou- sand. This is called the *geostrophic balance*, and the *geostrophic equations* are:

$$\frac{1}{\rho}\frac{\partial p}{\partial x} = fv; \quad \frac{1}{\rho}\frac{\partial p}{\partial y} = -fu; \quad \frac{1}{\rho}\frac{\partial p}{\partial z} = -g$$

This balance applies to oceanic flows with horizontal dimensions larger than roughly 50 km and times greater than a few days.

and, there is no friction:

$$fx = fy = fz = 0. \quad (10.3)$$

With these assumptions the momentum equation (7.12) becomes:

$$\frac{1}{\rho}\frac{\partial p}{\partial x} = 0; \qquad \frac{1}{\rho}\frac{\partial p}{\partial y} = 0; \qquad \frac{1}{\rho}\frac{\partial p}{\partial z} = -g(\varphi, z) \quad (10.4)$$

where I have explicitly noted that gravity *g* is a function of latitude ϕ and heightz. I will show later why I have kept this explicit.

Equations (10.4) require surfaces of constant pressure to be level surface (see page 30). A surface of constant pressure is an *isobaric surface*. The last equation can be integrated to obtain the pressure at any depth *h*. Recalling that ρ is a function of depth for an ocean at rest.

$$p = \int_{-h}^{0} g(\varphi, z)\, \rho(z)\, dz \quad (10.5)$$

For many purposes, *g* and ρ are constant, and $p = \rho g h$. Later, I will show that (10.5) applies with an accuracy of about one part per million even if the ocean is not at rest.

The SI unit for pressure is the pascal (Pa). A bar is another unit of pressure. One bar is exactly 10^5 Pa (table 10.1). Because the depth in meters and pressure in decibars are almost the same numerically, oceanographers prefer to state pressure in decibars.

Table 10.1 Units of Pressure

1 pascal (Pa) =	1 N/m^2 = 1 kg·s^{-2}·m^{-1}
1 bar =	10^5 Pa
1 decibar =	10^4 Pa
1 millibar =	100 Pa

10.2 GEOSTROPHIC EQUATIONS

The geostrophic balance requires that the Coriolis force balance the horizontal pressure gradient. The equations for geostrophic balance are derived from the equations of motion assuming the flow has no acceleration, $du/dt = dv/dt = dw/dt = 0$; that horizontal velocities are much larger than vertical, $w \ll u, v$; that the only external force is gravity; and that friction is small. With these assumptions (7.12) become

$$\frac{\partial p}{\partial x} = \rho f v; \quad \frac{\partial p}{\partial y} = -\rho f u; \quad \frac{\partial p}{\partial z} = -\rho g \quad (10.6)$$

where $f = 2\Omega \sin \phi$ is the Coriolis parameter. These are the *geostrophic equations.*

The equations can be written:

$$u = -\frac{1}{f\rho}\frac{\partial p}{\partial y}; \qquad v = \frac{1}{f\rho}\frac{\partial p}{\partial x} \quad (10.7a)$$

$$p = p_0 + \int_{-h}^{\zeta} g(\varphi, z)\rho(z)dz \quad (10.7b)$$

where $p0$ is atmospheric pressure at $z = 0$, and ζ is the height of the sea surface. Note that I have allowed for the sea surface to be above or below the surface $z = 0$; and the pressure gradient at the sea surface is balanced by a surface current us.

Substituting (10.7b) into (10.7a) gives:

$$u = -\frac{1}{f\rho}\frac{\partial}{\partial y}\int_{-h}^{0} g(\varphi, z)\,\rho(z)\,dz - \frac{g}{f}\frac{\partial \zeta}{\partial y} \quad (10.8a)$$

$$u = -\frac{1}{f\rho}\frac{\partial}{\partial y}\int_{-h}^{0} g(\varphi, z)\,\rho(z)\,dz - u_s$$

where I have used the Boussinesq approximation, retaining full accuracy for ρ only when calculating pressure. In a similar way, we can derive the equation for v.

$$v = \frac{1}{f\rho}\frac{\partial}{\partial x}\int_{-h}^{0} g(\varphi, z)\,\rho(z)\,dz + \frac{g}{f}\frac{\partial \zeta}{\partial x}$$

$$v = \frac{1}{f\rho}\frac{\partial}{\partial x}\int_{-h}^{0} g(\varphi, z)\,\rho(z)\,dz + v_s \quad (10.8b)$$

If the ocean is homogeneous and density and gravity are constant, the first term on the right-hand side of (10.8) is equal to zero; and the horizontal pressure gradients within the ocean are the same as the gradient at $z = 0$. This is barotropic flow described in §10.4.

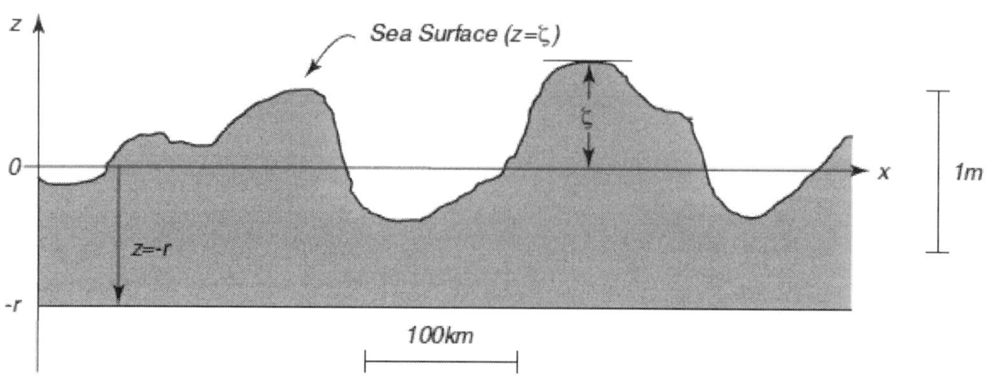

Figure 10.1 Sketch defining ζ and r, used for calculating pressure just below the sea surface.

10.3 SURFACE GEOSTROPHIC CURRENTS FROM ALTIMETRY

The geostrophic approximation applied at $z = 0$ leads to a very simple relation: surface geostrophic currents are proportional to surface slope. Consider a level surface slightly below the sea surface, say two meters below the sea surface, at $z = -r$ (figure 10.1).

The pressure on the level surface is:

$$p = \rho g (\zeta + r) \quad (10.9)$$

assuming ρ and g are essentially constant in the upper few meters of the ocean. Substituting this into (10.7a), gives the two components (*us, vs*) of the surface geostrophic current:

$$u_s = -\frac{g}{f}\frac{\partial \zeta}{\partial y}; \qquad v_s = \frac{g}{f}\frac{\partial \zeta}{\partial x} \quad (10.10)$$

where g is gravity, f is the Coriolis parameter, and ζ is the height of the sea surface above a level surface.

The Oceanic Topography: In §3.4 we define the topography of the sea surface ζ to be the height of the sea surface relative to a particular level surface, the geoid; and we defined the geoid to be the level surface that coincided with the surface of the ocean at rest. Thus, according to (10.10) the surface geostrophic currents are proportional to the slope of the topography (figure 10.2), a quantity that can be measured by satellite altimeters if the geoid is known.

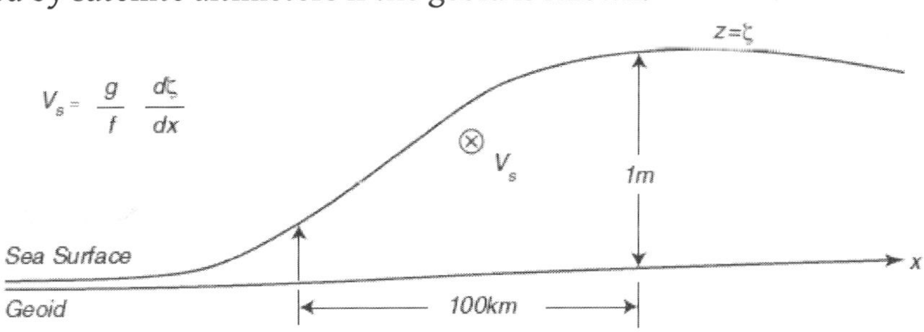

Figure 10.2 The slope of the sea surface relative to the geoid (∂ζ/∂x) is directly related to surface geostrophic currents *vs*. The slope of 1 meter per 100 kilometres (10 μrad) is typical of strong currents. *Vs* is into the paper in the northern Hemisphere.

Because the geoid is a level surface, it is a surface of constant geopotential. To see this, consider the work done in moving a mass m by a distance h perpendicular to a level surface. The work is $W = mgh$, and the change of potential energy per unit mass is gh. Thus level surfaces are surfaces of constant geopotential, where the *geopotential* $\Phi = gh$.

Topography is due to processes that cause the ocean to move: tides, ocean currents, and the changes in barometric pressure that produce the inverted barometer effect. Because the ocean's topography is due to dynamical processes, it is usually called *dynamic topography*. The topography is approximately one hundredth of the geoid undulations. Thus the shape of the sea surface is dominated by local variations of gravity. The influence of currents is much smaller.

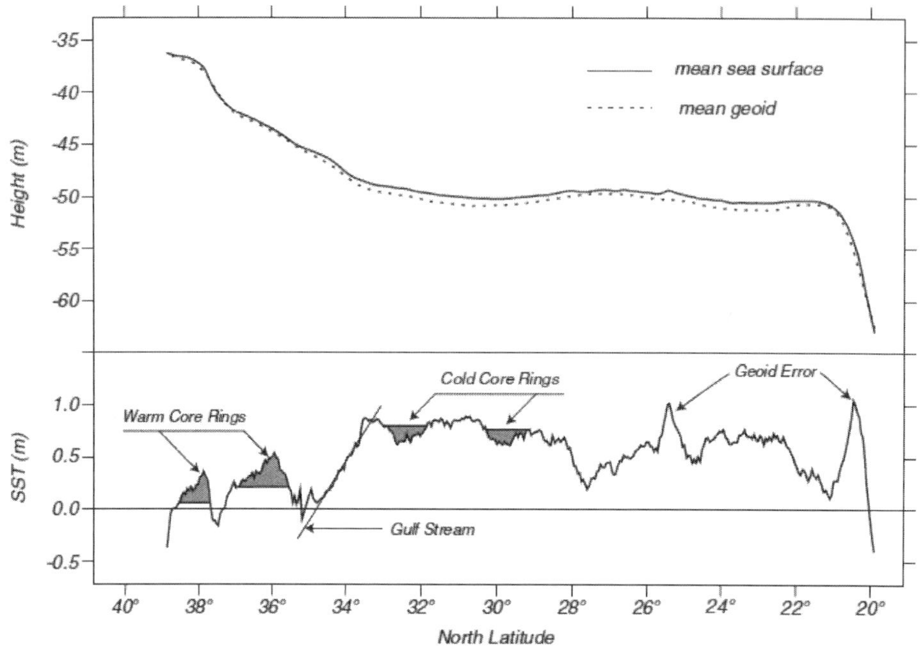

Figure 10.3 Topex/Poseidon altimeter observations of the Gulf Stream. When the altimeter observations are subtracted from the local geoid, they yield the oceanic topography, which is due primarily to ocean currents in this example. The gravimetric geoid was determined by the Ohio State University from ship and other surveys of gravity in the region. From Center for Space Research, University of Texas.

Typically, sea-surface topography has amplitude of ±1m (figure 10.3). Typical slopes are $\partial\zeta/\partial x \approx$ 1–10 micro radians for v = 0.1–1.0 m/s at mid latitude.

The height of the geoid, smoothed over horizontal distances greater than roughly 400 km, is known with an accuracy of ±1mm from data collected by the Gravity Recovery and Climate Experiment grace satellite mission.

Satellite Altimetry: Very accurate, satellite-altimeter systems are needed for measuring the oceanic topography. The first systems, carried on Seasat, Geosat, ERS–1, and ERS–2 were designed to measure week-to-week variability of currents. Topex/Poseidon, launched in 1992, was the first satellite designed to make the much more accurate measurements necessary for observing the permanent (time-averaged) surface circulation of the ocean, tides, and the variability of gyre-scale currents. It was followed in 2001 by Jason and in 2008 by Jason-2.

Because the geoid was not well known locally before about 2004, altimeters were usually flown in orbits that have an exactly repeating ground track. Thus Topex/Poseidon and Jason fly over the same ground track every 9.9156 days. By subtracting sea-surface height from one traverse of the ground track from height measured on a later traverse, changes in topography can be observed without knowing the geoid. The geoid is constant in time, and the subtraction removes the geoid, revealing changes due to changing currents, such as mesoscale eddies, assuming tides have been removed from the data (figure 10.4). Mesoscale variability includes eddies with diameters between roughly 20 and 500 km.

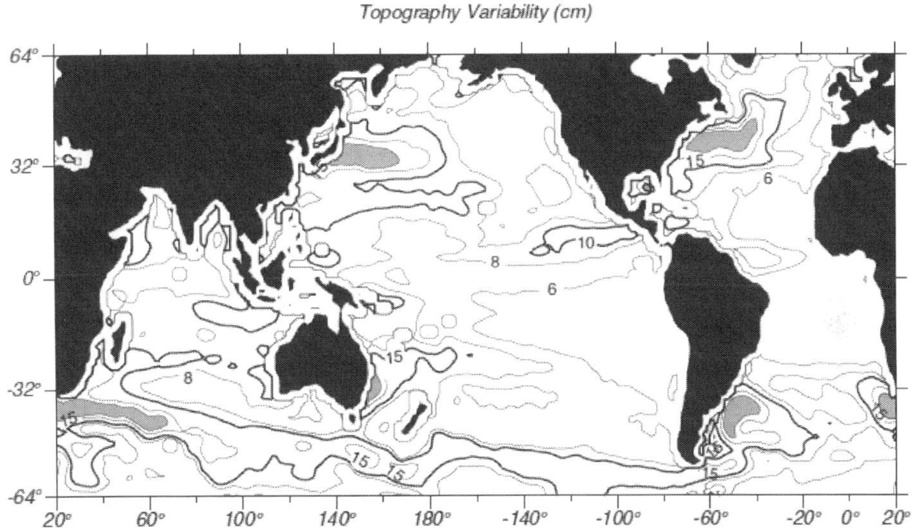

Figure 10.4 Global distribution of standard deviation of topography from Topex/Poseidon altimeter data from 10/3/92 to 10/6/94. The height variance is an indicator of variability of surface geostrophic currents. From Center for Space Research, University of Texas.

The great accuracy and precision of the Topex/Poseidon and Jason altimeter systems allow them to measure the oceanic topography over ocean basins with an accuracy of ±(2–5) cm (Chelton et al, 2001). This allows them to measure:

1. Changes in the mean volume of the ocean and sea-level rise with an accuracy of ±0.4 mm/yr since 1993 (Nerem et al, 2006);
2. Seasonal heating and cooling of the ocean (Chambers et al 1998);
3. Open ocean tides with an accuracy of ±(1–2) cm (Shum et al, 1997);
4. Tidal dissipation (Egbert and Ray, 1999; Rudnick et al, 2003);
5. The permanent surface geostrophic current system (figure 10.5);
6. Changes in surface geostrophic currents on all scales (figure 10.4); and
7. Variations in topography of equatorial current systems such as those associated with El Niño (figure 10.6).

Altimeter Errors (Topex/Poseidon and Jason): The most accurate observations of the sea-surface topography are from Topex/Poseidon and Jason. Errors for these satellite altimeter system are due to (Chelton et al, 2001):

1. Instrument noise, ocean waves, water vapour, free electrons in the ionosphere, and mass of the atmosphere. Both satellites carried a precise altimeter system able to observe the height of the satellite above the sea surface between ±66° latitude with a precision of ±(1–2) cm and an accuracy of ±(2–5) cm. The systems consist of a two-frequency radar altimeter to measure height above the sea, the influence of the ionosphere, and wave height, and a three-frequency microwave radiometer able to measure water vapour in the troposphere.

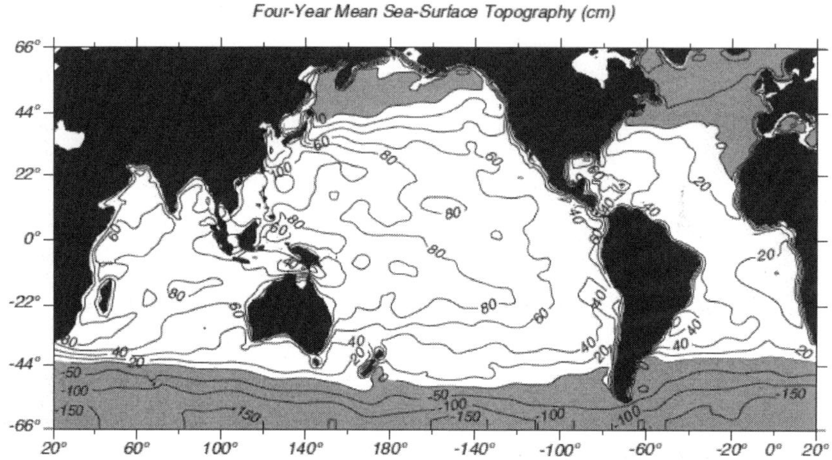

Figure 10.5 Global distribution of time-averaged topography of the ocean from Topex/Pos- eidon altimeter data from 10/3/92 to 10/6/99 relative to the jgm–3 geoid. Geostrophic currents at the ocean surface are parallel to the contours. Compare with figure 2.8 calculated from hydrographic data. From Center for Space Research, University of Texas.

2. Tracking errors. The satellites carried three tracking systems that enable their position in space, the ephemeris, to be determined with an accuracy of ±(1–3.5) cm.

3. Sampling error. The satellites measure height along a ground track that repeats within ±1 km every 9.9156 days. Each repeat is a cycle. Because currents are measured only along the sub-satellite track, there is a sampling error. The satellite cannot map the topography between ground tracks, nor can they observe changes with periods less than 2 × 9.9156 d (see §16.3).

4. Geoid error. The permanent topography is not well known over distances shorter than a hundred kilometres because geoid errors dominate for short distances. Maps of topography smoothed over greater distances are used to study the dominant features of the permanent geostrophic currents at the sea surface (figure 10.5). New satellite systems grace and champ are measuring earth's gravity accurately enough that the geoid error is now small enough to ignore over distances greater than 100 km.

Taken together, the measurements of height above the sea and the satellite position give sea-surface height in geocentric coordinates within ±(2–5) cm. Geoiderror adds further errors that depend on the size of the area being measured.

10.4 GEOSTROPHIC CURRENTS FROM HYDROGRAPHY

The geostrophic equations are widely used in oceanography to calculate currents at depth. The basic idea is to use hydrographic measurements of temperature, salinity or conductivity, and pressure to calculate the density field of the ocean using the equation of state of sea water. Density is used in (10.7b)to calculate the internal pressure field, from which the geostrophic currents are calculated using (10.8a, b). Usually, however, the constant of integration in (10.8) is not known, and only the relative velocity field can be calculated.

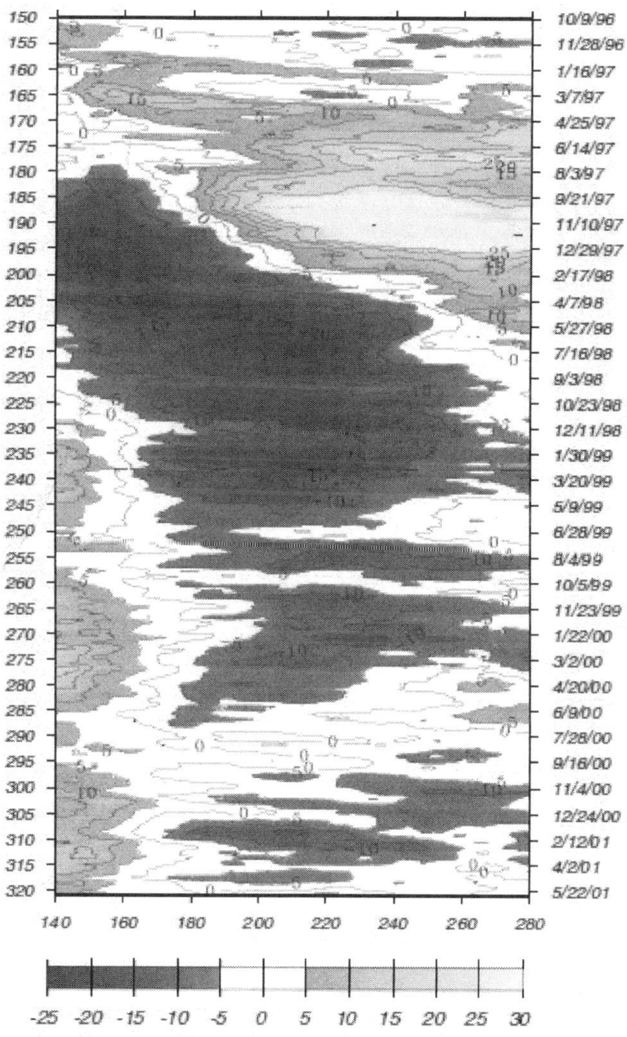

Figure 10.6 Time-longitude plot of sea-level anomalies in the Equatorial Pacific observed by Topex/Poseidon during the 1997–1998 El Niño. Warm anomalies are light gray, cold anomalies are dark gray. The anomalies are computed from 10-day deviations from a three-year mean surface from3 Oct 1992 to 8 Oct 1995. The data

are smoothed with a Gaussian weighted filter with a longitudinal span of 5° and a latitudinal span of 2°.The annotations on the left are cycles of satellite data. From Center for Space Research, University of Texas.

At this point, you may ask, why not just measure pressure directly as is done in meteorology, where direct measurements of pressure are used to calculate winds. And, aren't pressure measurements needed to calculate density from the equation of state? The answer is that very small changes in depth make large changes in pressure because water is so heavy. Errors in pressure caused by errors in determining the depth of a pressure gauge are much larger than the pressure due to currents. For example, using (10.7a), we calculate that the pressure gradient due to a 10 cm/s current at 30° latitude is 7.5×10^{-3} Pa/m, which is 750 Pa in 100 km. From the hydrostatic equation (10.5), 750 Pa is equivalent to a change of depth of 7.4 cm. Therefore, for this example, we must know the depth of a pressure gauge with an accuracy of much better than 7.4 cm. This is not possible.

Geopotential Surfaces within the Ocean: Calculation of pressure gradients within the ocean must be done along surfaces of constant geopotential just as we calculated surface pressure gradients relative to the geoid when we calculated surface geostrophic currents. As long ago as 1910, Vilhelm Bjerknes (Bjerknes and Sandstrom, 1910) realized that such surfaces are not at fixed heights in the atmosphere because g is not constant, and (10.4) must include the variability of gravity in both the horizontal and vertical directions (Saunders and Fofonoff, 1976) when calculating pressure in the ocean.

The *geopotential* Φ is:

$$\Phi = \int_0^z g dz \quad (10.11)$$

Because $\Phi/9.8$ in SI units has almost the same numerical value as height in meters, the meteorological community accepted Bjerknes' proposal that height be replaced by *dynamic meters* $D = \Phi/10$ to obtain a natural vertical coordinate. Later, this was replaced by the *geopotential meter* (gpm) $Z = \Phi/9.80$. The geopotential meter is a measure of the work required to lift a unit mass from sea level to a height z against the force of gravity. Harald Sverdrup, Bjerknes' student, carried the concept to oceanography, and depths in the ocean are often quoted in geopotential meters. The difference between depths of constant vertical distance and constant potential can be relatively large. For example, the geometric depth of the 1000 dynamic meter surface is 1017.40 m at the north pole and 1022.78 m at the equator, a difference of 5.38 m.

Note that depth in geopotential meters, depth in meters, and pressure in decibars are almost the same numerically. At a depth of 1 meter the pressure is approximately 1.007 decibars and the depth is 1.00 geopotential meters.

Equations for Geostrophic Currents within the Ocean: To calculate geostrophic currents, we need to calculate the horizontal pressure gradient within the ocean. This can be done using either of two approaches:

1. Calculate the slope of a constant pressure surface relative to a surface of constant geopotential. We used this approach when we used sea-surface slope from altimetry to calculate surface geostrophic currents. The sea surface is a constant-pressure surface. The constant geopotential surface was the geoid.
2. Calculate the change in pressure on a surface of constant geopotential. Such a surface is called a *geopotential surface.*

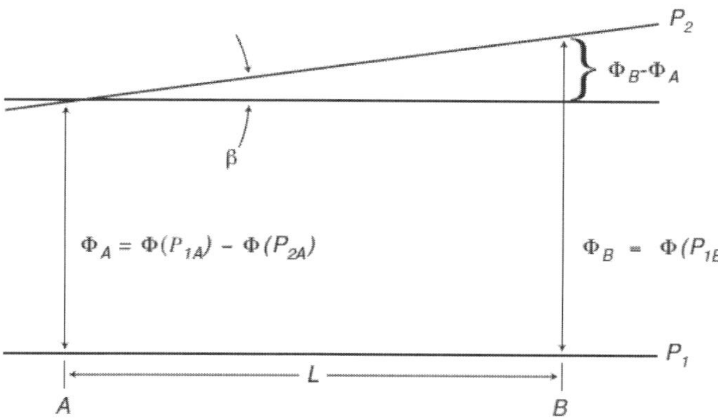

Figure 10.7. Sketch of geometry used for calculating geostrophic current from hydrography.

Oceanographers usually calculate the slope of constant-pressure surfaces.

The important steps are:

- Calculate differences in geopotential (ΦA - ΦB) between two constant-pressure surfaces ($P1$, $P2$) at hydrographic stations A and B (figure 10.7). This is similar to the calculation of ζ of the surface layer.
- Calculate the slope of the upper pressure surface relative to the lower.
- Calculate the geostrophic current at the upper surface relative to the current at the lower. This is the current shear.
- Integrate the current shear from some depth where currents are known to obtain currents as a function of depth. For example, from the surface downward, using surface geostrophic currents observed by satellite altimetry, or upward from an assumed level of no motion.

To calculate geostrophic currents oceanographers use a modified form of the hydrostatic equation. The vertical pressure gradient (10.6) is written

$$\frac{\delta p}{\rho} = \alpha\,\delta p = -g\,\delta z \quad (10.12a)$$

$$\alpha\,\delta p = \delta\Phi \quad (10.12b)$$

where $\alpha = \alpha(S, t, p)$ is the *specific volume*, and (10.12b) follows from (10.11). Differentiating (10.12b) with respect to horizontal distance x allows the geostrophic balance to be written in terms of the slope of the constant-pressure surface using (10.6) with $f = 2\Omega \sin \varphi$:

$$\alpha \frac{\partial p}{\partial x} = \frac{1}{\rho}\frac{\partial p}{\partial x} = -2\,\Omega\, v \sin \varphi \quad (10.13a)$$

$$\frac{\partial \Phi\,(p = p_0)}{\partial x} = -2\,\Omega\, v\, \sin \varphi \quad (10.13b)$$

where Φ is the geopotential at the constant-pressure surface.

Now let's see how hydrographic data are used for evaluating $\partial\Phi/\partial x$ on a constant-pressure surface. Integrating (10.12b) between two constant-pressure surfaces ($P1$, $P2$) in the ocean as shown in figure 10.7 gives the geopotential difference between two constant-pressure surfaces. At station A the integration gives:

$$\Phi\,(P_{1A}) - \Phi\,(P_{2A}) = \int_{P_{1A}}^{P_{2A}} \alpha\,(S, t, p)\, dp \quad (10.14)$$

The specific volume anomaly is written as the sum of two parts:

$$\alpha(S, t, p) = \alpha(35, 0, p) + \delta \quad (10.15)$$

where $\alpha(35, 0, p)$ is the specific volume of sea water with salinity of 35, temperature of 0°C, and pressure p. The second term δ is the *specific volume anomaly*. Using (10.15) in (10.14) gives:

$$\Phi(P_{1A}) - \Phi(P_{2A}) = \int_{P_{1A}}^{P_{2A}} \alpha(35, 0, p)\, dp + \int_{P_{1A}}^{P_{2A}} \delta\, dp$$

$$\Phi(P_{1A}) - \Phi(P_{2A}) = (\Phi_1 - \Phi_2)_{std} + \Delta\Phi_A$$

where $(\Phi1 - \Phi2)std$ is the *standard geopotential distance* between two constant-pressure surfaces $P1$ and $P2$, and

$$\Delta\Phi_A = \int_{P_{1A}}^{P_{2A}} \delta\, dp \quad (10.16)$$

is the anomaly of the geopotential distance between the surfaces. It is called the *geopotential anomaly*. The geometric distance between Φ_2 and Φ_1 is numerically approximately $(\Phi_2 - \Phi_1)/g$ where $g = 9.8 m/s^2$ is the approximate value of gravity. The geopotential anomaly is much smaller, being approximately 0.1% of the standard geopotential distance.

Consider now the geopotential anomaly between two pressure surfaces $P1$ and $P2$ calculated at two hydrographic stations A and B a distance L meters apart (figure 10.7). For simplicity we assume the lower constant-pressure surface is a level surface. Hence the constant-pressure and geopotential surfaces coincide, and there is no geostrophic velocity at this depth. The slope of the upper surface is

$$\frac{\Delta\Phi_B - \Delta\Phi_A}{L} = \text{slope of constant-pressure surface } P_2$$

because the standard geopotential distance is the same at stations A and B. The geostrophic velocity at the upper surface calculated from (10.13b) is:

$$V = \frac{(\Delta\Phi_B - \Delta\Phi_A)}{2\Omega\, L\, \sin\varphi} \quad (10.17)$$

where V is the velocity at the upper geopotential surface. The velocity V is perpendicular to the plane of the two hydrographic stations and directed into the plane of figure 10.7 if the flow is in the northern hemisphere. *A useful rule of thumb is that the flow is such that warmer, lighter water is to the right looking downstream in the northern hemisphere.*

Note that I could have calculated the slope of the constant-pressure surfaces using density ρ instead of specific volume α. I used α because it is the common practice in oceanography, and tables of specific volume anomalies and computer code to calculate the anomalies are widely available. The common practice follows from numerical methods developed before calculators and computers were available, when all calculations were done by hand or by mechanical calculators with the help of tables and nomograms. Because the computation must be done with an accuracy of a few parts per million, and because all scientific fields tend to be conservative, the common practice has continued to use specific volume anomalies rather than density anomalies.

Barotropic and Baroclinic Flow: If the ocean were homogeneous with constant density, then constant-pressure surfaces would always be parallel to the sea surface, and the geostrophic velocity would be independent of depth. In this case the relative velocity is zero, and hydrographic data cannot be used to measure the geostrophic current. If density varies with depth, but not with horizontal distance, the constant-pressure surfaces are always parallel to the sea surface and the levels of constant density, the *isopycnal surfaces*. In this case, the relative flow is also zero. Both cases are examples of *barotropic flow*.

Barotropic flow occurs when levels of constant pressure in the ocean are always parallel to the surfaces of constant density. Note, some authors call the vertically averaged flow the barotropic component of the flow. Wunsch (1996:74) points out that barotropic is used in so many different ways that the term is meaningless and should not be used.

Baroclinic flow occurs when levels of constant pressure are inclined to surfaces of constant density. In this case, density varies with depth and horizontal position. A good example is seen in figure 10.8 which shows levels of constant density changing depth by more than 1 km over horizontal distances of 100 km at the Gulf Stream. Baroclinic flow varies with depth, and the relative cur- rent can be calculated from hydrographic data. Note, constant-density surfaces cannot be inclined to constant-pressure surfaces for a fluid at rest.

In general, the variation of flow in the vertical can be decomposed into a barotropic component which is independent of depth, and a baroclinic component which varies with depth.

10.5 AN EXAMPLE USING HYDROGRAPHIC DATA

Let's now consider a specific numerical calculation of geostrophic velocity using generally accepted procedures from *Processing of Oceanographic Station Data* (jpots Editorial Panel, 1991). The book has worked examples using hydrographic data collected by the r/v *Endeavor* in the north Atlantic. Data were collected on Cruise 88 along 71°W across the Gulf Stream south of CapeCod, Massachusetts at stations 61 and 64. Station 61 is on the Sargasso Sea side of the Gulf Stream in water 4260 m deep. Station 64 is north of the Gulf Stream in water 3892 m deep. The measurements were made by a Conductivity- Temperature-Depth-Oxygen Profiler, Mark III CTD/02, made by Neil Brown Instruments Systems.

The CTD sampled temperature, salinity, and pressure 22 times per second, and the digital data were averaged over 2 dbar intervals as the CTD was lowered in the water. Data were tabulated at 2 dbar pressure intervals centered on odd values of pressure because the first observation is at the surface, and the first averaging interval extends to 2 dbar, and the center of the first interval is at 1 dbar. Data were further smoothed with a binomial filter and linearly interpolated to standard levels reported in the first three columns of tables 10.2 and 10.3. All processing was done by computer.

$\delta(S, t, p)$ in the fifth column of tables 10.2 and 10.3 is calculated from the values of t, S, p in the layer. $< \delta >$ is the average value of specific volume anomaly for the layer between standard pressure levels. It is the average of the values of $\delta\ (S, t, p)$ at the top and bottom of the layer (*cf.* the mean-value theorem of calculus). The last column ($10^{-5}\Delta\Phi$) is the product of the average specific volume anomaly of the layer times the thickness of the layer in decibars. Therefore, the last column is the geopotential anomaly $\Delta\Phi$ calculated by integrating (10.16) between $P1$ at the bottom of each layer and $P2$ at the top of each layer.

The distance between the stations is L = 110, 935 m; the average Coriolis parameter is $f = 0.88104\times10^{-4}$; and the denominator in (10.17) is 0.10231 s/m.

This was used to calculate the geostrophic currents relative to 2000 decibars reported in table 10.4 and plotted in figure 10.8.

Notice that there are no Ekman currents in figure 10.8. Ekman currents are not geostrophic, so they don't contribute directly to the topography. They contribute only indirectly through Ekman pumping.

10.6 REMARKS ON GEOSTROPHIC CURRENTS

Now that we know how to calculate geostrophic currents from hydrographic data, let's consider some of the limitations of the theory and techniques.

Converting Relative Velocity to Velocity: Hydrographic data give geostrophic currents relative to geostrophic currents at some reference level. How can we convert the relative geostrophic velocities to velocities relative to the earth?

1. *Assume a Level of no Motion*: Traditionally, oceanographers assume there is a *level of no motion*, sometimes called a *reference surface*, roughly 2,000 m below the surface. This is the assumption used to derive the currents in table 10.4. Currents are assumed to be zero at this depth, and relative currents are integrated up to the surface and down to the bottom to obtain current velocity as a function of depth. There is some experimental evidence that such a level exists on average for mean currents (see for example, Defant, 1961: 492).

Table 10.2 Computation of Relative Geostrophic Currents.

Data from Endeavor Cruise 88, Station 61 (36°40.03'N, 70°59.59'W; 23 August 1982; 1102Z)

Pressure decibar	t °C	S	σ(θ) kg/m³	δ(S, t, p) 10⁻8m³/kg	< δ > 10⁻8m³/kg	10⁻5ΔΦ m²/s²
0	25.698	35.221	23.296	457.24	457.26	0.046
1	25.698	35.221	23.296	457.28	440.22	0.396
10	26.763	36.106	23.658	423.15	423.41	0.423
20	26.678	36.106	23.658	423.66	423.82	0.424
30	26.676	36.107	23.659	423.98	376.23	0.752
50	24.528	36.561	24.670	328.48	302.07	0.755
75	22.753	36.614	25.236	275.66	257.41	0.644
100	21.427	36.637	25.630	239.15	229.61	0.574
125	20.633	36.627	25.841	220.06	208.84	0.522
150	19.522	36.558	26.086	197.62	189.65	0.948
200	18.798	36.555	26.273	181.67	178.72	0.894
250	18.431	36.537	26.354	175.77	174.12	0.871
300	18.189	36.526	26.408	172.46	170.38	1.704
400	17.726	36.477	26.489	168.30	166.76	1.668
500	17.165	36.381	26.557	165.22	158.78	1.588
600	15.952	36.105	26.714	152.33	143.18	1.432
700	13.458	35.776	26.914	134.03	124.20	1.242

Pressure decibar	t °C	S	σ(θ) kg/m³	δ(S, t, p) $10^{-8} m^3/kg$	< δ > $10^{-8} m^3/kg$	$10^{-5}\Delta\Phi$ m^2/s^2
800	11.109	35.437	27.115	114.36	104.48	1.045
900	8.798	35.178	27.306	94.60	80.84	0.808
1000	6.292	35.044	27.562	67.07	61.89	0.619
1100	5.249	35.004	27.660	56.70	54.64	0.546
1200	4.813	34.995	27.705	52.58	51.74	0.517
1300	4.554	34.986	27.727	50.90	50.40	0.504
1400	4.357	34.977	27.743	49.89	49.73	0.497
1500	4.245	34.975	27.753	49.56	49.30	1.232
1750	4.028	34.973	27.777	49.03	48.83	1.221
2000	3.852	34.975	27.799	48.62	47.77	2.389
2500	3.424	34.968	27.839	46.92	45.94	2.297
3000	2.963	34.946	27.868	44.96	43.40	2.170
3500	2.462	34.920	27.894	41.84	41.93	2.097
4000	2.259	34.904	27.901	42.02		

Defant recommends choosing a reference level where the current shear in the vertical is smallest. This is usually near 2 km. This leads to useful maps of surface currents because surface currents tend to be faster than deeper currents. Figure 10.9 shows the geopotential anomaly and surface currents in the Pacific relative to the 1,000 dbar pressure level.

2. *Use known currents:* The known currents could be measured by current meters or by satellite altimetry. Problems arise if the currents are not measured at the same time as the hydrographic data. For example, the hydrographic data may have been collected over a period of months to decades, while the currents may have been measured over a period of only a few months. Hence, the hydrography may not be consistent with the current measurements. Sometimes currents and hydrographic data are measured at nearly the same time (figure 10.10). In this example, currents were measured continuously by moored current meters (points) in a deep western boundary current and calculated from CTD data taken just after the current meters were deployed and just before they were recovered (smooth curves). The solid line is the current assuming a level of no motion at 2,000 m, the dotted line is the current adjusted using the current meter observations smoothed for various intervals before or after the CTD casts.

Table 10.3 Computation of Relative Geostrophic Currents.
Data from Endeavor Cruise 88, Station 64 (37°39.93'N, 71°0.00'W; 24 August 1982; 0203Z)

Pressure decibar	t °C	S	σ(θ) kg/m^3	δ(S, t, p) $10^{-8} m^3/kg$	< δ > $10^{-8} m^3/kg$	$10^{-5}\Delta\Phi$ m^2/s^2
0	26.148	34.646	22.722	512.09	512.15	0.051
1	26.148	34.646	22.722	512.21	512.61	0.461
10	26.163	34.645	22.717	513.01	512.89	0.513
20	26.167	34.655	22.724	512.76	466.29	0.466
30	25.640	35.733	23.703	419.82	322.38	0.645
50	18.967	35.944	25.755	224.93	185.56	0.464
75	15.371	35.904	26.590	146.19	136.18	0.340
100	14.356	35.897	26.809	126.16	120.91	0.302
125	13.059	35.696	26.925	115.66	111.93	0.280
150	12.134	35.567	27.008	108.20	100.19	0.501
200	10.307	35.360	27.185	92.17	87.41	0.437
250	8.783	35.168	27.290	82.64	79.40	0.397
300	8.046	35.117	27.364	76.16	66.68	0.667
400	6.235	35.052	27.568	57.19	52.71	0.527
500	5.230	35.018	27.667	48.23	46.76	0.468
600	5.005	35.044	27.710	45.29	44.67	0.447
700	4.756	35.027	27.731	44.04	43.69	0.437
800	4.399	34.992	27.744	43.33	43.22	0.432
900	4.291	34.991	27.756	43.11	43.12	0.431
1000	4.179	34.986	27.764	43.12	43.10	0.431
1100	4.077	34.982	27.773	43.07	43.12	0.431
1200	3.969	34.975	27.779	43.17	43.28	0.433
1300	3.909	34.974	27.786	43.39	43.38	0.434
1400	3.831	34.973	27.793	43.36	43.31	0.433
1500	3.767	34.975	27.802	43.26	43.20	1.080
1750	3.600	34.975	27.821	43.13	43.00	1.075
2000	3.401	34.968	27.837	42.86	42.13	2.106
2500	2.942	34.948	27.867	41.39	40.33	2.016
3000	2.475	34.923	27.891	39.26	39.22	1.961
3500	2.219	34.904	27.900	39.17	40.08	2.004
4000	2.177	34.896	27.901	40.98		

Table 10.4 Computation of Relative Geostrophic Currents. Data from Endeavor Cruise 88, Station 61 and 64

Pressure decibar	$10^{-5}\Delta\Phi_{61}$ m^2/s^2	$\Sigma\Delta\Phi$ at 61*	$10^{-5}\Delta\Phi_{64}$ m^2/s^2	$\Sigma\Delta\Phi$ at 64*	V (m/s)
0		2.1872	0.051	1.2583	0.95
1	0.046	2.1826	0.461	1.2532	0.95
10	0.396	2.1430	0.513	1.2070	0.96
20	0.423	2.1006	0.466	1.1557	0.97
30	0.424	2.0583	0.645	1.1091	0.97
50	0.752	1.9830	0.464	1.0446	0.96
75	0.755	1.9075	0.340	0.9982	0.93
100	0.644	1.8431	0.302	0.9642	0.90
125	0.574	1.7857	0.280	0.9340	0.87
150	0.522	1.7335	0.501	0.9060	0.85
200	0.948	1.6387	0.437	0.8559	0.80
250	0.894	1.5493	0.397	0.8122	0.75
300	0.871	1.4623	0.667	0.7725	0.71
400	1.704	1.2919	0.527	0.7058	0.60
500	1.668	1.1252	0.468	0.6531	0.48
600	1.588	0.9664	0.447	0.6063	0.37
700	1.432	0.8232	0.437	0.5617	0.27
800	1.242	0.6990	0.432	0.5180	0.19
900	1.045	0.5945	0.431	0.4748	0.12
1000	0.808	0.5137	0.431	0.4317	0.08
1100	0.619	0.4518	0.431	0.3886	0.06
1200	0.546	0.3972	0.433	0.3454	0.05
1300	0.517	0.3454	0.434	0.3022	0.04
1400	0.504	0.2950	0.433	0.2588	0.04
1500	0.497	0.2453	1.080	0.2155	0.03
1750	1.232	0.1221	1.075	0.1075	0.01
2000	1.221	0.0000	2.106	0.0000	0.00
2500	2.389	-0.2389	2.016	-0.2106	-0.03
3000	2.297	-0.4686	1.961	-0.4123	-0.06
3500	2.170	-0.6856	2.004	-0.6083	-0.08
4000	2.097	-0.8952		-0.8087	-0.09

* Geopotential anomaly integrated from 2000 dbar level. Velocity is calculated from (10.17)

3. *Use Conservation Equations:* Lines of hydrographic stations across a strait or an ocean basin may be used with conservation of mass and salt to calculate currents. This is an example of an inverse problem (Wunsch,1996 describes the application of inverse methods in oceanography). See Mercier et al. (2003) for a description of how they determined the circulation in the upper layers of the eastern basins of the south Atlantic using hydrographic data from the World Ocean Circulation Experiment and direct measurements of current in a box model constrained by inverse theory.

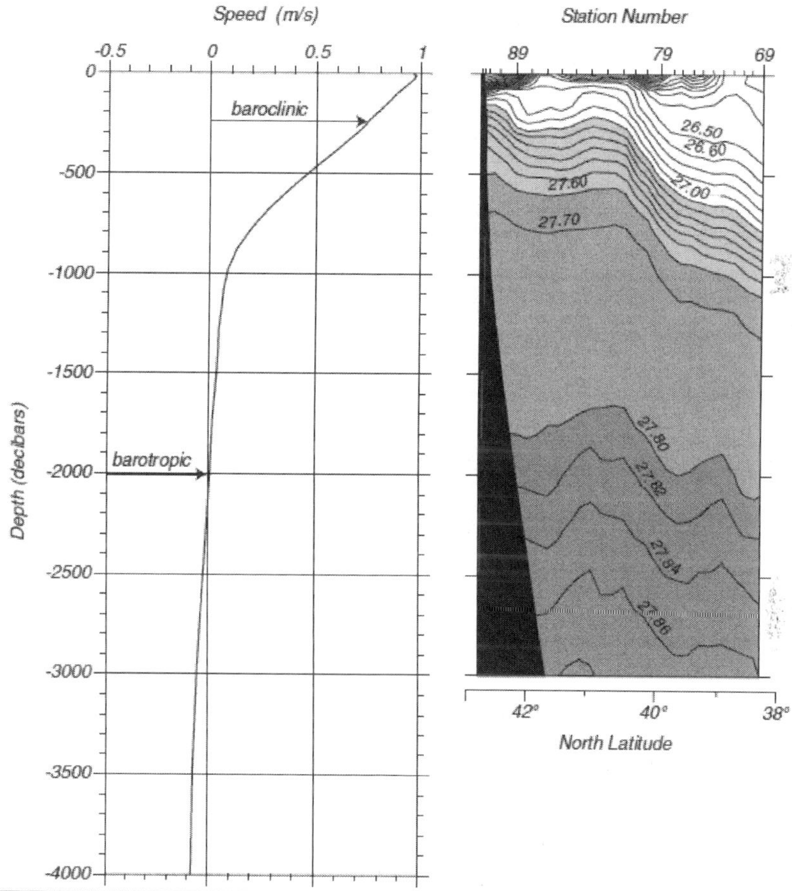

Figure 10.8 Left Relative current as a function of depth calculated from hydrographic data collected by the *Endeavor* cruise south of Cape Cod in August 1982. The Gulf Stream is the fast current shallower than 1000 decibars. The assumed depth of no motion is at 2000decibars. Right Cross section of potential density σθ across the Gulf Stream along 63.66° W calculated from CTD data collected from *Endeavor* on 25–28 April 1986. The Gulf Stream is centered on the steeply sloping contours shallower than 1000m between 40° and 41°. Notice that the vertical scale is 425 times the horizontal scale. (Data contoured by Lynn Talley, Scripps Institution of Oceanography).

Disadvantage of Calculating Currents from Hydrographic Data: Currents calculated from hydrographic data have been used to make maps of ocean currents since the early 20th century. Nevertheless, it is important to review the limitations of the technique.

1. Hydrographic data can be used to calculate only the current relative to a current at another level.

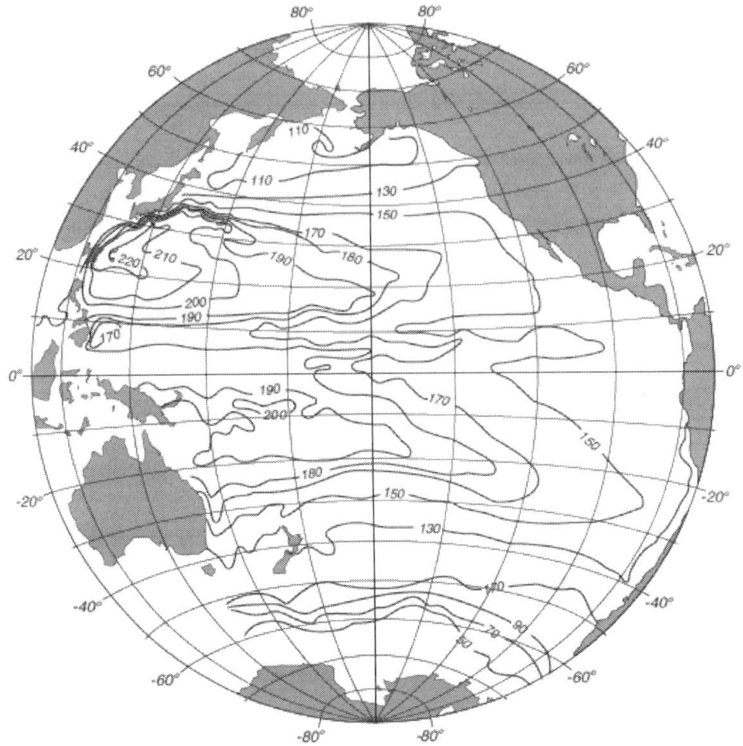

Figure 10.9. Mean geopotential anomaly relative to the 1,000 dbar surface in the Pacific based on 36,356 observations. Height of the anomaly is in geopotential centimetres. If the velocity at 1,000 dbar were zero, the map would be the surface topography of the Pacific. After Wyrtki (1979).

2. The assumption of a level of no motion may be suitable in the deep ocean, but it is usually not a useful assumption when the water is shallow such as over the continental shelf.

3. Geostrophic currents cannot be calculated from hydrographic stations that are close together. Stations must be tens of kilometres apart.

Limitations of the Geostrophic Equations: I began this section by showing that the geostrophic balance applies with good accuracy to flows that exceed a few tens of kilometres in extent and with periods greater than a few days. The balance cannot, however, be perfect. If it were, the flow in the ocean would never change because the balance ignores any acceleration of the flow. The important limitations of the geostrophic assumption are:

4. Geostrophic currents cannot evolve with time because the balance ignores acceleration of the flow. Acceleration dominates if the horizontal dimensions are less than roughly 50 km and times are less than a few days. Acceleration is negligible, but not zero, over longer times and distances.

1. The geostrophic balance does not apply within about 2° of the equator where the Coriolis force goes to zero because $\sin \phi \rightarrow 0$.
2. The geostrophic balance ignores the influence of friction.

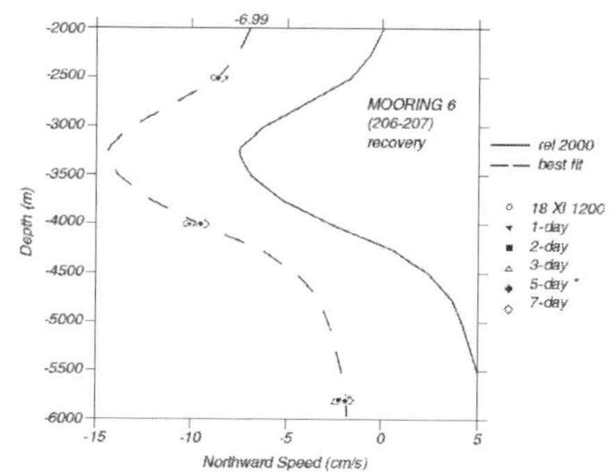

Figure 10.10 Current meter measurements can be used with CTD measurements to determine current as a function of depth avoiding the need for assuming a depth of no motion. Solid line: profile assuming a depth of no motion at 2000 decibars. Dashed line: profile adjusted to agree with currents measured by current meters 1–7 days before the CTD measurements. (Plots from Tom Whitworth, Texas A&M University)

Accuracy: Strub et al. (1997) showed that currents calculated from satellite altimeter measurements of sea-surface slope have an accuracy of ±3–5 cm/s.

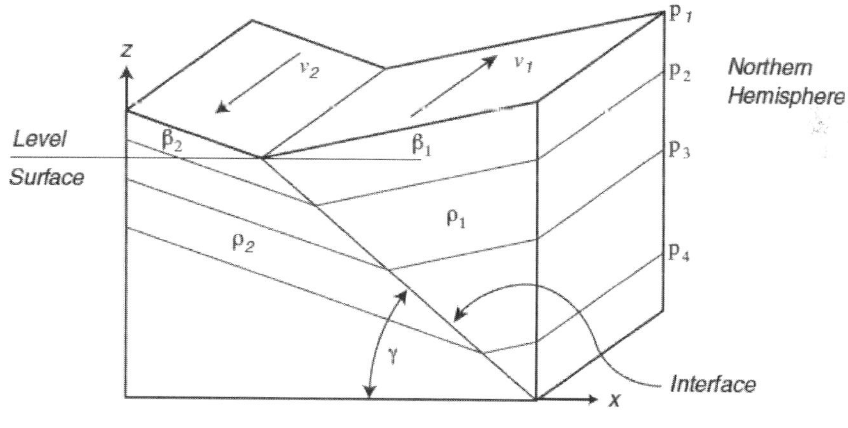

Figure 10.11 Slopes β of the sea surface and the slope γ of the interface between two homogeneous, moving layers, with density ρ1 and ρ2 in the northern hemisphere. After Neumann and Pierson (1966: 166)

10.7 CURRENTS FROM HYDROGRAPHIC SECTIONS

Lines of hydrographic data along ship tracks are often used to produce con- tour plots of density in a vertical section along the track. Cross-sections of currents sometimes show sharply dipping density surfaces with a large contrast in density on either side of the current. The baroclinic currents in the section can be estimated using a technique first proposed by Margules (1906) and de- scribed by Defant (1961: 453). The technique allows oceanographers to estimate the speed and direction of currents perpendicular to the section by a quick look at the section.

To derive Margules' equation, consider the slope ∂z/∂x of a stationary interface between two water masses with densities ρ1 and ρ2 (see figure 10.11). To calculate the change in velocity across the interface we assume homogeneous layers of density ρ1 < ρ2 both of which are in geostrophic equilibrium. Although the ocean does not have an idealized interface that we assumed, and the water masses do not have uniform density, and the interface between the water masses is not sharp, the concept is still useful in practice.

The change in pressure on the interface is:

$$\delta p = \frac{\partial p}{\partial x}\,\delta x + \frac{\partial p}{\partial z}\,\delta z, \quad (10.18)$$

and the vertical and horizontal pressure gradients are obtained from (10.6):

Therefore:

$$\frac{\partial p}{\partial z} = -\rho_1 g + \rho_1 f v_1 \quad (10.19)$$

$$\delta p_1 = -\rho_1 f v_1\,\delta x + \rho_1 g\,\delta z \quad (10.20a)$$

$$\delta p_2 = -\rho_2 f v_2\,\delta x + \rho_2 g\,\delta z \quad (10.20b)$$

The boundary conditions require $\delta_{p1} = \delta_{p2}$ on the interface if the interface is not moving. Equating (10.20a) with (10.20b), dividing by δx, and solving for δz/δx gives:

$$\frac{\delta z}{\delta x} \equiv \tan\gamma = \frac{f}{g}\left(\frac{\rho_2 v_2 - \rho_1 v_1}{\rho_2 - \rho_1}\right)$$

Because ρ1 ≈ ρ2, and for small β and γ,

$$\tan\gamma \approx \frac{f}{g}\left(\frac{\rho_1}{\rho_2 - \rho_1}\right)(v_2 - v_1) \quad (10.21a)$$

$$\tan\beta_1 = -\frac{f}{g}\,v_1 \quad (10.21b)$$

$$\tan\beta_2 = -\frac{f}{g}\,v_2 \quad (10.21c)$$

where β is the slope of the sea surface, and γ is the slope of the interface between the two water masses. Because the internal differences in density are small, the slope is approximately 1000 times larger than the slope of the constant pressure surfaces.

Consider the application of the technique to the Gulf Stream (figure 10.8). From the figure: $\phi = 36^\circ$, $\rho_1 = 1026.7$ kg/m^3, and $\rho 2 = 1027.5$ kg/m^3 at a depth of 500 decibars. If we use the $\sigma t = 27.1$ surface to estimate the slope between the two water masses, we see that the surface changes from a depth of 350 m to a depth of 650 m over a distance of 70 km. Therefore, $\tan \gamma = 4300 \times 10^{-6} = 0.0043$, and $\Delta v = v_2 - v_1 = -0.38$ m/s. Assuming $v2 = 0$, then $v1 = 0.38$ m/s. This rough estimate of the velocity of the Gulf Stream compares well with velocity at a depth of 500m calculated from hydrographic data (table 10.4) assuming a level of no motion at 2,000 decibars.

The slope of the constant-density surfaces are clearly seen in figure 10.8. And plots of constant-density surfaces can be used to quickly estimate current directions and a rough value for the speed. In contrast, the slope of the sea surface is 8.4×10^{-6} or 0.84 m in 100 km if we use data from table 10.4.

Note that constant-density surfaces in the Gulf Stream slope downward to the east, and that sea-surface topography slopes upward to the east. Constant pressure and constant density surfaces have opposite slope.

If the sharp interface between two water masses reaches the surface, it is an oceanic front, which has properties that are very similar to atmospheric fronts. Eddies in the vicinity of the Gulf Stream can have warm or cold cores (figure 10.12). Application of Margules' method to these mesoscale eddies gives the direction of the flow. Anticyclonic eddies (clockwise rotation in the northern hemisphere) have warm cores ($\rho 1$ is deeper in the center of the eddy than elsewhere) and the constant-pressure surfaces bow upward. In particular, the sea surface is higher at the center of the ring. Cyclonic eddies are the reverse.

10.8 LAGRANGIAN MEASUREMENTS OF CURRENTS

Oceanography and fluid mechanics distinguish between two techniques for measuring currents: Lagrangian and Eulerian. Lagrangian techniques follow a water particle. Eulerian techniques measure the velocity of water at a fixed position.

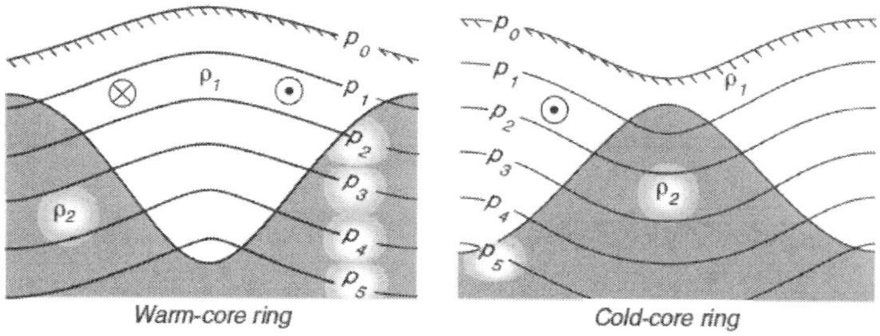

Figure 10.12 Shape of constant-pressure surfaces *pi* and the interface between two water masses of density ρ1, ρ2 if the upper is rotating faster than the lower. Left: Anti-cyclonic motion, warm-core eddy. Right: Cyclonic, cold-core eddy. Note that the sea surface *p0* slopes up toward the center of the warm-core ring, and the constant-density surfaces slope down toward the center. Circle with dot is current toward the reader, circle with cross is current away from the reader. After Defant (1961: 466).

Basic Technique: Lagrangian techniques track the position of a drifter designed to follow a water parcel either on the surface or deeper within the water column. The mean velocity over some period is calculated from the distance between positions at the beginning and end of the period divided by the period. Errors are due to:

1. The failure of the drifter to follow a parcel of water. We assume the drifter stays in a parcel of water, but wind blowing on the surface float of a surface drifter can cause the drifter to move relative to the water.
2. Errors in determining the position of the drifter.
3. Sampling errors. Drifters go only where drifters want to go. And drifters want to go to convergent zones. Hence drifters tend to avoid areas of divergent flow.

Satellite Tracked Surface Drifters: Surface drifters consist of a drogue plus a float. Its position is determined by the Argos system on meteorological satellites (Swenson and Shaw, 1990) or calculated from GPS data recorded continuously by the buoy and relayed to shore.

Argos-tracked buoys carry a radio transmitter with a very stable frequency *F0*. A receiver on the satellite receives the signal and determines the Doppler shift *F* as a function of time *t* (figure 10.13). The Doppler frequency is

$$F = \frac{dR}{dt}\frac{F_0}{c} + F_0$$

where *R* is the distance to the buoy, *c* is the velocity of light. The closer the buoy to the satellite the more rapidly the frequency changes. When *F* = *F0* the range is a minimum. This is the time of closest approach, and the satellite's velocity vector is perpendicular to the line from the satellite to the buoy. The time of closest approach and the time rate of change of Doppler frequency at that time gives the buoy's position relative to the orbit with a 180° ambiguity(B and BB in the figure). Because the orbit is accurately known, and because the buoy can be observed many times, its position can be determined without ambiguity.

The accuracy of the calculated position depends on the stability of the frequency transmitted by the buoy. The Argos system tracks buoys with an accuracy of ±(1–2) km, collecting 1–8 positions per day depending on latitude. Because 1 cm/s ≈ 1 km/day, and because typical values of currents in the ocean range from one to two hundred centimetres per second, this is an very useful accuracy.

Holey-Sock Drifters: The most widely used, satellite-tracked drifter is the holey-sock drifter. It consists of a cylindrical drogue of cloth 1 m in diameter by 15 m long with 14 large holes cut in the sides. The weight of the drogue is supported by a float set 3 m below the surface. The submerged float is tethered to a partially submerged surface float carrying the Argos transmitter.

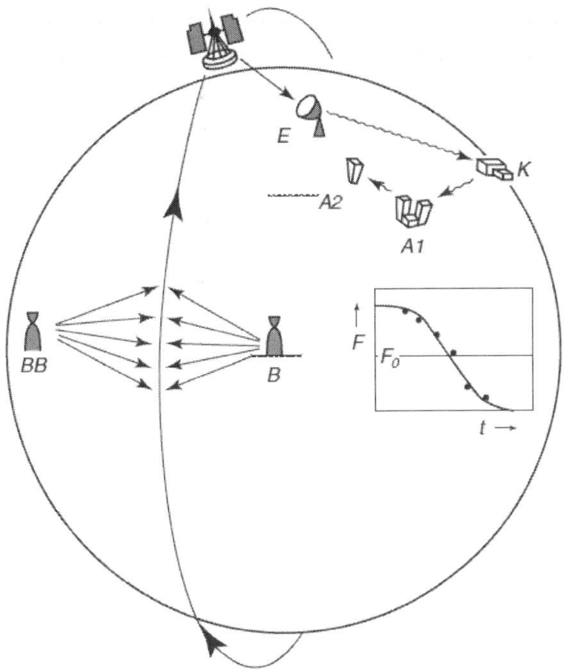

Figure 10.13 System Argos uses radio signals transmitted from surface buoys to determine the position of the buoy. A satellite receives the signal from the buoy B. The time rate of change of the signal, the Doppler shift F , is a function of buoy position and distance from the satellite's track. Note that a buoy at BB would produce the same Doppler shift as the buoy at B. The recorded Doppler signal is transmitted to ground stations E, which relays the information to processing centers A via control stations K. After Dietrich et al. (1980: 149).

The buoy was designed for the Surface Velocity Program and extensively tested. Niiler et al. (1995) carefully measured the rate at which wind blowing on the surface float pulls the drogue through the water, and they found that the buoy moves 12 ± 9° to the right of the wind at a speed

$$U_s = (4.32 \pm 0.67\times)\, 10^{-2} \frac{U_{10}}{DAR} + (11.04 \pm 1.63) \frac{D}{DAR} \quad (10.22)$$

where *DAR* is the drag area ratio defined as the drogue's drag area divided by the sum of the tether's drag area and the surface float's drag area, and *D* is the difference in velocity of the water between the top of the cylindrical drogue and the bottom. Drifters typically have a *DAR* of 40, and the drift *Us* < 1 cm/s for *U10* < 10 m/s.

Argo Floats: The most widely used subsurface floats are the Argo floats. The floats (figure 10.14) are designed to cycle between the surface and some pre-determined depth. Most floats drift for 10 days at a depth of 1 km, sink to 2 km, then rise to the surface. While rising, they profile temperature and salinity as a function of pressure (depth). The floats remains on the surface for a few hours, relays data to shore via the Argos system, then sink again to 1 km. Each float carries enough power to repeat this cycle for several years. The float thus measures currents at 1 km depth and density distribution in the upper ocean. Three thousand Argo floats are being deployed in all parts of the ocean for the Global Ocean Data Assimilation Experiment GODAE.

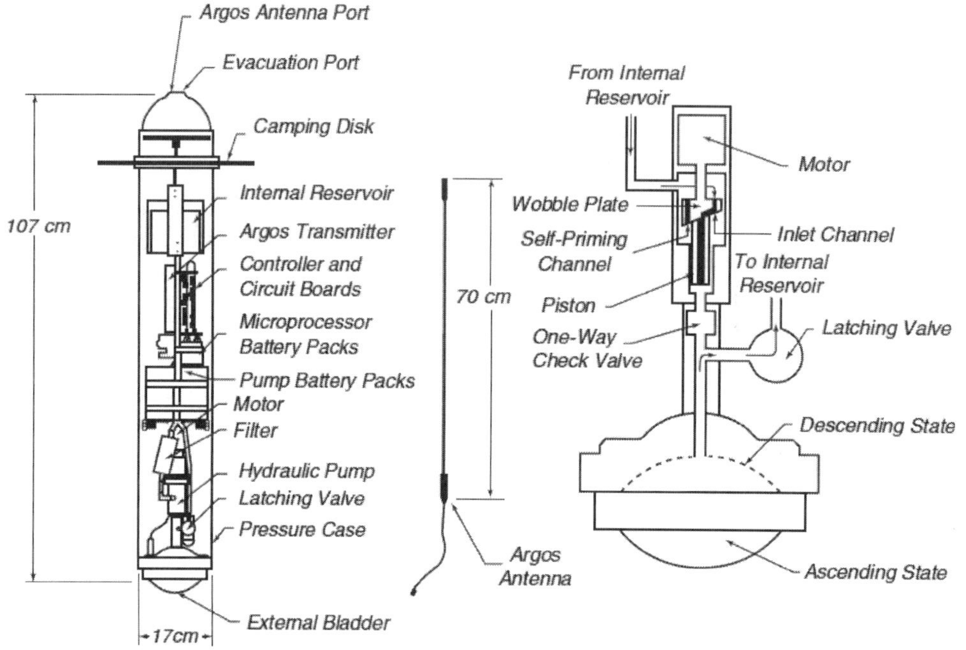

Figure 10.14 The Autonomous Lagrangian Circulation Explorer (ALACE) floats is the prototype for the Argos floats. It measures currents at a depth of 1 km. Left: Schematic of the drifter. To ascend, the hydraulic pump moves oil from an internal reservoir to an external bladder, reducing the drifter's density. To descend, the latching valve is opened to allow oil to flow back into the internal reservoir. The antenna is mounted to the end cap. Right: Expanded schematic of the hydraulic system. The motor rotates the wobble plate actuating the piston which pumps hydraulic oil. After Davis et al. (1992).

Lagrangian Measurements Using Tracers: The most common method for measuring the flow in the deep ocean is to track parcels of water containing molecules not normally found in the ocean. Thanks to atomic bomb tests in the 1950s and the recent exponential increase of chlorofluorocarbons in the atmosphere, such tracers have been introduced into the ocean in large quantities.

See §13.4 for a list of tracers used in oceanography. The distribution of trace molecules is used to infer the movement of the water. The technique is especially useful for calculating velocity of deep water masses averaged over decades and for measuring turbulent mixing discussed in §8.4.

The distribution of trace molecules is calculated from the concentration of the molecules in water samples collected on hydrographic sections and surveys. Because the collection of data is expensive and slow, there are few repeated sections. Figure 10.15 shows two maps of the distribution of tritium in the north

Atlantic collected in 1972–1973 by the Geosecs Program and in 1981, a decade later. The sections show that tritium, introduced into the atmosphere during the atomic bomb tests in the atmosphere in the 1950s to 1972, penetrated to depths below 4 km only north of 40°N by 1971 and to 35°N by 1981. This shows that deep currents are very slow, about 1.6 mm/s in this example.

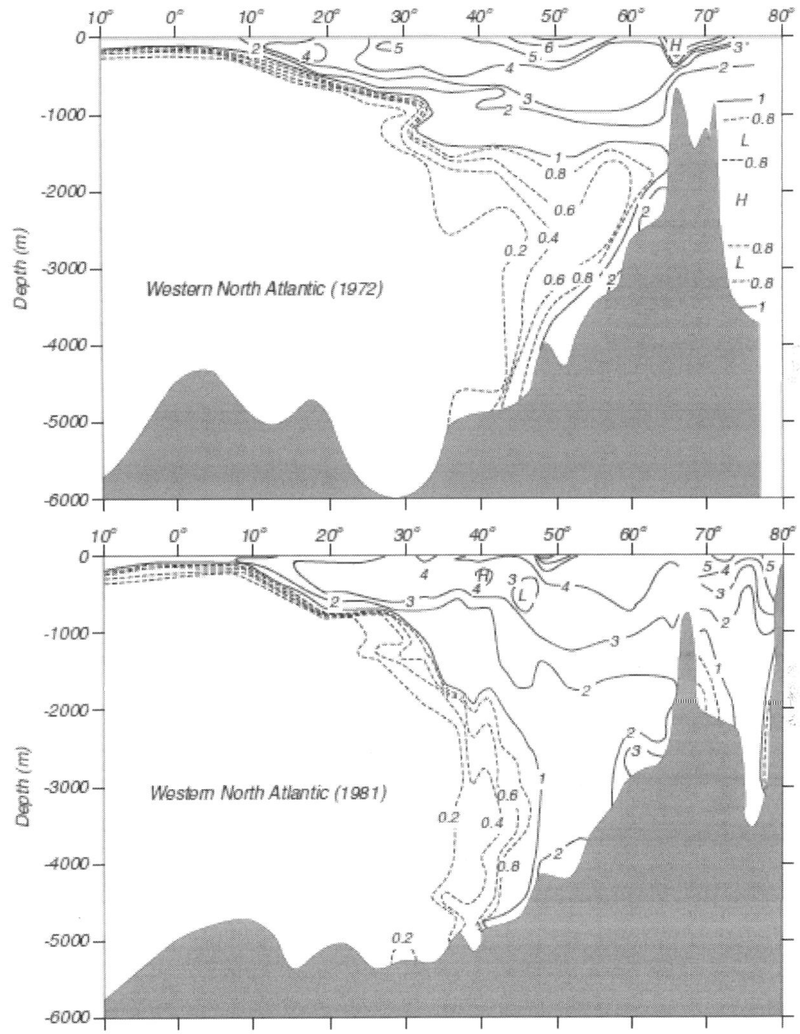

Figure 10.15 Distribution of tritium along a section through the western basins in the north Atlantic, measured in 1972 (Top) and remeasured in 1981 (Bottom). Units are tritium units, where one tritium unit is 10^{18} (tritium atoms)/(hydrogen atoms) corrected to the activity levels that would have been observed on 1 January 1981. Compare this figure to the density in the ocean shown in figure 13.10. After Toggweiler (1994).

Because the deep currents are so small, we can question what process are responsible for the observed distribution of tracers. Both turbulent diffusion and advection by currents can fit the observations. Hence, does figure 10.15 give mean currents in the deep Atlantic, or the turbulent diffusion of tritium?

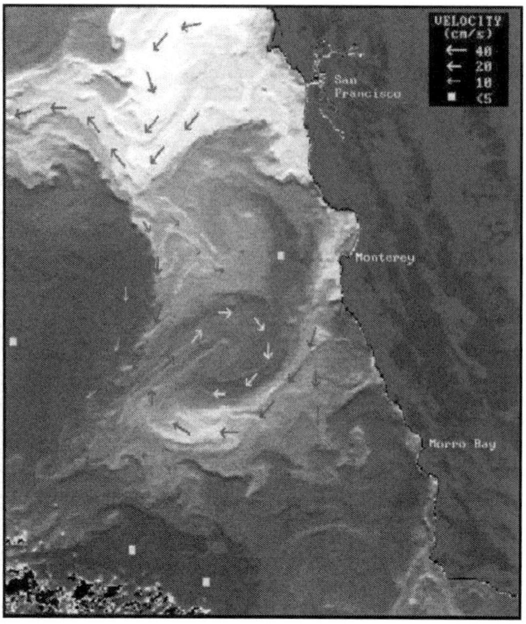

Figure 10.16 Ocean temperature and current patterns are combined in this AVHRR analysis. Surface currents were computed by tracking the displacement of small thermal or sediment features between a pair of images. A directional edge-enhancement filter was applied here to define better the different water masses. Warm water is shaded darker. From Ocean Imaging, Solana Beach, California, with permission.

Another useful tracer is the temperature and salinity of the water. I will consider these observations in §13.4 where I describe the core method for studying deep circulation. Here, I note that AVHRR observations of surface temperature of the ocean are an additional source of information about currents.

Sequential infrared images of surface temperature are used to calculate the displacement of features in the images (figure 10.16). The technique is especially useful for surveying the variability of currents near shore. Land provides reference points from which displacement can be calculated accurately, and large temperature contrasts can be found in many regions in some seasons.

There are two important limitations.

- Many regions have extensive cloud cover, and the ocean cannot be seen.
- Flow is primarily parallel to temperature fronts, and strong currents can exist along fronts even though the front may not move. It is therefore essential to track the motion of small eddies embedded in the flow near the front and not the position of the front.

The Rubber Duckie Spill: On January 10, 1992 a 12.2-m container with 29,000 bathtub toys, including rubber ducks (called rubber duckies by children)washed overboard from a container ship at 44.7°N, 178.1°E (figure 10.17). Ten months later the toys began washing ashore near Sitka, Alaska. A similar accident on May

27, 1990 released 80,000 Nike-brand shoes at 48°N, 161°W when waves washed containers from the *Hansa Carrier*.

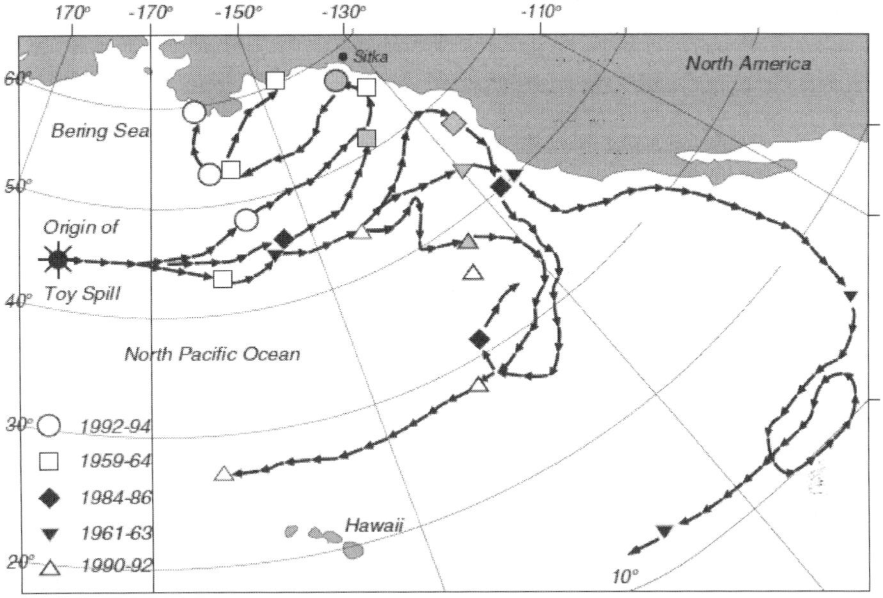

Figure 10.17 Trajectories that spilled rubber duckies would have followed had they been spilled on January 10 of different years. Five trajectories were selected from a set of 48 simulations of the spill each year between 1946 and 1993. The trajectories begin on January 10 and end two years later (solid symbols). Grey symbols indicate positions on November 16 of the year of the spill. The grey circle gives the location where rubber ducks first came ashore near Sitka in 1992. The code at lower left gives the dates of the trajectories. After Ebbesmeyer and Ingraham (1994).

The spills and eventual recovery of the toys and shoes proved to be good tests of a numerical model for calculating the trajectories of oil spills developed by Ebbesmeyer and Ingraham (1992, 1994). They calculated the possible trajectories of the spilled toys using the Ocean Surface Current Simulations oscurs numerical model driven by winds calculated from the Fleet Numerical Oceanography Center's daily sea-level pressure data. After modifying their calculations by increasing the wind age coefficient by 50% for the toys and by decreasing their angle of deflection function by 5°, their calculations accurately predicted the arrival of the toys near Sitka, Alaska on November 16, 1992, ten months after the spill.

10.9 MEASUREMENTS CONDUCTED USING THE EULERIAN METHOD

Eulerian measurements are made by many different types of instruments on ships and moorings.

Moorings (figure 10.18) are placed on the sea floor by ships. The moorings may last for months to longer than a year. Because the mooring must be deployed and recovered by deep-sea research ships, the technique is expensive and few moorings are now being deployed. The subsurface mooring shown on the right in the figure is preferred for several reasons: it does not have a surface float that is forced by high frequency, strong, surface currents; the mooring is out of sight and it does not attract the attention of fishermen; and the floatation is usually deep enough to avoid being caught by fishing nets. Measurements made from moorings have errors due to:

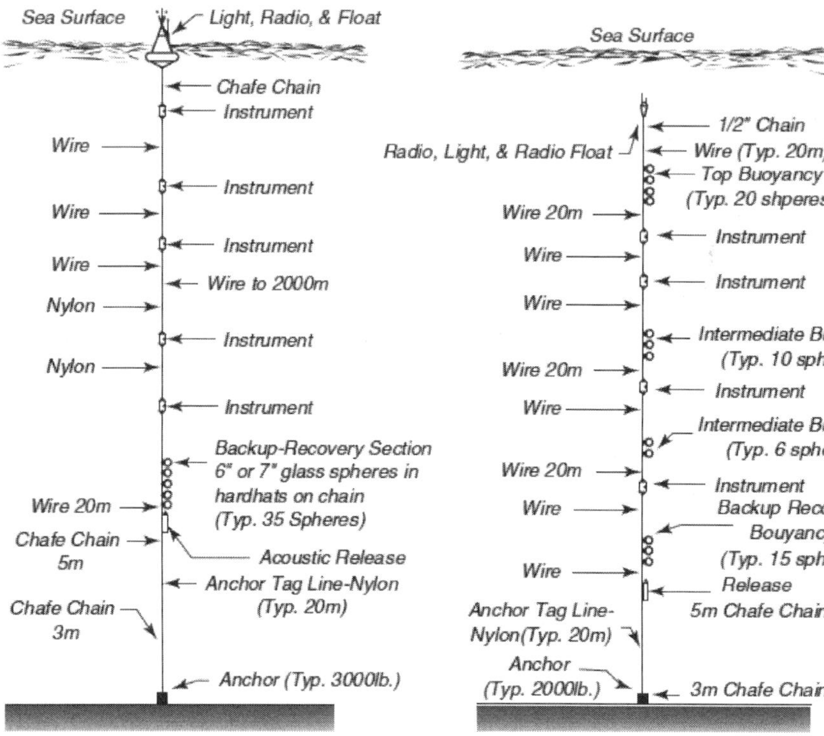

Figure 10.18 Left: An example of a surface mooring of the type deployed by the Woods Hole Oceanographic Institution's Buoy Group. Right: An example of a subsurface mooring deployed by the same group. After Baker (1981: 410–411).

- Mooring motion. Subsurface moorings move least. Surface moorings in strong currents move most, and are seldom used.
- Inadequate Sampling. Moorings tend not to last long enough to give accurate estimates of mean velocity or inter annual variability of the velocity.
- Fouling of the sensors by marine organisms, especially instruments deployed for more than a few weeks close to the surface.

Acoustic-Doppler Current Meters and Profilers: The most common Eulerian measurements of currents are made using sound. Typically, the current meter or profiler transmits sound in three or four narrow beams pointed in different directions. Plankton and tiny bubbles reflect the sound back to the instrument. The Doppler shift of the reflected sound is proportional to the radial component of the velocity of whatever reflects the sound. By combining data from three or four beams, the horizontal velocity of the current is calculated assuming the bubbles and plankton do not move very fast relative to the water. Two types of acoustic current meters are widely used. The Acoustic-Doppler Current Profiler, called the ADCP, measures the Doppler shift of sound reflected from water at various distances from the instrument using sound beams projected into the water just as a radar measures radio scatter as a function of range using radio beams projected into the air. Data from the beams are combined to give profiles of current velocity as a function of distance from the instrument. On ships, the beams are pointed diagonally downward at 3–4 horizontal angles relative to the ship's bow. Bottom-mounted meters use beams pointed diagonally upward.

Ship-board instruments are widely used to profile currents within 200 to 300 m of the sea surface while the ship steams between hydrographic stations. Because a ship moves relative to the bottom, the ship's velocity and orientation must be accurately known. GPS data have provided this information since the early 1990s.

Acoustic-Doppler current meters are much simpler than the ADCP. They transmit continuous beams of sound to measure current velocity close to the meter, not as a function of distance from the meter. They are placed on moorings and sometimes on a CTD. Instruments on moorings record velocity as a function of time for many days or months. The Aanderaa current meter (figure 10.19) in the figure is an example of this type. Instruments on CTDs profile currents from the surface to the bottom at hydrographic stations.

10.10 CONCISE SUMMARY

- Pressure distribution is almost precisely the hydrostatic pressure obtained by assuming the ocean is at rest. Pressure is therefore calculated very accurately from measurements of temperature and conductivity as a function of pressure using the equation of state of seawater. Hydrographic data give the relative, internal pressure field of the ocean.
- Flow in the ocean is in almost exact geostrophic balance except for flow in the upper and lower boundary layers. Coriolis force almost exactly balances the horizontal pressure gradient.
- Satellite altimetric observations of the oceanic topography give the surface geostrophic current. The calculation of topography requires an accurate geoid. If the geoid is not known, altimeters can measure the change in topography as a function of time, which gives the change in surface geostrophic currents.
- Topex/Poseidon and Jason are the most accurate altimeter systems, and they can measure the topography or changes in topography with an accuracy of ±4 cm.

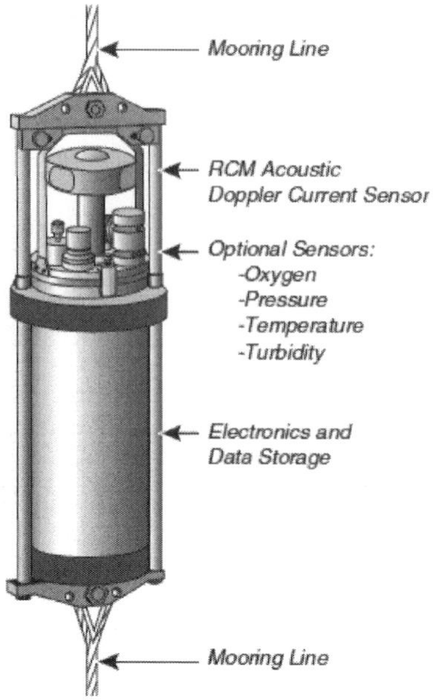

Figure 10.19 An example of a moored acoustic current meter, the rcm 9 produced by Aanderaa Instruments. Two components of horizontal velocity are measured by an acoustic system, and the directions are referenced to north using an internal Hall-effect compass. The electronics, data recorder, and battery are in the pressure-resistant housing. Accuracy is ±0.15 cm/s and ±5°. (Courtesy Aanderaa Instruments)

- Hydrographic data are used to calculate the internal geostrophic currents in the ocean relative to known currents at some level. The level can be surface currents measured by altimetry or an assumed level of no motion at depths below 1–2 km.
- Flow in the ocean that is independent of depth is called barotropic flow, flow that depends on depth is called baroclinic flow. Hydrographic data give only the baroclinic flow.
- Geostrophic flow cannot change with time, so the flow in the ocean is not exactly geostrophic. The geostrophic method does not apply to flows at the equator where the Coriolis force vanishes.
- Slopes of constant density or temperature surfaces seen in a cross-section of the ocean can be used to estimate the speed of flow through the section.
- Lagrangian techniques measure the position of a parcel of water in the ocean. The position can be determined using surface drifters or subsurface floats, or chemical tracers such as tritium.
- Eulerian techniques measure the velocity of flow past a point in the ocean. The velocity of the flow can be measured using moored current meters or acoustic velocity profilers on ships, CTDs or moorings.

11 OCEAN CIRCULATION PROPELLED BY WIND FORCES

What drives the ocean currents? At first, we might answer, the winds. But if we think more carefully about the question, we might not be so sure. We might notice, for example, that strong currents, such as the North Equatorial Counter currents in the Atlantic and Pacific Ocean go upwind. Spanish navigators in the 16th century noticed strong northward currents along the Florida coast that seemed to be unrelated to the wind. How can this happen? And, why are strong currents found offshore of east coasts but not offshore of west coasts?

Answers to the questions can be found in a series of three remarkable papers published from 1947 to 1951. In the first, Harald Sverdrup (1947) showed that the circulation in the upper kilometre or so of the ocean is directly related to the curl of the wind stress if the Coriolis force varies with latitude. Henry Stommel (1948) showed that the circulation in oceanic gyres is asymmetric also because the Coriolis force varies with latitude. Finally, Walter Munk (1950) added eddy viscosity and calculated the circulation of the upper layers of the Pacific. Together the three oceanographers laid the foundations for a modern theory of ocean circulation.

11.1 SVERDRUP'S THEORY OF THE OCEANIC CIRCULATION

While Sverdrup was analyzing observations of equatorial currents, he came upon (11.6) below relating the curl of the wind stress to mass transport within the upper ocean. To derive the relationship, Sverdrup assumed that the flow is stationary, that lateral friction and molecular viscosity are small, that non-linear terms such as $u\,\partial u/\partial x$ are small, and that turbulence near the sea surface can be described using a vertical eddy viscosity. He also assumed that the wind-driven circulation vanishes at some depth of no motion. With these assumptions, the horizontal components of the momentum equation from 8.9 and 8.12 become:

$$\frac{\partial p}{\partial x} = f\,\rho\,v + \frac{\partial T_{xz}}{\partial z} \quad (11.1a)$$

$$\frac{\partial p}{\partial y} = -f\,\rho\,u + \frac{\partial T_{yz}}{\partial z} \quad (11.1b)$$

Sverdrup integrated these equations from the surface to a depth *-D* equal to or greater than the depth at which the horizontal pressure gradient becomes zero. He defined:

$$\frac{\partial P}{\partial x} = \int_{-D}^{0} \frac{\partial p}{\partial x}\, dz, \qquad \frac{\partial P}{\partial y} = \int_{-D}^{0} \frac{\partial p}{\partial y}\, dz, \quad (11.2a)$$

$$M_x \equiv \int_{-D}^{0} \rho u(z)\, dz, \qquad M_y \equiv \int_{-D}^{0} \rho v(z)\, dz, (11.2b)$$

where *Mx, My* are the mass transports in the wind-driven layer extending down to an assumed depth of no motion.

The horizontal boundary condition at the sea surface is the wind stress. At depth *-D* the stress is zero because the currents go to zero:

Txz (0) = *Tx Txz* (*-D*) = 0

Tyz (0) = *Ty Tyz* (*-D*) = 0 (11.3)

where *Tx* and *Ty* are the components of the wind stress.

Using these definitions and boundary conditions, (11.1) become:

$$\frac{\partial P}{\partial x} = f\, M_y + T_x \quad (11.4a)$$

$$\frac{\partial P}{\partial y} = -f\, M_x + T_y \quad (11.4b)$$

In a similar way, Sverdrup integrated the continuity equation (7.19) over the same vertical depth, assuming the vertical velocity at the surface and at depth-*D* are zero, to obtain:

$$\frac{\partial M_x}{\partial x} + \frac{\partial M_y}{\partial y} = 0 \quad (11.5)$$

Differentiating (11.4a) with respect to *y* and (11.4b) with respect to *x*, subtracting, and using (11.5) gives:

$$\beta\, M_y = \frac{\partial T_y}{\partial x} - \frac{\partial T_x}{\partial y} \quad (11.6)$$
$$\beta\, M_y = \text{curl}_z(T)$$

where $\beta \equiv \partial f/\partial y$ is the rate of change of Coriolis parameter with latitude, and where curlz (*T*) is the vertical component of the curl of the wind stress.

This is an important and fundamental result—the northward mass transport of wind driven currents is equal to the curl of the wind stress. Note that Sverdrup allowed *f* to vary with latitude. We will see later that this is essential.

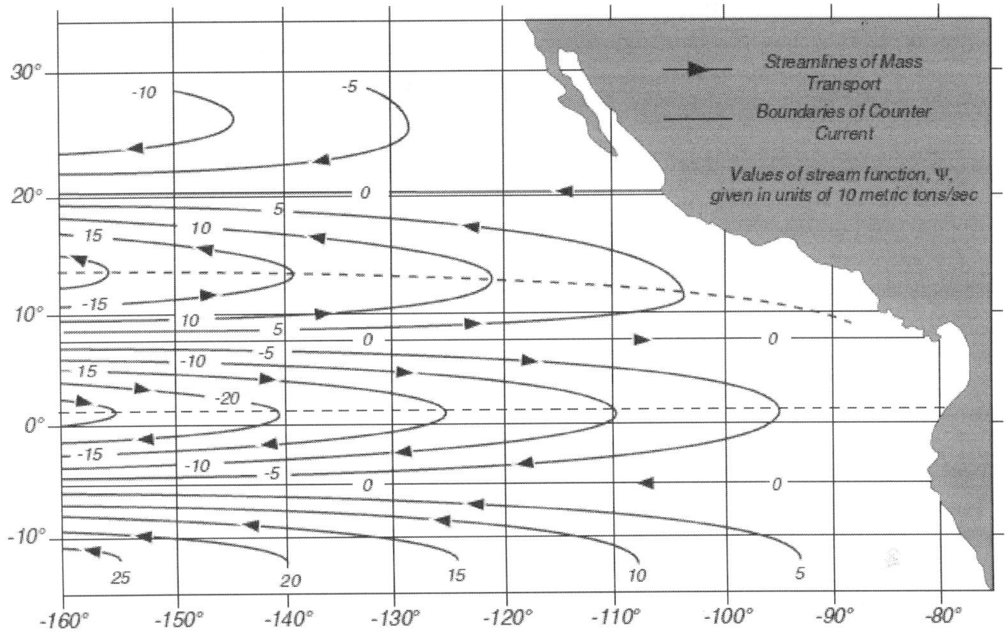

Figure 11.1 Streamlines of mass transport in the eastern Pacific calculated from Sverdrup's theory using mean annual wind stress. After Reid (1948).

We calculate β from

$$\beta \equiv \frac{\partial f}{\partial y} = \frac{2\Omega\cos\varphi}{R} \quad (11.7)$$

where R is earth's radius and ϕ is latitude.

Over much of the open ocean, especially in the tropics, the wind is zonal and $\partial Ty/\partial x$ is sufficiently small that

$$M_y \approx -\frac{1}{\beta}\frac{\partial T_x}{\partial y} \quad (11.8)$$

Substituting (11.8) into (11.5), assuming β varies with latitude, Sverdrup obtained:

$$\frac{\partial M_x}{\partial x} = -\frac{1}{2\Omega\cos\varphi}\left(\frac{\partial T_x}{\partial y}\tan\varphi + \frac{\partial^2 T_x}{\partial y^2}R\right) \quad (11.9)$$

Sverdrup integrated this equation from a north-south eastern boundary at $x = 0$, assuming no flow into the boundary. This requires $Mx = 0$ at $x = 0$. Then

$$M_x = -\frac{\Delta x}{2\Omega\cos\varphi}\left[\left\langle\frac{\partial T_x}{\partial y}\right\rangle\tan\varphi + \left\langle\frac{\partial^2 T_x}{\partial y^2}\right\rangle R\right] \quad (11.10)$$

where Δx is the distance from the eastern boundary of the ocean basin, and brackets indicate zonal averages of the wind stress (figure 11.1).

To test his theory, Sverdrup compared transports calculated from known winds in the eastern tropical Pacific with transports calculated from hydrographic data collected by the *Carnegie* and *Bushnell* in October and November 1928, 1929, and 1939 between 34°N and 10°S and between 80°W and 160°W.

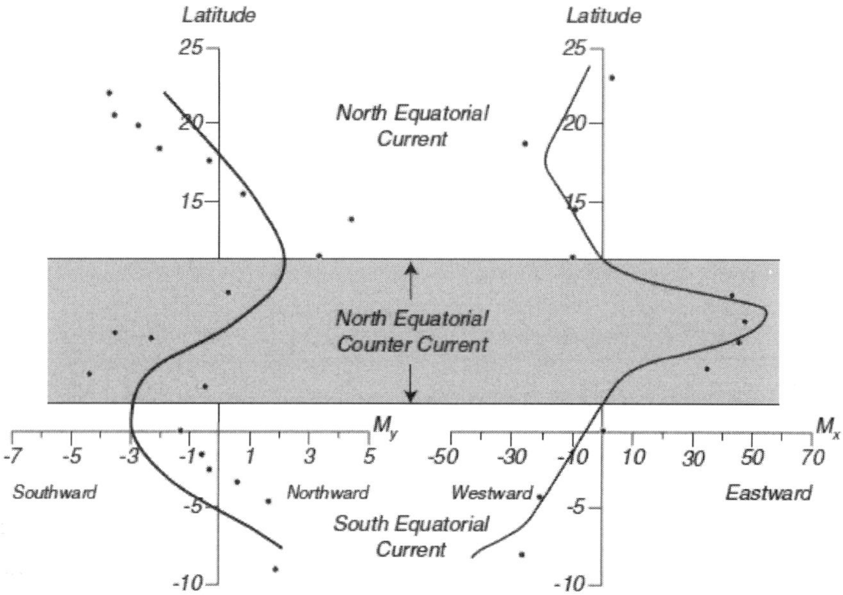

Figure 11.2 Mass transport in the eastern Pacific calculated from Sverdrup's theory using observed winds with 11.8 and 11.10 (solid lines) and pressure calculated from hydrographic data from ships with 11.4 (dots). Transport is in tons per second through a section one meter wide extending from the sea surface to a depth of one kilometer. Note the difference in scale between *My* and *Mx*. After Reid (١٩٤٨).

The hydrographic data were used to compute *P* by integrating from a depth of *D* = -1000 m. The comparison, figures 11.2, showed not only that thetransports can be accurately calculated from the wind, but also that the theory predicts wind-driven currents going upwind.

Comments on Sverdrup's Solutions:

- Sverdrup assumed i) The internal flow in the ocean is geostrophic; ii) there is a uniform depth of no motion; and iii) Ekman's transport is correct. I examined Ekman's theory in Chapter 9, and the geostrophic balance in Chapter 10. We know little about the depth of no motion in the tropical Pacific.

- The solutions are limited to the east side of the ocean because *Mx* grows with *x*. The result comes from neglecting friction which would eventually balance the wind-driven flow. Nevertheless, Sverdrup solutions have been used for describing the global system of surface currents. The solutions are applied throughout each basin all the way to the western limit of the basin. There, conservation of mass is forced by including north-south currents confined to a thin, horizontal boundary layer (figure 11.3).

- Only one boundary condition can be satisfied, no flow through the eastern boundary. More complete descriptions of the flow require more complete equations.
- The solutions do not give the vertical distribution of the current.
- Results were based on data from two cruises plus average wind data assuming a steady state. Later calculations by Leetmaa, McCreary, and Moore (1981) using more recent wind data produces solutions with sea- sonal variability that agrees well with observations provided the level of no motion is at 500 m. If another depth were chosen, the results are not as good.

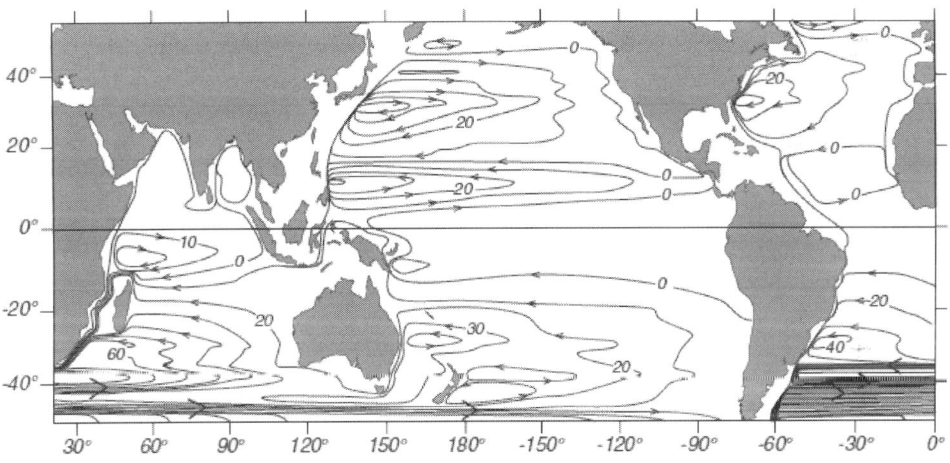

Figure 11.3 Depth-integrated Sverdrup transport applied globally using the wind stress from Hellerman and Rosenstein (1983). Contour interval is 10 Sverdrups. After Tomczak and Godfrey (1994: 46).

- Wunsch (1996: §2.2.3) after carefully examining the evidence for a Sverdrup balance in the ocean concluded we do not have sufficient information to test the theory. He writes

The purpose of this extended discussion has not been to disapprove the validity of Sverdrup balance. Rather, it was to emphasize the gap commonly existing in oceanography between a plausible and attrac- tive theoretical idea and the ability to demonstrate its quantitative applicability to actual oceanic flow fields.—Wunsch (1996).

Wunsch, however, notes

Sverdrup's relationship is so central to theories of the ocean circulation that almost all discussions assume it to be valid without any comment at all and proceed to calculate its consequences for higher- order dynamics. . . it is difficult to overestimate the importance of Sverdrup balance.—Wunsch (1996).

But the gap is shrinking. Measurements of mean stress in the equatorial Pacific (Yu and McPhaden, 1999) show that the flow there is in Sverdrup balance.

Stream Lines, Path Lines, and the Stream Function: Before discussing more about the ocean's wind-driven circulation, we need to introduce the concept of stream lines and the stream function (see Kundu, 1990: 51 & 66).

At each instant in time, we can represent a flow field by a vector velocity at each point in space. The instantaneous curves that are everywhere tangent tothe direction of the vectors are called the *stream lines* of the flow. If the flow is unsteady, the pattern of stream lines change with time.

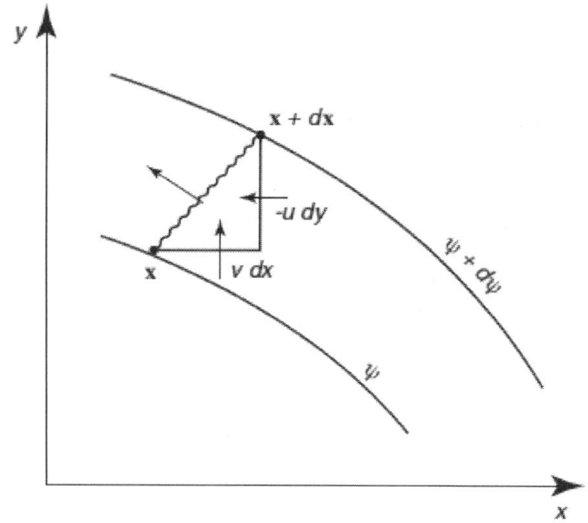

Figure 11.4 Volume transport between stream lines in a two-dimensional, steady flow. After Kundu (1990: 68).

The trajectory of a fluid particle, the path followed by a Lagrangian drifter, is called the *path line* in fluid mechanics. The path line is the same as the stream line for steady flow, and they are different for an unsteady flow.

We can simplify the description of two-dimensional, incompressible flows by using the *stream function* ψ defined by:

$$u \equiv \frac{\partial \psi}{\partial y}, \qquad v \equiv -\frac{\partial \psi}{\partial x}, \quad (11.11)$$

The stream function is often used because it is a scalar from which the vector velocity field can be calculated. This leads to simpler equations for some flows. Stream functions are also useful for visualizing the flow. At each instant, the flow is parallel to lines of constant ψ. Thus if the flow is steady, the lines of constant stream function are the paths followed by water parcels.

The volume rate of flow between any two stream lines of a steady flow is$d\psi$, and the volume rate of flow between two stream lines $\psi 1$ and $\psi 2$ is equal to $\psi 1$ - $\psi 2$. To see this, consider an arbitrary line *dx* = (*dx, dy*) between twostream lines (figure 11.4). The volume rate of flow between the stream lines is:

$$v\,dx + (-u)\,dy = -\frac{\partial\psi}{\partial x}\,dx - \frac{\partial\psi}{\partial y}\,dy = -d\psi \quad (11.12)$$

and the volume rate of flow between the two stream lines is numerically equal to the difference in their values of ψ.

Now, lets apply the concepts to satellite-altimeter maps of the oceanic topography. In §10.3 I wrote (10.10)

$$u_s = -\frac{g}{f}\frac{\partial\zeta}{\partial y} \quad (11.13)$$
$$v_s = \frac{g}{f}\frac{\partial\zeta}{\partial x}$$

Comparing (11.13) with (11.11) it is clear that and the sea surface is a stream function scaled by *g/f*. Turning to figure 10.5, the lines of constant height are stream lines, and flow is along the lines. The surface geostrophic transport is proportional to the difference in height, independent of distance between the stream lines. The same statements apply to figure 10.9, except that the transport is relative to transport at the 1000 decibars surface, which is roughly one kilometer deep.

$$\psi = -\frac{g}{f}\zeta \quad (11.14)$$

In addition to the stream function, oceanographers use the mass-transport stream function Ψ defined by:

$$M_x \equiv \frac{\partial\Psi}{\partial y}, \qquad M_y \equiv -\frac{\partial\Psi}{\partial x} \quad (11.15)$$

This is the function shown in figures 11.2 and 11.3.

11.2 STOMMEL'S THEORY OF WESTERN BOUNDARY CURRENTS

At the same time Sverdrup was beginning to understand circulation in the eastern Pacific, Stommel was beginning to understand why western boundary currents occur in ocean basins. To study the circulation in the north Atlantic, Stommel (1948) used essentially the same equations used by Sverdrup (11.1, 11.2, and 11.3) but he added a bottom stress proportional to velocity to (11.3):

$$\left(A_z\frac{\partial u}{\partial z}\right)_0 = -T_x = -F\cos(\pi\, y/b) \qquad \left(A_z\frac{\partial u}{\partial z}\right)_D = -Ru \quad (11.16a)$$

$$\left(A_z\frac{\partial v}{\partial z}\right)_0 = -T_y = 0 \qquad \left(A_z\frac{\partial v}{\partial z}\right)_D = -Rv \quad (11.16b)$$

where F and R are constants.

Stommel calculated steady-state solutions for flow in a rectangular basin $0 \le y \le b$, $0 \le x \le \lambda$ of constant depth D filled with water of constant density.

His first solution was for a non-rotating earth. This solution had a symmetric flow pattern with no western boundary current (figure 11.5, left). Next, Stommel assumed a constant rotation, which again led to a symmetric solution with no western boundary current. Finally, he assumed that the Coriolis force varies with latitude. This led to a solution with western intensification (figure 11.5, right). Stommel suggested that the crowding of stream lines in the west indicated that the variation of Coriolis force with latitude may explain why the Gulf Stream is found in the ocean. We now know that the variation of Coriolis force with latitude is required for the existence of the western boundary current, and that other models for the flow which use different formulations for friction, lead to western boundary currents with different structure. Pedlosky (1987, Chapter 5) gives a very useful, succinct, and mathematically clear description of the various theories for western boundary currents.

In the next chapter, we will see that Stommel's results can also be explained in terms of vorticity—wind produces clockwise torque (vorticity), which must be balanced by a counterclockwise torque produced at the western boundary.

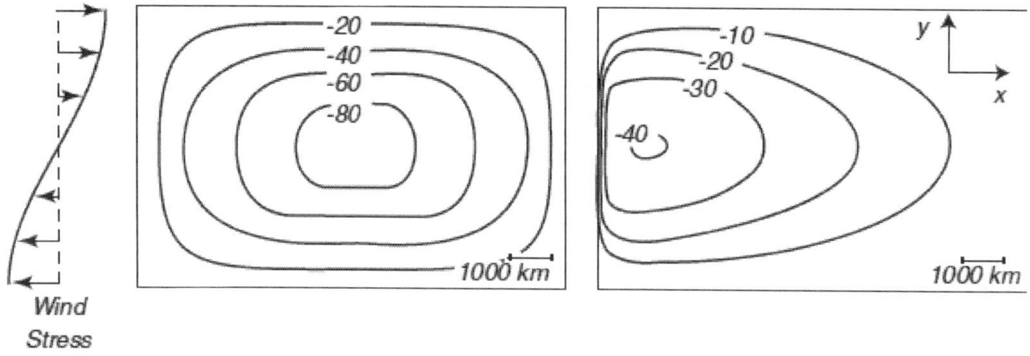

Figure 11.5 Stream function for flow in a basin as calculated by Stommel (1948). Left: Flow for non-rotating basin or flow for a basin with constant rotation. Right: Flow when rotation varies linearly with y.

11.3 MUNK'S SOLUTION

Sverdrup's and Stommel's work suggested the dominant processes producing a basin-wide, wind-driven circulation. Munk (1950) built upon this foundation, adding information from Rossby (1936) on lateral eddy viscosity, to obtain a solution for the circulation within an ocean basin. Munk used Sverdrup's idea of a vertically integrated mass transport flowing over a motionless deeper layer. This simplified the mathematical problem, and it is more realistic. The ocean currents are concentrated in the upper kilometer of the ocean, they are not barotropic and independent of depth. To include friction, Munk used lateral eddy friction with constant *AH* = *Ax* = *Ay*. Equations (11.1) become:

$$\frac{1}{\rho}\frac{\partial p}{\partial x} = f\,v + \frac{\partial}{\partial z}\left(A_z\frac{\partial u}{\partial z}\right) + A_H\frac{\partial^2 u}{\partial x^2} + A_H\frac{\partial^2 u}{\partial y^2} \quad (11.17a)$$

$$\frac{1}{\rho}\frac{\partial p}{\partial y} = -f\,u + \frac{\partial}{\partial z}\left(A_z\frac{\partial v}{\partial z}\right) + A_H\frac{\partial^2 v}{\partial x^2} + A_H\frac{\partial^2 v}{\partial y^2} \quad (11.17b)$$

Munk integrated the equations from a depth -*D* to the surface at $z = z_0$ which is similar to Sverdrup's integration except that the surface is not at *z* = 0. Munk assumed that currents at the depth -*D* vanish, that (11.3) apply atthe horizontal boundaries at the top and bottom of the layer, and that *AH* is constant.

To simplify the equations, Munk used the mass-transport stream function (11.15), and he proceeded along the lines of Sverdrup. He eliminated the pressure term by taking the *y* derivative of (11.17a) and the *x* derivative of (11.17b) to obtain the equation for mass transport:

$$\underbrace{A_H\nabla^4\Psi}_{\text{Friction}} - \underbrace{\beta\frac{\partial \Psi}{\partial x} = -\,\mathrm{curl}_z T}_{\text{Sverdrup Balance}} \quad (11.18)$$

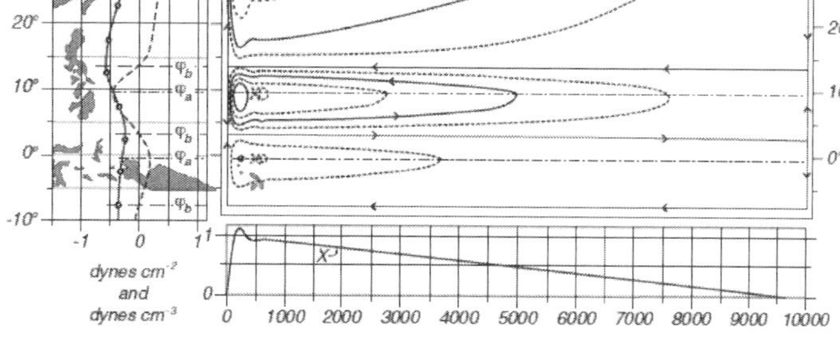

Figure 11.6 Left: Mean annual wind stress *Tx*(*y*) **over the Pacific and the curl of the wind stress. ϕ*b* are the northern and southern boundaries of the gyres, where** *My* **= 0 and curl τ = 0. ϕ0 is the center of the gyre. Upper Right: The mass transport stream function for a rectangular basin calculated by Munk (1950) using observed wind stress for the Pacific. Contour interval is 10 Sverdrups. The total transport between the coast and any point** *x*, *y* **is ψ(*x*, *y*).The transport in the relatively narrow northern section is greatly exaggerated. Lower Right: North-South component of the mass transport. After Munk (1950).**

where

$$\nabla^4 = \frac{\partial^4}{\partial x^4} + 2\frac{\partial^4}{\partial x^2\,\partial y^2} + \frac{\partial^4}{\partial y^4} \quad (11.19)$$

is the biharmonic operator. Equation (11.18) is the same as (11.6) with the addition of the lateral friction term AH. The friction term is large close to a lateral boundary where the horizontal derivatives of the velocity field are large, and it is small in the interior of the ocean basin. Thus in the interior, the balance of forces is the same as that in Sverdrup's solution.

Equation (11.18) is a fourth-order partial differential equation, and four boundary conditions are needed. Munk assumed the flow at a boundary is parallel to a boundary and that there is no slip at the boundary:

$$\Psi_{bdry} = 0, \qquad \left(\frac{\partial \Psi}{\partial n}\right)_{bdry} = 0 \quad (11.20)$$

where n is normal to the boundary. Munk then solved (11.18) with (11.20) assuming the flow was in a rectangular basin extending from $x = 0$ to $x = r$, and from $y = -s$ to $y = +s$. He further assumed that the wind stress was zonaland in the form:

$$T = a \cos ny + b \sin ny + c$$

$$n = j\,\pi/s, j = 1, 2, \ldots \quad (11.21)$$

Munk's solution (figure 11.6) shows the dominant features of the gyre-scale circulation in an ocean basin. It has a circulation similar to Sverdrup's in the eastern parts of the ocean basin and a strong western boundary current in thewest. Using $AH = 5 \times 10^3\ m^2/s$ gives a boundary current roughly 225 km widewith a shape similar to the flow observed in the Gulf Stream and the Kuroshio.

The transport in western boundary currents is independent of AH , and it depends only on (11.6) integrated across the width of the ocean basin. Hence, it depends on the width of the ocean, the curl of the wind stress, and β. Using the best available estimates of the wind stress, Munk calculated that the Gulf Stream should have a transport of 36 Sv and that the Kuroshio should have a transport of 39 Sv. The values are about one half of the measured values of the flow available to Munk. This is very good agreement considering the wind stress was not well known.

Recent recalculations show good agreement except for the region offshore ofCape Hatteras where there is a strong recirculation. Munk's solution was based on wind stress averaged aver 5°squares. This underestimated the curl of the stress. Leetmaa and Bunker (1978) used modern drag coefficient and $2^\circ \times 5^\circ$averages of stress to obtain 32 Sv transport in the Gulf Stream, a value very close to that calculated by Munk.

11.4 OBSERVED SURFACE CIRCULATION IN THE ATLANTIC

The theories by Sverdrup, Munk, and Stommel describe an idealized flow. But the ocean is much more complicated. To see just how complicated the flow is at the surface, let's look at a whole ocean basin, the north Atlantic. I have chosen this region because it is the best observed, and because mid-latitude processes in the Atlantic are similar to mid-latitude processes in the other ocean. Thus, for example, I use the Gulf Stream as an example of a western boundary current.

Let's begin with the Gulf Stream to see how our understanding of ocean surface currents has evolved. Of course, we can't look at all aspects of theflow.You can find out much more by reading Tomczak and Godfrey (1994) book on*Regional Oceanography: An Introduction*.

North Atlantic Circulation: The north Atlantic is the most thoroughly studied ocean basin. There is an extensive body of theory to describe most aspects of the circulation, including flow at the surface, in the thermocline, and at depth, together with an extensive body of field observations. By looking at figures showing the circulation, we can learn more about the circulation, and by looking at the figures produced over the past few decades we can trace an ever more complete understanding of the circulation.

Let's begin with the traditional view of the time-averaged surface flow in the north Atlantic based mostly on hydrographic observations of the density field(figure 2.7). It is a contemporary view of the mean circulation of the entire ocean based on a century of more of observations. Because the figure includes all the ocean, perhaps it is overly simplified. So, let's look then at a similar view of the mean circulation of just the north Atlantic (figure 11.7).

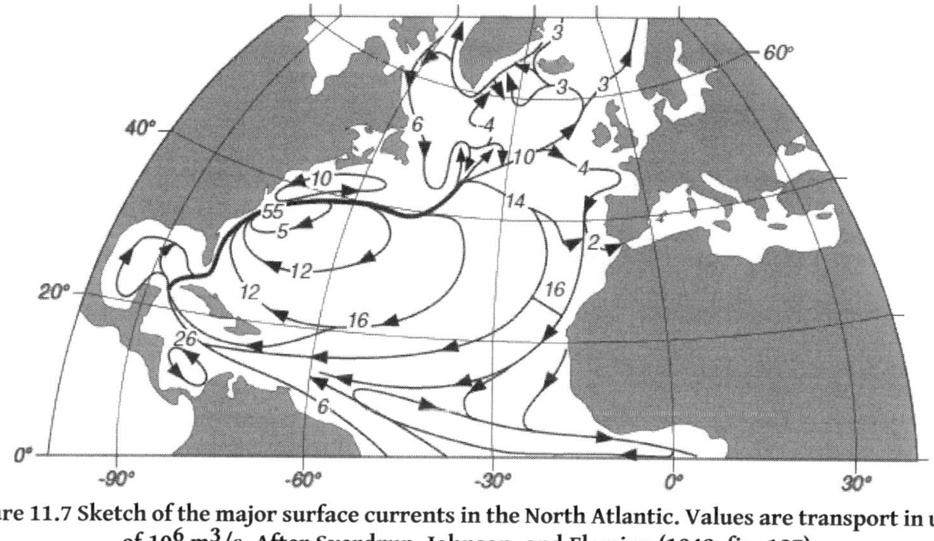

Figure 11.7 Sketch of the major surface currents in the North Atlantic. Values are transport in units of 10^6 m^3/s. After Sverdrup, Johnson, and Fleming (1942: fig. 187).

The figure shows a broad, basin-wide, mid latitude gyre as we expect from Sverdrup's theory described in §11.1. In the west, a western boundary current,the Gulf Stream, completes the gyre. In the north a subpolar gyre includes the Labrador current. An equatorial current system and counter current are found at low latitudes with flow similar to that in the Pacific. Note, however, the strong cross equatorial flow in the west which flows along the northeast coast of Brazil toward the Caribbean.

If we look closer at the flow in the far north Atlantic (figure 11.8) we see that the flow is still more complex. This figure includes much more detail of a region important for fisheries and commerce. Because it is based on an extensive base of hydrographic observations, is this reality? For example, if we were to drop a Lagrangian float into the Atlantic would it follow the streamline shown in the figure?

To answer the question, let's look at the tracks of a 110 buoys drifting on the sea surface compiled by Phil Richardson (figure 11.9 top). The tracks give a very different view of the currents in the north Atlantic. It is hard to distinguish the flow from the jumble of lines, sometimes called spaghetti tracks. Clearly, the flow is very turbulent, especially in the Gulf Stream, a fast, western-boundary current. Furthermore, the turbulent eddies seem to have a diameter of a few degrees. This is much different than turbulence in the atmosphere. In the air,the large eddies are called storms, and storms have diameters of 10°–20°. Thus oceanic "storms" are much smaller than atmospheric storms.

Perhaps we can see the mean flow if we average the drifter tracks. What happens when Richardson averages the tracks through $2^\circ \times 2^\circ$ boxes? The averages (figure 11.9 bottom) begin to show some trends, but note that in some regions, such as east of the Gulf Stream, adjacent boxes have very different means, some having currents going in different directions. This indicates the flow is so variable, that the average is not stable. Forty or more observations do not yields a stable mean value. Overall, Richardson finds that the kinetic energy of the eddies is 8 to 37 times larger than the kinetic energy of the mean flow. Thus oceanic turbulence is very different than laboratory turbulence. In the lab, the mean flow is typically much faster than the eddies.

Further work by Richardson (1993) based on subsurface buoys freely drifting at depths between 500 and 3,500 m, shows that the current extends deep below the surface, and that typical eddy diameter is 80 km.

Gulf Stream Recirculation Region: If we look closely at figure 11.7 we see that the transport in the Gulf Stream increases from 26 Sv in the Florida Strait (between Florida and Cuba) to 55 Sv offshore of Cape Hatteras. Later measurements showed the transport increases from 30 Sv in the Florida Straitto 150 Sv near 40°N.

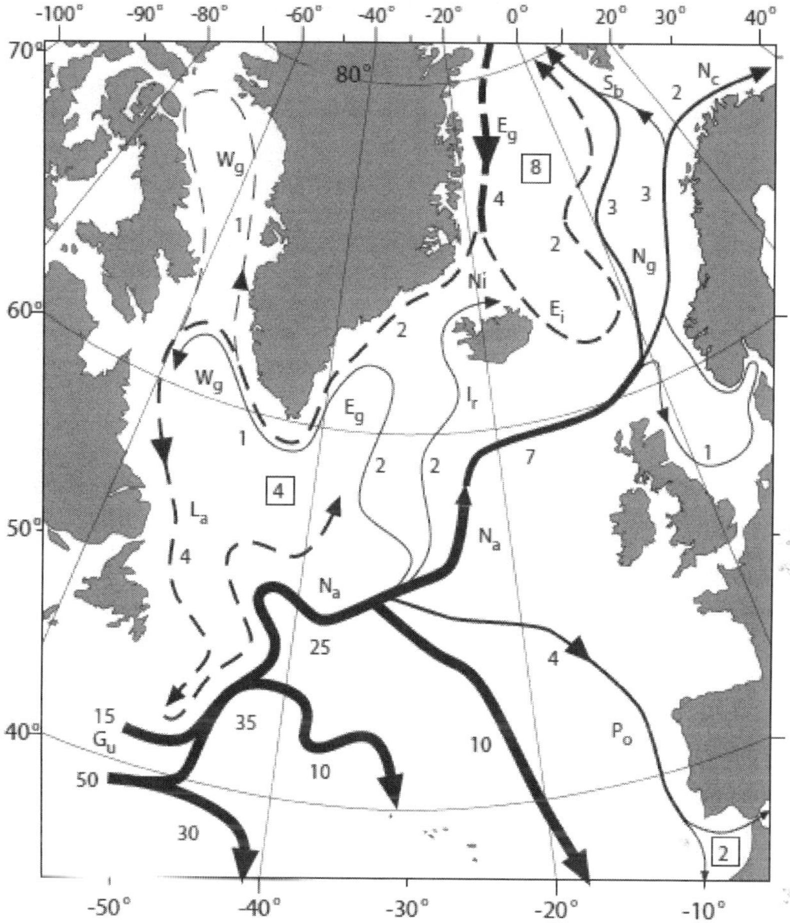

Figure 11.8 Detailed schematic of named currents in the north Atlantic. The numbers give the transport in units on 106m³/s from the surface to a depth of 1 km. Eg: East Greenland Current; Ei: East Iceland Current; Gu: Gulf Stream; Ir: Irminger Current; La: Labrador Current; Na: North Atlantic Current; Nc: North Cape Current; Ng: Norwegian Current; Ni: North Iceland Current; Po: Portugal Current; Sb: Spitsbergen Current; Wg: West Greenland Current. Numbers with in squares give sinking water in units on 106m3/s. Solid Lines: Warmer currents. Broken Lines: Colder currents. After Dietrich et al. (1980: 542).

The observed increase, and the large transport off Hatteras, disagree with transports calculated from Sverdrup's theory. Theory predicts a much smaller maximum transport of 30 Sv, and that the maximum ought to be near 28°N. Now we have a problem: What causes the high transports near 40°N?

Niiler (1987) summarizes the theory and observations. First, there is nohydrographic evidence for a large influx of water from the Antilles Current that flows north of the Bahamas and into the Gulf Stream. This rules out the possibility that the Sverdrup flow is larger than the calculated value, and that the flow bypasses the Gulf of Mexico. The flow seems to come primarily fromthe Gulf Stream itself. The flow between 60°W and 55°W is to the south. Thewater then flows south and west, and rejoins the Stream between 65°W and 75°W. Thus, there are two subtropical gyres: a small gyre directly south of the Stream centered on 65°W, called the Gulf Stream recirculation region, and the broad, wind-driven gyre near the surface seen in figure 11.7 that extends all the way to Europe.

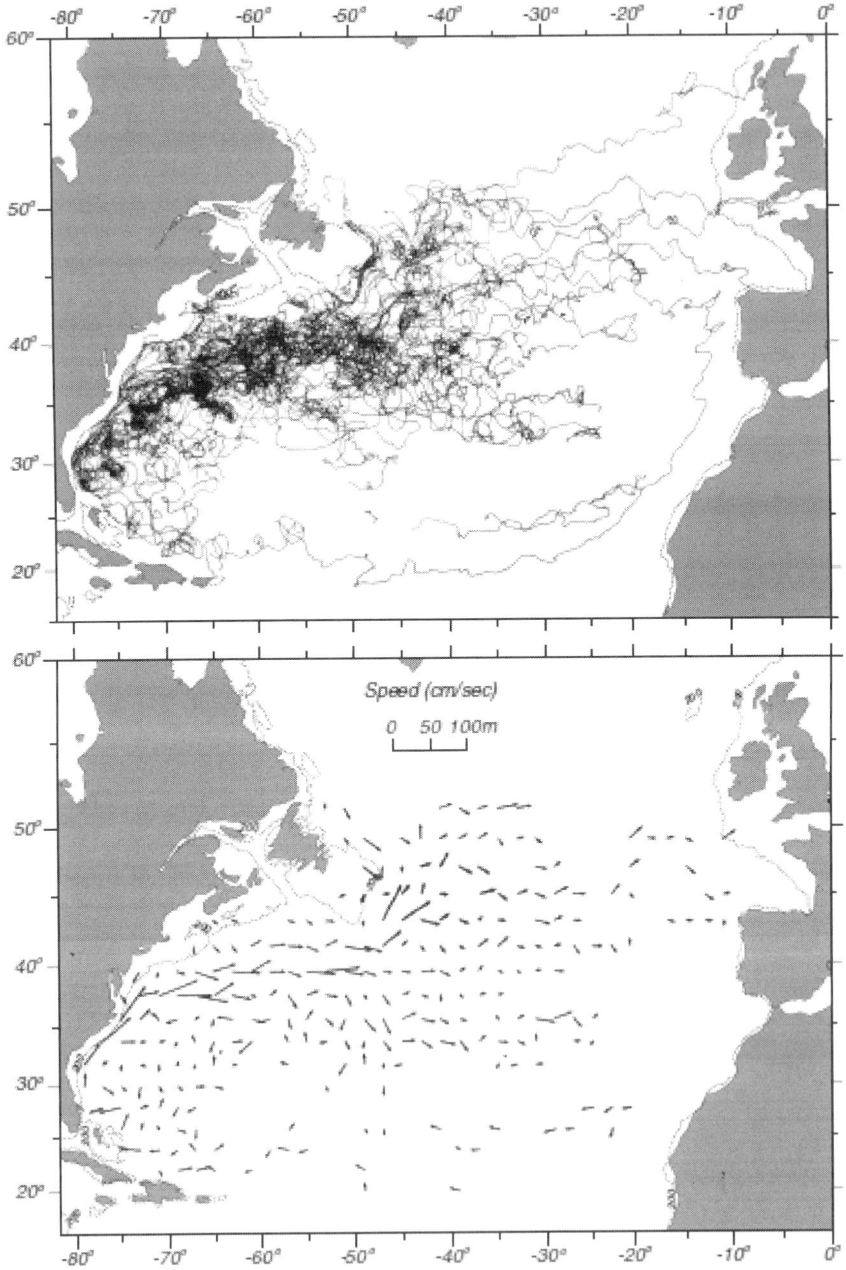

Figure 11.9 Top Tracks of 110 drifting buoys deployed in the western north Atlantic. Bottom Mean velocity of currents in 2° × 2° boxes calculated from tracks above. Boxes with fewer than 40 observations were omitted. Length of arrow is proportional to speed. Maximum values are near 0.6 m/s in the Gulf Stream near 37°N 71°W. After Richardson(1981).

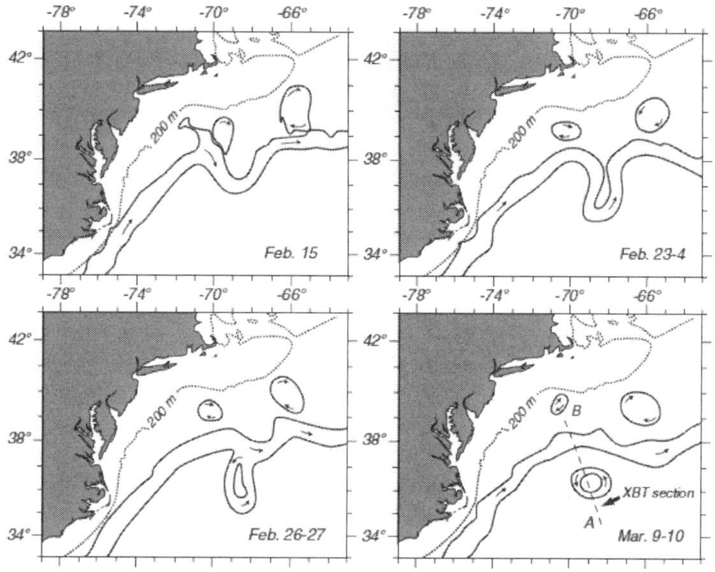

Figure 11.10 Gulf Stream meanders lead to the formation of a spinning eddy, a ring. Notice that rings have a diameter of about 1°. After Ring Group (1981).

The Gulf Stream recirculation carries two to three times the mass of the broader gyre. Current meters deployed in the recirculation region show that the flow extends to the bottom. This explains why the recirculation is weak in the maps calculated from hydrographic data. Currents calculated from the density distribution give only the baroclinic component of the flow, and they miss the component that is independent of depth, the barotropic component.

The Gulf Stream recirculation is driven by the potential energy of the steeplysloping thermocline at the Gulf Stream. The depth of the 27.00 sigma-theta ($\sigma\theta$) surface drops from 250 meters near 41°N in figure 10.8 to 800 m near 38°N south of the Stream. Eddies in the Stream convert the potential energyto kinetic energy through baroclinic instability. The instability leads to an interesting phenomena: negative viscosity. The Gulf Stream accelerates not decelerates. It acts as though it were under the influence of a negative viscosity. The same process drives the jet stream in the atmosphere. The steeply sloping density surface separating the polar air mass from mid-latitude air masses atthe atmosphere's polar front also leads to baroclinic instability. For more on this topic see Starr's (1968) book on *Physics of Negative Viscosity Phenomena*. Let's look at this process in the Gulf Stream (figure 11.10). The strong current shear in the Stream causes the flow to begin to meander. The meander intensifies, and eventually the Stream throws off a ring. Those on the south side drift southwest, and eventually merge with the stream several months later (figure 11.11). The process occurs all along the recirculation region, and satellite images show nearly a dozen or so rings occur north and south of the stream (figure 11.11).

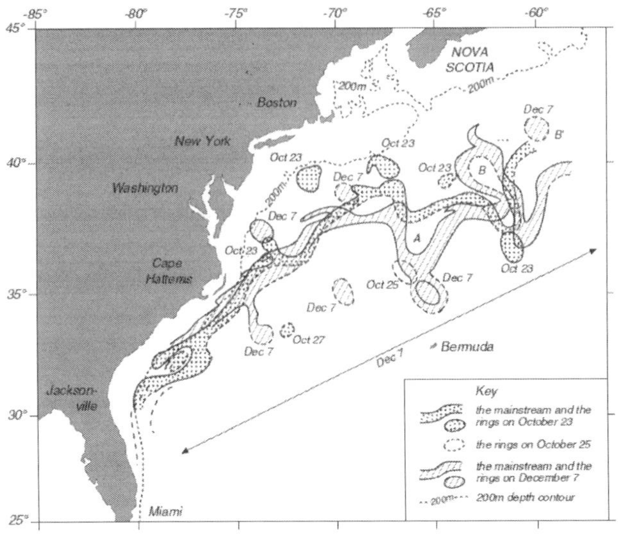

Figure 11.11 Sketch of the position of the Gulf Stream, warm core, and cold core eddies observed in infrared images of the sea surface collected by the infrared radiometer on NOAA-5 in October and December 1978. After Tolmazin (1985: 91).

11.5 CONCISE SUMMARY

The theory for wind-driven, geostrophic currents was first outlined in a series of papers by Sverdrup, Stommel, and Munk between 1947 and 1951.

- They showed that realistic currents can be calculated only if the Coriolis parameter varies with latitude.
- Sverdrup showed that the curl of the wind stress drives a northward masstransport, and that this can be used to calculate currents in the ocean away from western boundary currents.
- Stommel showed that western boundary currents are required for flow to circulate around an ocean basin when the Coriolis parameter varies with latitude.
- Munk showed how to combine the two solutions to calculate the wind-driven geostrophic circulation in an ocean basin. In all cases, the current is driven by the curl of the wind stress.
- The observed circulation in the ocean is very turbulent. Many years of observations may need to be averaged together to obtain a stable map of the mean flow.
- The Gulf Stream is a region of baroclinic instability in which turbulence accelerates the stream. This creates a Gulf Stream recirculation. Transports in the recirculation region are much larger than transports calculated from the Sverdrup-Munk theory.

12 VORTICITY IN THE OCEAN: AT A GLANCE

Most of the fluid flows with which we are familiar, from bathtubs to swimming pools, are not rotating, or they are rotating so slowly that rotation is not important except maybe at the drain of a bathtub as water is let out. As a result, we do not have a good intuitive understanding of rotating flows. In the ocean, rotation and the conservation of vorticity strongly influence flow over distances exceeding a few tens of kilometers. The consequences of the rotation leads to results we have not seen before in our day-to-day dealings with fluids. For example, did you ask yourself why the curl of the wind stress leads to a mass transport in the north-south direction and not in the east-west direction? What is special about north-south motion? In this chapter, I will explore some of the consequences of rotation for flow in the ocean.

12.1 VORTICITY

In simple words, vorticity is the rotation of the fluid. The rate of rotation can be defined various ways. Consider a bowl of water sitting on a table in a laboratory. The water may be spinning in the bowl. In addition to the spinning of the water, the bowl and the laboratory are rotating because they are on a rotating earth. The two processes are separate and lead to two types of vorticity.

Planetary Vorticity Everything on earth, including the ocean, the atmosphere, and bowls of water, rotates with the earth. This rotation is the *planetary vorticity* f. It is twice the local rate of rotation of earth:

$$\boxed{f \equiv 2\Omega \sin \varphi \text{ (radians/s)} = 2 \sin \varphi \text{ (cycles/day)}} \qquad (12.1)$$

Planetary vorticity is the Coriolis parameter I used earlier to discuss flow in the ocean. It is greatest at the poles where it is twice the rotation rate of earth. Note that the vorticity vanishes at the equator and that the vorticity in the southern hemisphere is negative because ϕ is negative.

Relative Vorticity: The ocean and atmosphere do not rotate at exactly the same rate as earth. They have some rotation relative to earth due to currents and winds. *Relative vorticity* ζ is the vorticity due to currents in the ocean. Mathematically it is:

$$\zeta \equiv \text{curl}_z \mathbf{V} = \frac{\partial v}{\partial x} - \frac{\partial u}{\partial y} \qquad (12.2)$$

where **V** = (u, v) is the horizontal velocity vector, and where we have assumed that the flow is two-dimensional. This is true if the flow extends over distances greater than a few tens of kilometers. ζ is the vertical component of the three-dimensional vorticity vector ω, and it is sometimes written ωz . ζ is positive for counter-clockwise rotation viewed from above. This is the same sense as earth's rotation in the northern hemisphere.

Note on Symbols Symbols commonly used in one part of oceanography often have very different meaning in another part. Here I use ζ for vorticity, but in Chapter 10, I used ζ to mean the height of the sea surface. I could use ωz for relative vorticity, but ω is also commonly used to mean frequency in radians per second. I have tried to eliminate most confusing uses, but the dual use of ζ is one we will have to live with. Fortunately, it shouldn't cause much confusion.

For a rigid body rotating at rate Ω, curl**V** = 2 Ω. Of course, the flow does not need to rotate as a rigid body to have relative vorticity. Vorticity can also result from shear. For example, at a north/south western boundary in the ocean, $u = 0$, $v = v(x)$ and $\zeta = \partial v(x)/\partial x$.

ζ is usually much smaller than f, and it is greatest at the edge of fast currents such as the Gulf Stream. To obtain some understanding of the size of ζ, consider the edge of the Gulf Stream off Cape Hatteras where the velocity decreases by 1 m/s in 100 km at the boundary. The curl of the current is approximately (1 m/s)/(100 km) = 0.14 cycles/day = 1 cycle/week. Hence even this large relative vorticity is still almost seven times smaller than f. A more typical values of relative vorticity, such as the vorticity of eddies, is a cycle per month.

Absolute Vorticity: The sum of the planetary and relative vorticity is called*absolute vorticity*:

$$\text{Absolute Vorticity} \equiv (\zeta + f) \quad (12.3)$$

We can obtain an equation for absolute vorticity in the ocean by manipulating the equations of motion for frictionless flow. We begin with:

$$\frac{Du}{Dt} - f\,v = -\frac{1}{\rho}\frac{\partial p}{\partial x} \quad (12.4a)$$

$$\frac{Dv}{Dt} + f\,u = -\frac{1}{\rho}\frac{\partial p}{\partial y} \quad (12.4b)$$

If we expand the substantial derivative, and if we subtract ∂ /∂y of (12.4a) from∂ /∂x of (12.4b) to eliminate the pressure terms, we obtain after some algebraic manipulations:

$$\frac{D}{Dt}(\zeta + f) + (\zeta + f)\left(\frac{\partial u}{\partial x} + \frac{\partial v}{\partial y}\right) = 0 \quad (12.5)$$

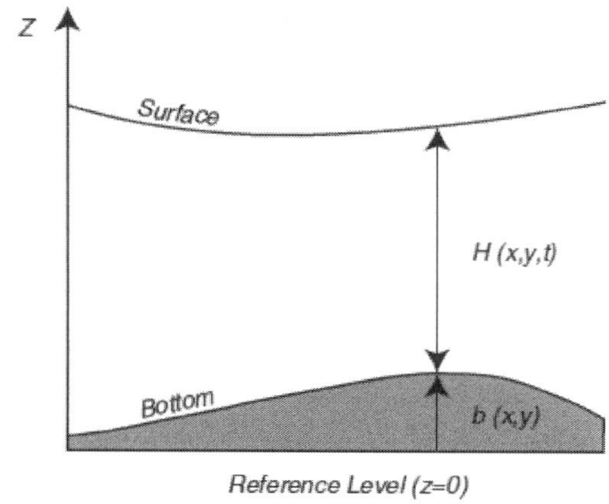

Figure 12.1 Sketch of fluid flow used for deriving conservation of potential vorticity. After Cushman-Roisin (1994: 55).

In deriving (12.5) we used:

$$\frac{Df}{Dt} = \frac{\partial f}{\partial t} + u\frac{\partial f}{\partial x} + v\frac{\partial f}{\partial y} = \beta v$$

because f is independent of time t and eastward distance x.

Potential Vorticity: The rotation rate of a column of fluid changes as the column is expanded or contracted. This changes the vorticity through changes in ζ. To see how this happens, consider barotropic, geostrophic flow in an ocean with depth $H(x, y, t)$, where H is the distance from the sea surface to the bottom. That is, we allow the surface to have topography (figure 12.1).

Integrating the continuity equation (7.19) from the bottom to the top of the ocean gives (Cushman-Roisin, 1994):

$$\left(\frac{\partial u}{\partial x} + \frac{\partial v}{\partial y}\right)\int_{b}^{b+H} dz + w\Big|_{b}^{b+H} = 0 \quad (12.6)$$

where b is the topography of the bottom, and H is the depth of the water. Notice that $\partial u/\partial x$ and $\partial v/\partial y$ are independent of z because they are barotropic, and the terms can be taken outside the integral.

The boundary conditions require that flow at the surface and the bottom be along the surface and the bottom. Thus the vertical velocities at the top and the bottom are:

$$w(b+H) = \frac{\partial(b+H)}{\partial t} + u\frac{\partial(b+H)}{\partial x} + v\frac{\partial(b+H)}{\partial y} \quad (12.7)$$

$$w(b) = u\frac{\partial(b)}{\partial x} + v\frac{\partial(b)}{\partial y} \quad (12.8)$$

where we used $\partial b/\partial t = 0$ because the bottom does not move, and $\partial H/\partial z = 0$. Substituting (12.7) and (12.8) into (12.6) we obtain

$$\left(\frac{\partial u}{\partial x}+\frac{\partial v}{\partial y}\right)+\frac{1}{H}\frac{DH}{Dt}=0$$

Substituting this into (12.5) gives:

$$\frac{D}{Dt}(\zeta+f)-\frac{(\zeta+f)}{H}\frac{DH}{Dt}=0$$

which can be written:

$$\frac{D}{Dt}\left(\frac{\zeta+f}{H}\right)=0$$

The quantity within the parentheses must be constant. It is called *potential vorticity* Π. Potential vorticity is conserved along a fluid trajectory:

$$\text{Potential Vorticity} = \Pi \equiv \frac{\zeta+f}{H} \quad (12.9)$$

For baroclinic flow in a continuously stratified fluid, the potential vorticity can be written (Pedlosky, 1987: §2.5):

$$\Pi = \frac{\zeta+f}{\rho}\cdot\nabla\lambda \quad (12.10)$$

where λ is any conserved quantity for each fluid element. In, particular, if $\lambda = \rho$then:

$$\Pi = \frac{\zeta+f}{\rho}\frac{\partial\rho}{\partial z} \quad (12.11)$$

assuming the horizontal gradients of density are small compared with the vertical gradients, a good assumption in the thermocline. In most of the interior of the ocean, $f \gg \zeta$ and (12.11) is written (Pedlosky, 1996, eq 3.11.2):

$$\Pi = \frac{f}{\rho}\frac{\partial\rho}{\partial z} \quad (12.12)$$

This allows the potential vorticity of various layers of the ocean to be determined directly from hydrographic data without knowledge of the velocity field.

12.2 CONSERVATION OF VORTICITY

The angular momentum of any isolated spinning body is conserved. The spinning body can be an eddy in the ocean or the earth in space. If the spinning body is not isolated, that is, if it is linked to another body, then angular momentum can be transferred between the bodies. The two bodies need not be in physical contact. Gravitational forces can transfer momentum between bodies in space. I will return to this topic in Chapter 17 when I discuss tides in the ocean. Here, let's look at conservation of vorticity in a spinning ocean.

Friction is essential for the transfer of momentum in a fluid. Friction transfers momentum from the atmosphere to the ocean through the thin, frictional, Ekman layer at the sea surface. Friction transfers momentum from the ocean to the solid earth through the Ekman layer at the sea floor. Friction along the sides of sub-sea mountains leads to pressure differences on either side of themounta in which causes another kind of drag called *form drag*. This is the same drag that causes wind force on cars moving at high speed. In the vast interior of the ocean, however, the flow is frictionless, and vorticity is conserved. Such a flow is said to be *conservative.*

Conservation of Potential Vorticity: The conservation of potential vorticity couples changes in depth, relative vorticity, and changes in latitude. All three interact.

1. Changes in the depth H of the flow changes in the relative vorticity. The concept is analogous with the way figure skaters decreases their spin by extending their arms and legs. The action increases their moment of inertia and decreases their rate of spin (figure 12.2).

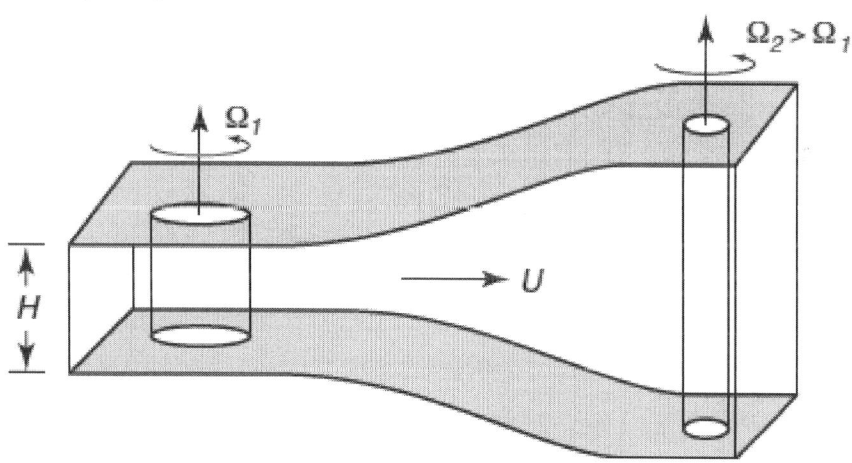

Figure 12.2 Sketch of the production of relative vorticity by the changes in the height of a fluid column. As the vertical fluid column moves from left to right, vertical stretching reduces the moment of inertia of the column, causing it to spin faster.

2. Changes in latitude require a corresponding change in ζ. As a column of water moves equatorward, f decreases, and ζ must increase (figure 12.3). If this seems somewhat mysterious, von Arx (1962) suggests we consider a barrel of water at rest at the north pole. If the barrel is moved southward, the water in it retains the rotation it had at the pole, and it will appear to rotate counter-clockwise at the new latitude where f is smaller.

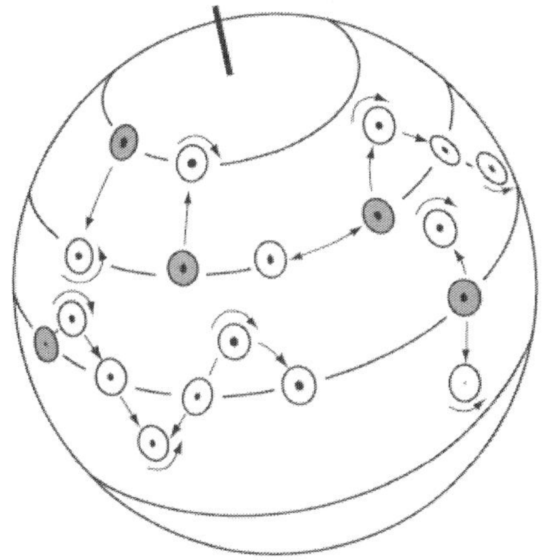

Figure 12.3 Angular momentum tends to be conserved as columns of water change latitude.

This changes the relative vorticity of the columns. After von Arx (1962: 110).

12.3 INFLUENCE OF VORTICITY

The concept of conservation of potential vorticity has far reaching consequences, and its application to fluid flow in the ocean gives a deeper under- standing of ocean currents.

Flow Tends to be Zonal: In the ocean f tends to be much larger than ζ and thus f/H = constant. This requires that the flow in an ocean of constant depth be zonal. Of course, depth is not constant, but in general, currents tend to be east-west rather than north south. Wind makes small changes in ζ, leading to a small meridional component to the flow (see figure 11.3).

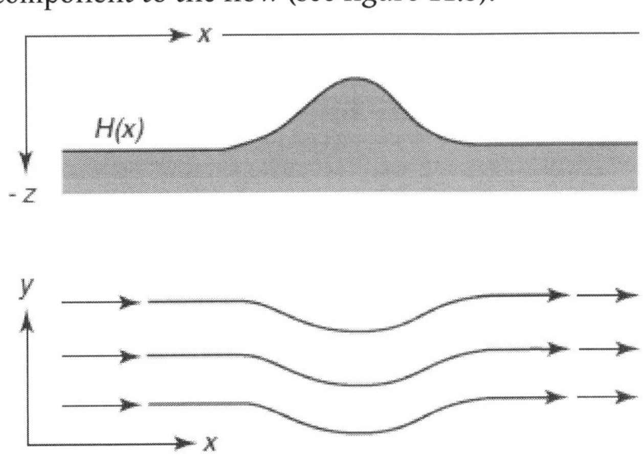

Figure 12.4 Barotropic flow over a sub-sea ridge is turned equatorward to conserve potential vorticity. After Dietrich et al. (1980: 333).

Topographic Steering: Barotropic flows are diverted by sea floor features. Consider what happens when a flow that extends from the surface to the bottom encounters a sub-sea ridge (figure 12.4). As the depth decreases, $\zeta + f$ must also decrease, which requires that f decrease, and the flow is turned toward theequator. This is called *topographic steering*. If the change in depth is sufficientlylarge, no change in latitude will be sufficient to conserve potential vorticity, and the flow will be unable to cross the ridge. This is called *topographic blocking*.

Western Boundary Currents: The balance of vorticity provides an alternate explanation for the existence of western boundary currents. Consider the gyre-scale flow in an ocean basin (figure 12.5), say in the north Atlantic from 10°Nto 50°N. The wind blowing over the Atlantic adds negative vorticity ζ_T . Asthe water flows around the gyre, the vorticity of the gyre must remain nearly constant, else the flow would spin faster or slower. Overall, the negative vorticity input by the wind must be balanced by a source of positive vorticity.

Throughout most of the basin the negative vorticity input by the wind is balanced by an increase in relative vorticity. As the flow moves southward throughout the basin, f decreases and ζ must increase according to (12.9) be- cause H, the depth of the wind-driven circulation, does not change much.

The balance breaks down, however, in the west where the flow returns north- ward. In the west, f increases, ζ decreases, and a source of positive vorticity is needed. The positive vorticity ζb is produced by the western boundary boundary current.

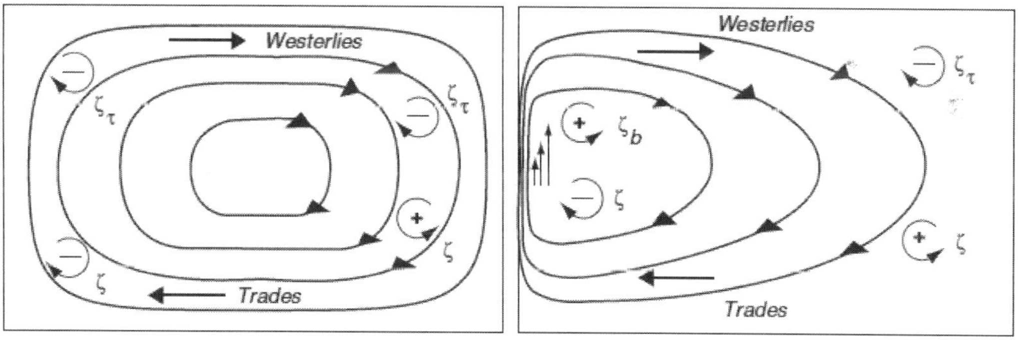

Figure 12.5 The balance of potential vorticity can clarify why western boundary currents are necessary. Left: Vorticity input by the wind ζ_T balances the change in relative vorticity ζ in the east as the flow moves southward and f decreases. The two do not balance in the west where ζ must decrease as the flow moves northward and f increases. Right: Vorticity in the west is balanced by relative vorticity ζb generated by shear in the western boundary current.

12.4 VORTICITY AND EKMAN PUMPING

Rotation places another very interesting constraint on the geostrophic flow field. To help understand the constraints, let's first consider flow in a fluid with constant rotation. Then we will look into how vorticity constrains the flow of a fluid with rotation that varies with latitude. An understanding of the constraints leads to a deeper understanding of Sverdrup's and Stommel's results discussed in the last chapter.

Fluid dynamics on the *f* Plane: the Taylor-Proudman Theorem: The influence of vorticity due to earth's rotation is most striking for geostrophic flow of a fluid with constant density ρ0 on a plane with constant rotation $f = f0$. From Chapter 10, the three components of the geostrophic equations (10.4) are:

$$f\,v = \frac{1}{\rho_0}\frac{\partial p}{\partial x} \quad (12.13a)$$

$$f\,u = -\frac{1}{\rho_0}\frac{\partial p}{\partial y} \quad (12.13b)$$

$$g = -\frac{1}{\rho_0}\frac{\partial p}{\partial z} \quad (12.13c)$$

and the continuity equations (7.19) is:

$$0 = \frac{\partial u}{\partial x} + \frac{\partial v}{\partial y} + \frac{\partial w}{\partial z} \quad (12.13d)$$

Taking the z derivative of (12.13a) and using (12.13c) gives:

$$-f_0\frac{\partial v}{\partial z} = -\frac{1}{\rho_0}\frac{\partial}{\partial z}\left(\frac{\partial p}{\partial x}\right) = \frac{\partial}{\partial x}\left(-\frac{1}{\rho_0}\frac{\partial p}{\partial z}\right) = \frac{\partial g}{\partial x} = 0$$

$$f_0\frac{\partial v}{\partial z} = 0$$

$$\therefore \quad \frac{\partial v}{\partial z} = 0$$

Similarly, for the u-component of velocity (12.13b). Thus, the vertical derivative of the horizontal velocity field must be zero.

$$\frac{\partial u}{\partial z} = \frac{\partial v}{\partial z} = 0 (12.14)$$

This is the *Taylor-Proudman Theorem*, which applies to slowly varying flows in a homogeneous, rotating, inviscid fluid. The theorem places strong constraints on the flow:

If therefore any small motion be communicated to a rotating fluid the resulting motion of the fluid must be one in which any two particles originally in a line parallel to the axis of rotation must remain so, except for possible small oscillations about that position—Taylor (1921).

Hence, rotation greatly stiffens the flow! Geostrophic flow cannot go over a seamount, it must go around it. Taylor (1921) explicitly derived (12.14) and (12.16) below. Proudman (1916) independently derived the same theorem but not as explicitly.

Further consequences of the theorem can be obtained by eliminating the pressure terms from (12.13a & 12.13b) to obtain:

$$\frac{\partial u}{\partial x} + \frac{\partial v}{\partial y} = -\frac{\partial}{\partial x}\left(\frac{1}{f_0\,\rho_0}\frac{\partial p}{\partial y}\right) + \frac{\partial}{\partial y}\left(\frac{1}{f_0\,\rho_0}\frac{\partial p}{\partial x}\right) \text{(12.15a)}$$

$$\frac{\partial u}{\partial x} + \frac{\partial v}{\partial y} = \frac{1}{f_0\,\rho_0}\left(-\frac{\partial^2 p}{\partial x\,\partial y} + \frac{\partial^2 p}{\partial x\,\partial y}\right) \text{(12.15b)}$$

$$\frac{\partial u}{\partial x} + \frac{\partial v}{\partial y} = 0 \text{ (12.15c)}$$

Because the fluid is incompressible, the continuity equation (12.13d) requires

$$\frac{\partial w}{\partial z} = 0 \text{ (12.16)}$$

Furthermore, because $w = 0$ at the sea surface and at the sea floor, if the bottom is level, there can be no vertical velocity on an f-plane. Note that the derivation of (12.16) did not require that density be constant. It requires only slow motion in a frictionless, rotating fluid.

Fluid Dynamics on the Beta Plane: Ekman Pumping: If (12.16) is true, the flow cannot expand or contract in the vertical direction, and it is indeed as rigid as a steel bar. There can be no gradient of vertical velocity in an ocean with constant planetary vorticity. How then can the divergence of the Ekman transport at the sea surface lead to vertical velocities at the surface or at the base of the Ekman layer? The answer can only be that one of the constraints used in deriving (12.16) must be violated. One constraint that can be relaxed is the requirement that $f = f0$.

Consider then flow on a beta plane. If $f = f0 + \beta\, y$, then (12.15a) becomes:

$$\frac{\partial u}{\partial x} + \frac{\partial v}{\partial y} = -\frac{1}{f\,\rho_0}\frac{\partial^2 p}{\partial x\,\partial y} + \frac{1}{f\,\rho_0}\frac{\partial^2 p}{\partial x\,\partial y} - \frac{\beta}{f}\frac{1}{f\,\rho_0}\frac{\partial p}{\partial x} \text{ (12.17)}$$

$$f\left(\frac{\partial u}{\partial x} + \frac{\partial v}{\partial y}\right) = -\beta\, v \text{ (12.18)}$$

where we have used (12.13a) to obtain v in the right-hand side of (12.18).

Using the continuity equation, and recalling that $\beta\, y \ll f0$

$$f_0\frac{\partial w_G}{\partial z} = \beta\, v \text{ (12.19)}$$

where we have used the subscript G to emphasize that (12.19) applies to the ocean's interior, geostrophic flow. Thus the variation of Coriolis force with latitude allows vertical velocity gradients in the geostrophic interior of the ocean, and the vertical velocity leads to north-south currents. This explains why Sverdrup and Stommel both needed to do their calculations on a β-plane.

Ekman Pumping in the Ocean: In Chapter 9, we saw that the curl of the wind stress **T** produced a divergence of the Ekman transports leading to a vertical velocity *wE* (0) at the top of the Ekman layer. In Chapter 9 we derived

$$w_E(0) = -\text{curl}\left(\frac{\mathbf{T}}{\rho f}\right) \quad (12.20)$$

which is (9.30b) where ρ is density and *f* is the Coriolis parameter. Because the vertical velocity at the sea surface must be zero, the Ekman vertical velocity must be balanced by a vertical geostrophic velocity *wG*(0).

$$w_E(0) = -w_G(0) = -\text{curl}\left(\frac{\mathbf{T}}{\rho f}\right) \quad (12.21)$$

Ekman pumping (*wE* (0)) drives a vertical geostrophic current (-*wG*(0)) inthe ocean's interior. But why does this produce the northward current calculated by Sverdrup (11.6)? Peter Niiler (1987: 16) gives an explanation.

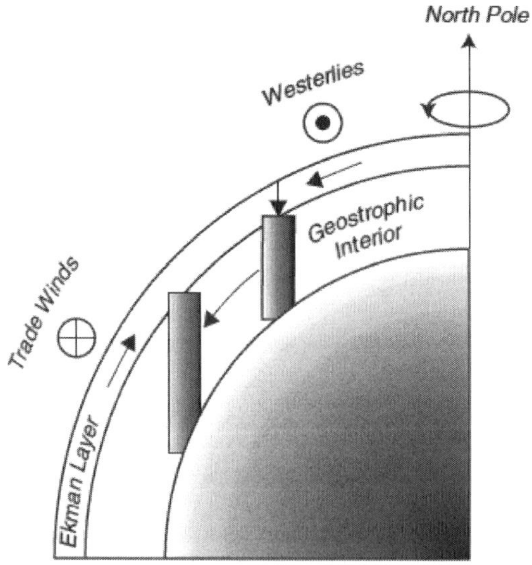

Figure 12.6 Ekman pumping that produces a downward velocity at the base of the Ekman layer forces the fluid in the interior of the ocean to move southward. See text for why this happens. After Niiler (1987).

Let us postulate there exists a deep level where horizontal and vertical motion of the water is much reduced from what it is just below the mixed layer [figure 12.6]... Also let us assume that vorticity is conserved there (or mixing is small) and the flow is so slow that accelerations over the earth's surface are much smaller than Coriolis accelerations. In such a situation a column of water of depth *H* will conserve its spin per unit volume, f/H (relative to the sun, parallel to the earth's axis of rotation). A vortex column which is compressed from the top by wind-forced sinking (H decreases) and whose bottom is in relatively quiescent water would tend to shorten and slow its spin. Thus because of the curved ocean surface it has to move southward (or extend its column) to regain its spin. Therefore, there should be a massive flow of water at some depthbelow the surface to the south in areas where the surface layers produce a sinking motion and to the north where rising motion is produced. This phenomenon was first modeled correctly by Sverdrup (1947) (after he wrote "ocean") and gives a dynamically plausible explanation of how wind produces deeper circulation in the ocean.

Peter Rhines (1984) points out that the rigid column of water trying to es- cape the squeezing imposed by the atmosphere escapes by moving southward. The southward velocity is about 5,000 times greater than the vertical Ekman velocity.

Ekman Pumping: An Example: Now let's see how Ekman pumping drives geostrophic flow in say the central north Pacific (figure 12.7) where the curl of the wind stress is negative. Westerlies in the north drive a southward transport, the trades in the south drive a northward transport. The converging Ekman transports must be balanced by downward geostrophic velocity (12.21).

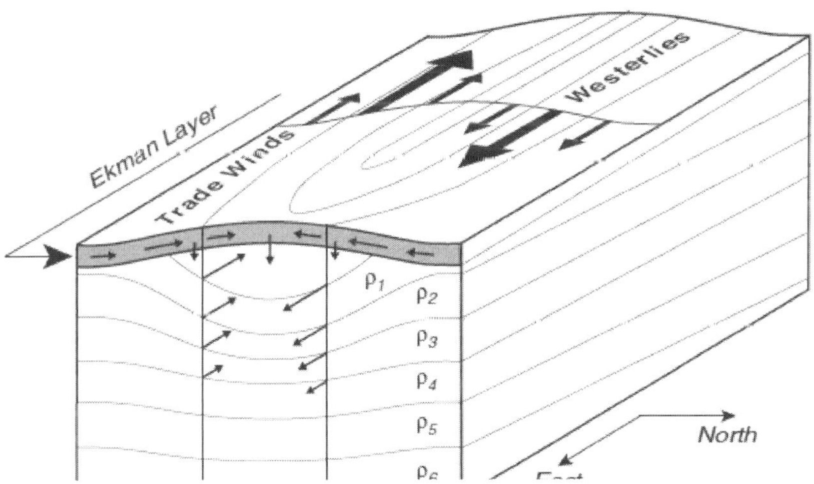

Figure 12.7 Winds at the sea surface drive Ekman transports to the right of the wind in this northern hemisphere example (bold arrows in shaded Ekman layer). The converging Ekman transports driven by the trades and westerlies drives a downward geostrophic flow just below the Ekman layer (bold vertical arrows), leading to downward bowing constant density surfaces ρ_i. Geostrophic currents associated with the warm water are shown by bold arrows. After Tolmazin (1985: 64).

Because the water near the surface is warmer than the deeper water, the vertical velocity produces a pool of warm water. Much deeper in the ocean, the wind-driven geostrophic current must go to zero (Sverdrup's hypothesis) and the deep pressure gradients must be zero. As a result, the surface must dome upward because a column of warm water is longer than a column of cold water having the same weight (they must have the same weight, otherwise, the deep pressure would not be constant, and there would be a deep horizontal pressure gradient). Such a density distribution produces north-south pressure gradients at mid depths that must be balanced by east-west geostrophic currents. In short, the divergence of the Ekman transports redistributes mass within the frictionless interior of the ocean leading to the wind-driven geostrophic currents.

Now let's continue the idea to include the entire north Pacific to see how winds produce currents flowing upwind. The example will give a deeper understanding of Sverdrup's results we discussed in §11.1.

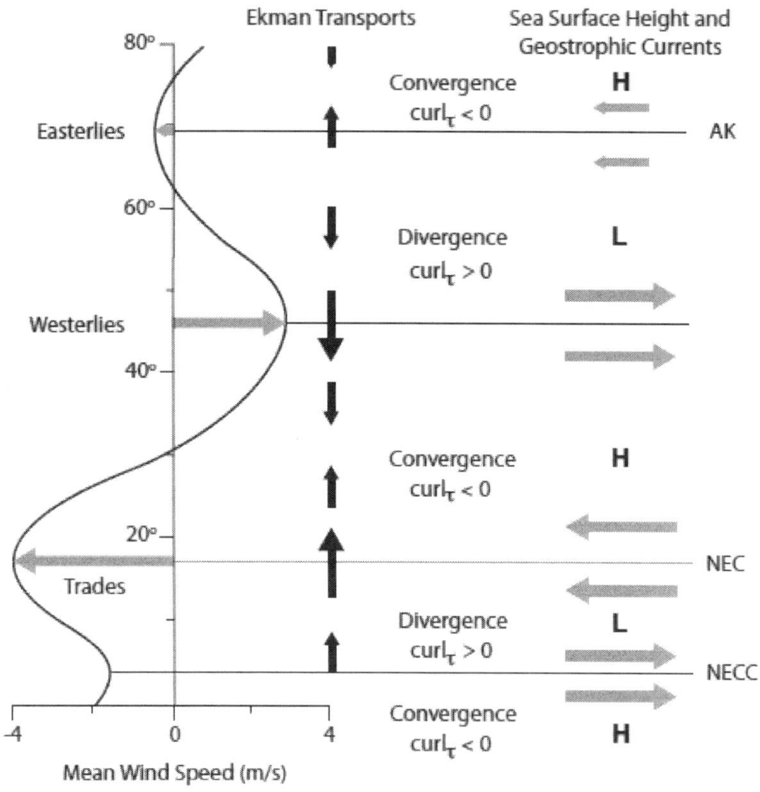

Figure 12.8 An example of how winds produce geostrophic currents running upwind. Ekman transports due to winds in the north Pacific (Left) lead to Ekman pumping (Center), which sets up north-south pressure gradients in the upper ocean. The pressure gradients are balanced by the Coriolis force due to east-west geostrophic currents (Right). Horizontal lines indicate regions where the curl of the zonal wind stress changes sign. AK: Alaskan Current, NEC: North Equatorial Current, NECC: North Equatorial Counter Current.

Figure 12.8 shows shows the mean zonal winds in the Pacific, together withthe north-south Ekman transports driven by the zonal winds. Notice that con- vergence of transport leads to downwelling, which produces a thick layer of warm water in the upper kilometer of the water column, and high sea level.Figure 12.6 is a sketch of the cross section of the region between 10°N and60°N, and it shows the pool of warm water in the upper kilometer centered on 30°N. Conversely, divergent transports leads to low sea level. The mean north-south pressure gradients associated with the highs and lows are balanced by the Coriolis force of east-west geostrophic currents in the upper ocean (shown at the right in the figure).

12.5 CONCISE SUMMARY

- Vorticity strongly constrains ocean dynamics.
- Vorticity due to earth's rotation is much greater than other sources of vorticity.
- Taylor and Proudman showed that vertical velocity is impossible in a uniformly rotating flow. The ocean is rigid in the direction parallel to the rotation axis. Hence Ekman pumping requires that planetary vorticity vary with latitude. This explains why Sverdrup and Stommel found that realistic oceanic circulation, which is driven by Ekman pumping, requires that f vary with latitude.
- The curl of the wind stress adds relative vorticity to central gyres of each ocean basin. For steady state circulation in the gyre, the ocean must lose vorticity in western boundary currents.
- Positive wind stress curl leads to divergent flow in the Ekman layer. The ocean's interior geostrophic circulation adjusts through a northward mass transport.
- Conservation of absolute vorticity in an ocean with constant density leads to the conservation of potential vorticity. Thus changes in depth in an ocean of constant density requires changes of latitude of the current.

13 OCEANIC DEEP CIRCULATION

The direct forcing of the oceanic circulation by wind discussed in the last few chapters is strongest in the upper kilometer of the water column. Below a kilometer lies the vast water masses of the ocean extending to depths of 4–5 km.

The water is everywhere cold, with a potential temperature less than 4°C. The water mass is formed when cold, dense water sinks from the surface to great depths at high latitudes. It spreads out from these regions to fill the ocean basins. Deep mixing eventually pulls the water up through the thermocline over large areas of the ocean. It is this upward mixing that drives the deep circulation. The vast deep ocean is usually referred to as the *abyss*, and the circulation as the *abyssal circulation*.

The densest water at the sea surface, water that is dense enough to sink to the bottom, is formed when frigid air blows across the ocean at high latitudes in winter in the Atlantic between Norway and Greenland and near Antarctica. The wind cools and evaporates water. If the wind is cold enough, sea ice forms, further increasing the salinity of the water because ice is fresher than sea water. Bottom water is produced only in these two regions. Cold, dense water is formed in the North Pacific, but it is not salty enough to sink to the bottom.

At mid and low latitudes, the density, even in winter, is sufficiently low that the water cannot sink more than a few hundred meters into the ocean. The only exception are some seas, such as the Mediterranean Sea, where evaporation is so great that the salinity of the water is sufficiently great for the water to sink to the bottom of these seas. If these seas can exchange water with the open ocean, the waters formed in winter in the seas mixes with the water in the open ocean and it spreads out along intermediate depths in the open ocean.

13.1 DEEP CIRCULATION

Many terms have been used to describe the deep circulation. They include:

1) *abyssal circulation*, 2) *thermohaline circulation* 3) *meridional overturning circulation*, and 4) *global conveyor*. The term thermohaline circulation was once widely used, but it has almost entirely disappeared from the oceanographic literature (Toggweiler and Russell, 2008). It is no longer used because it is now clear that the flow is not density driven, and because the concept has not been clearly defined (Wunsch, 2002b).

The meridional overturning circulation is better defined. It is the zonal average of the flow plotted as a function of depth and latitude. Plots of the circulation show where vertical flow is important, but they show no information about how circulation in the gyres influences the flow.

Following Wunsch (2002b), I define the deep circulation as the circulation of mass. Of course, the mass circulation also carries heat, salt, oxygen, and other properties. But the circulation of the other properties is not the same as the mass transport. For example, Wunsch points out that the north Atlantic imports heat but exports oxygen.

13.2 IMPORTANCE OF THE DEEP CIRCULATION

The deep circulation carries heat, salinity, oxygen, CO2, and other properties from high latitudes in winter to lower latitudes throughout the world. This has very important consequences.

- The contrast between the cold deep water and the warm surface waters determines the stratification of the ocean, which strongly influences ocean dynamics.
- The volume of deep water is far larger than the volume of surface water. Although currents in the deep ocean are relatively weak, they have transports comparable to the surface transports.
- The fluxes of heat and other variables carried by the deep circulation influences earth's heat budget and climate. The fluxes vary from decades to centuries to millennia, and this variability modulates climate over such time intervals. The ocean may be the primary cause of variability over times ranging from years to decades, and it may have helped modulate ice-age climate.

Two aspects of the deep circulation are especially important for understanding earth's climate and its possible response to increased carbon dioxide CO2 in the atmosphere: i) the ability of cold water to store CO2 and heat absorbed from the atmosphere, and ii) the ability of deep currents to modulate the heat transported from the tropics to high latitudes.

The ocean as a Reservoir of Carbon Dioxide: The ocean are the primary reservoir of readily available CO2, an important greenhouse gas. The ocean contain 40,000 GtC of dissolved, particulate, and living forms of carbon. The land contains 2,200 GtC, and the atmosphere contains only 750 GtC. Thus the ocean hold 50 times more carbon than the air. Furthermore, the amount of new carbon put into the atmosphere since the industrial revolution, 150 GtC, is less than the amount of carbon cycled through the marine ecosystem in fiveyears. (1 GtC = 1 gigaton of carbon = 10^{12} kilograms of carbon.) Carbonate rocks such as limestone, the shells of marine animals, and coral are other, much larger, reservoirs. But this carbon is locked up. It cannot be easily exchanged with carbon in other reservoirs.

More CO_2 dissolves in cold water than in warm water. Just imagine shaking and opening a hot can of Coke™. The CO_2 from a hot can will spew out far faster than from a cold can. Thus the cold deep water in the ocean is the major reservoir of dissolved CO_2 in the ocean.

New CO2 is released into the atmosphere when fossil fuels and trees are burned. Very quickly, 48% of the CO2 released into the atmosphere dissolves into the ocean (Sabine et al, 2004), much of which ends up deep in the ocean.

Forecasts of future climate change depend strongly on how much CO2 is stored in the ocean and for how long. If little is stored, or if it is stored and later released into the atmosphere, the concentration in the atmosphere will change, modulating earth's long-wave radiation balance. How much and how long CO2 is stored in the ocean depends on the transport of CO2 by the deep circulation and on the net flux of carbon deposited on the sea floor. The amount that dissolves depends on the temperature of the deep water, the storage time in the deep ocean depends on the rate at which deep water is replenished, and the deposition depends on whether the dead plants and animals that drop to the sea floor are oxidized. Increased ventilation of deep layers, and warming of the deep layers could release large quantities of the gas to the atmosphere.

The storage of carbon in the ocean also depends on the dynamics of marine ecosystems, upwelling, and the amount of dead plants and animals stored in sediments. But I won't consider these processes.

Oceanic Transport of Heat: The ocean carry about half the heat out of the tropics needed to maintain earth's temperature. Heat carried by the Gulf Stream and the north Atlantic drift keeps the far north Atlantic ice free, and ithelps warm Europe. Norway, at 60°N is far warmer than southern Greenland or northern Labrador at the same latitude. Palm trees grow on the west coast of Ireland, but not in Newfoundland which is further south.

Wally Broecker (1987), working at Lamont-Doherty Geophysical Observatory of Columbia University, calls the oceanic component of the heat-transport system the *Global Conveyor Belt*. The basic idea is that surface currents carry heat to the far north Atlantic (figure 13.1). There the surface water releasesheat and water to the atmosphere, and it becomes sufficiently cold, salty, and dense that it sinks to the bottom in the Norwegian and Greenland Seas. It then flows southward in cold, bottom currents. Some of the water remains on the surface and returns to the south in cool surface currents such as the Labrador Current and Portugal Current (see figure 11.8). Richardson (2008) has written a very useful paper surveying our understanding of the global conveyor belt.

The deep bottom water from the north Atlantic is mixed upward in other regions and ocean, and eventually it makes its way back to the Gulf Stream and the North Atlantic. Thus most of the water that sinks in the north Atlantic must be replaced by water from the far south Atlantic. As this surface watermoves northward across the equator and eventually into the Gulf Stream, it carries heat out of the south Atlantic.

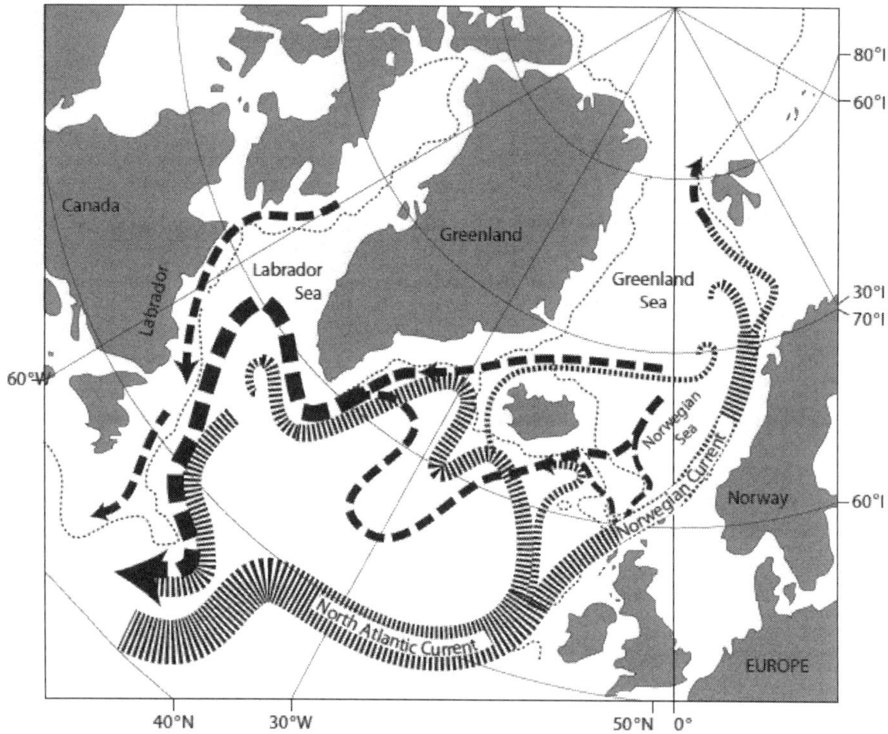

Figure 13.1 The surface (narrow dashes) and deep (wide dashes) currents in the north Atlantic. The North Atlantic Current brings warm water northward where it cools. Some sinks and returns southward as a cold, deep, western-boundary current. Some returns southward at the surface. From Woods Hole Oceanographic Institution.

So much heat is pulled northward by the formation of north Atlantic bottom water in winter that heat transport in the Atlantic is entirely northward, even in the southern hemisphere (figure 5.11). Much of the solar heat absorbed by the tropical Atlantic is shipped north to warm Europe and the northern hemisphere. Imagine then what might happen if the supply of heat is shut off. I will get back to that topic in the next section.

We can make a crude estimate of the importance of the north Atlantic surface and deep circulation from a calculation based on what we know about waters in the Atlantic compiled by Bill Schmitz (1996) in his wonderful summary of hislife's work. The Gulf Stream carries 40 Sv of 18°C water northward. Of this,14 Sv return southward in the deep western boundary current at a temperature of 2°C. The water must therefore lose 0.9 petawatts (1 petawatt = 10^{15} watt) in the north Atlantic north of 24°N. Although the calculation is very crude, it is remarkably close to the value of 1.2 ± 0.2 petawatts estimated much morecarefully by Rintoul and Wunsch (1991).

Note that if the water remained on the surface and returned as an eastern boundary current, it would be far warmer than the deep current when it returned southward. Hence, the heat transport would be much reduced and it would probably not keep the far north Atlantic ice free.

The production of bottom water is influenced by the surface salinity and winds in the north Atlantic (Toggweiler and Russell, 2008). It is also influenced by the rate of upwelling due to mixing in other oceanic areas. First, let's look at the influence of salinity.

Saltier surface waters form denser water in winter than less salty water. At first you may think that temperature is also important, but at high latitudeswater in all ocean basins gets cold enough to freeze, so all ocean produce -2°Cwater at the surface. Of this, only the most salty will sink, and the saltiestwater is in the Atlantic and under the ice on the continental shelves around Antarctica.

The production of bottom water is remarkably sensitive to small changesin salinity. Rahmstorf (1995), using a numerical model of the meridional over- turning circulation, showed that a ±0.1Sv variation of the flow of fresh waterinto the north Atlantic can switch on or off the deep circulation of 14 Sv. If the deep-water production is shut off during times of low salinity, the 1 petawatt of heat may also be shut off. Weaver and Hillaire-Marcel (2004) point out that the shutdown of the production of bottom water is unlikely, and if it did happen, it would lead to a colder Europe, not a new ice age, because of the higher concentrations of CO2 now in the atmosphere.

I write *may be shut off* because the ocean is a very complex system. Wedon't know if other processes will increase heat transport if the deep circulationis disturbed. For example, the circulation at intermediate depths may increase when deep circulation is reduced.

The production of bottom water is also remarkably sensitive to small changes in mixing in the deep ocean. Munk and Wunsch (1998) calculate that 2.1 TW (terawatts = 10^{12} watts) are required to drive the deep circulation, and that this small source of mechanical mixing drives a poleward heat flux of 2000 TW. Some of the energy for mixing comes from winds which can produce turbulent mixing throughout the ocean. Some energy comes from the dissipation of tidal currents, which depend on the distribution of the continents. Some of the energy comes from the flow of deep water past the mid-ocean ridge system. Thus during the last ice age, when sea level was much lower, tides, tidal currents, tidal dissipation, winds, and deep circulation all differed from present values.

Role of the Ocean in Ice-Age Climate Fluctuations: What might happen if the production of deep water in the Atlantic is shut off? Information contained in Greenland and Antarctic ice sheets, in north Atlantic sediments, and in lake sediments provide important clues.

Several ice cores through the Greenland and Antarctic ice sheets provide a continuous record of atmospheric conditions over Greenland and Antarctica extending back more than 700,000 years before the present in some cores. Annual layers in the core are counted to get age. Deeper in the core, where annual layers are hard to see, age is calculated from depth and from dust layers from well-dated volcanic eruptions. Oxygen-isotope ratios of the ice give air temperature at the glacier surface when the ice was formed. Deuterium concentrations give ocean-surface temperature at the moisture source region. Bubbles in the ice give atmospheric CO2 and methane concentration. Pollen, chemical com- position, and particles give information about volcanic eruptions, wind speed, and direction. Thickness of annual layers gives snow accumulation rates. And isotopes of some elements give solar and cosmic ray activity (Alley, 2000).

Cores through deep-sea sediments in the north Atlantic made by the Ocean Drilling Program give information about i) surface and deep temperatures and salinity at the location of above the core, ii) the production of north Atlantic deep water, iii) ice volume in glaciers, and iv) production of icebergs. Ice-sheet and deep-sea cores have allowed reconstructions of climate for the past few hundred thousand years.

- The oxygen-isotope and deuterium records in the ice cores show abrupt climate variability many times over the past 700,000 years. Many times during the last ice age temperatures near Greenland warmed rapidly over periods of 1–100 years, followed by gradual cooling over longer periods.
- (Dansgaard et al, 1993). For example, ≈ 11, 500 years ago, temperatures over Greenland warmed by ≈ 8°C in 40 years in three steps, each spanning 5 years (Alley, 2000). Such abrupt warming is called a Dansgaard/Oeschger event. Other studies have shown that much of the north- ern hemisphere warmed and cooled in phase with temperatures calculated from the ice core.
- The climate of the past 8,000 years was constant with very little variability. Our perception of climate change is thus based on highly unusual circumstances. All of recorded history has been during a period of warm and stable climate.
- Hartmut Heinrich and colleagues (Bond et al. 1992), studying the sediments in the north Atlantic found periods when coarse material was de- posited

on the bottom in mid ocean. Only icebergs can carry such material out to sea, and the find indicated times when large numbers of ice- bergs were released into the north Atlantic. These are now called Heinrich events.

- The correlation of Greenland temperature with iceberg production is related to the deep circulation. When icebergs melted, the surge of fresh water increased the stability of the water column shutting off the production of north Atlantic Deep Water. The shut-off of deep-water formation greatly reduced the northward transport of warm water into the north Atlantic, producing very cold northern hemisphere climate (figure 13.2). The melting of the ice pushed the polar front, the boundary between cold and warm water in the north Atlantic further south than its present position. The location of the front, and the time it was at different positions can be determined from analysis of bottom sediments.
- When the meridional overturning circulation shuts down, heat normally carried from the south Atlantic to the north Atlantic becomes available to warm the southern hemisphere. This results in a climate 'sea-saw' between northern and southern hemispheres.

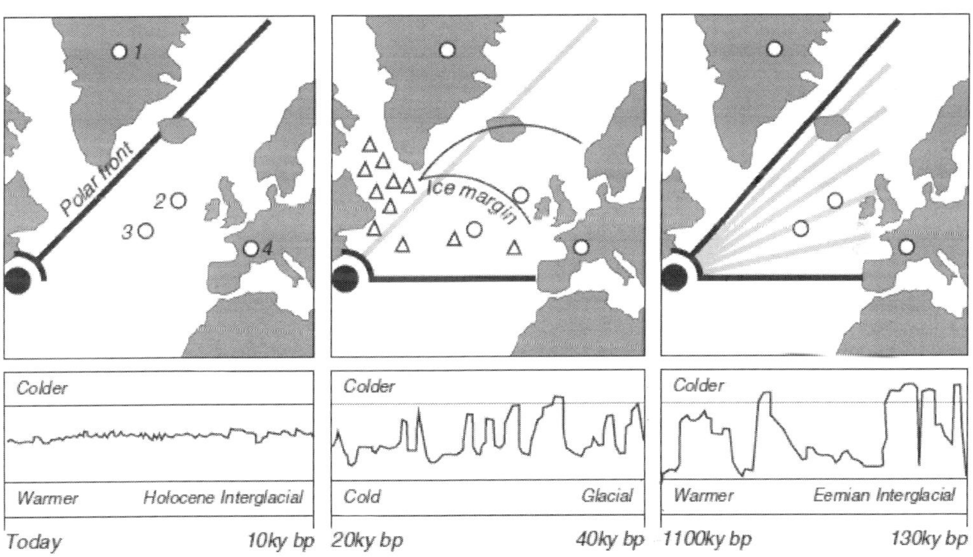

Figure 13.2 Periodic surges of icebergs during the last ice age appear to have modulated temperatures of the northern hemisphere by lowering the salinity of the far north Atlantic and reducing the meridional overturning circulation. Data from cores through the Greenland ice sheet (1), deep-sea sediments (2,3), and alpine-lake sediments (4) indicate that: Left: During recent times the circulation has been stable, and the polar front which separates warm and cold water masses has allowed warm water to penetrate beyond Norway. Center: During the last ice age, periodic surges of icebergs reduced salinity and reduced the meridional overturning circulation, causing the polar front to move southward and keeping warm water south of Spain. Right: Similar fluctuations during the last interglacial appear to have caused rapid, large changes in climate. The Bottom plot is a rough indication of temperature in the region, but the scales are not the same. After Zahn (1994).

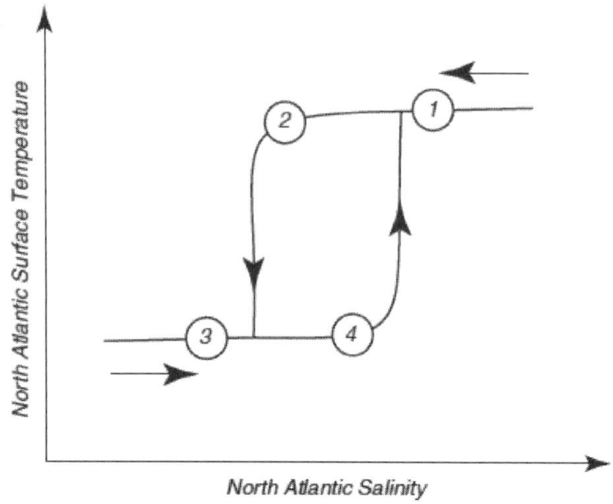

Figure 13.3 The meridional-overturning circulation in the north Atlantic may be stable near *2* and *4*. But, the switching from a warm, salty regime to a cold, fresh regime and back has hysteresis. This means that as the warm salty ocean in an initial state *1* freshens, and becomes more fresh than *2* it quickly switches to a cold, fresh state *3*. When the area again becomes salty, it must move past state *4* before it can switch back to *1*.

- The switching on and off of the deep circulation has large hysteresis (figure 13.3). The circulation has two stable states. The first is the present circulation. In the second, deep water is produced mostly near Antarctica, and upwelling occurs in the far north Pacific (as it does today) and in the far north Atlantic. Once the circulation is shut off, the system switches to the second stable state. The return to normal salinity does not cause the circulation to turn on. Surface waters must become saltier than average for the first state to return (Rahmstorf, 1995).
- Heinrich events seem to precede the largest Dansgaard/Oeschger events (Stocker and Marchal, 2000). Here's what seems to happen. The Heinrich event shuts off the Atlantic deep circulation which leads to a very cold north Atlantic (Martrat et al, 2007). This is followed about 1000 years later by a Dansgaard/Oeschger event with rapid warming.
- Dansgaard/Oeschger–Heinrich tandem events have global influence, and they are related to warming events seen in Antarctic ice cores. Temperatures changes in the two hemispheres are out of phase. When Greenland warms, Antarctica cools. Recent data from the European Project for Ice Coring in Antarctica (epica) shows that in the period between 20,000 and 90,000 years ago, 40% of the variance in the Greenland temperature data can be explained by Antarctic temperature data (Steig, 2006).

- A weakened version of this process with a period of about 1000 years may be modulating present-day climate in the north Atlantic, and it may have been responsible for the Little Ice Age from 1100 to 1800.

The relationship between variations in salinity, air temperature, deep-water formation, and the atmospheric circulation is not yet understood. For example, we don't know if changes in the atmospheric circulation trigger changes in the meridional overturning circulation, or if changes in the meridional overturning circulation trigger changes in the atmospheric circulation (Brauer et al, 2008). Furthermore, surges may result from warmer temperatures caused by increased water vapor from the tropics (a greenhouse gas) or from an internal instability of a large ice sheet. We do know, however, that climate can change very abruptly, and that circulation in the northern hemisphere has a very sensitive threshold, that when crossed, causes large changes in the circulation pattern.

For example, Steffensen (2008) found that 11,704, 12,896, and 14,694 years before 2000 ad the temperature of the source water for Greenland precipitation warmed 2–4°C in 1–3 years. This indicates a very rapid reorganization of the atmospheric circulation at high latitudes in the northern hemisphere and a shift in the location of the source region. During the earliest event air temperature over Greenland warmed by ≈ 10°C in 3 years. At the later events, air temper-=ature over Greenland changed more slowly, over 60 to 200 years. Brauer et al (2008) found an abrupt change in storminess over Germany at almost exactly the same time, 12,679 years ago.

13.3 THEORY FOR THE DEEP CIRCULATION

Stommel, Arons, and Faller in a series of papers from 1958 to 1960 described a simple theory of the abyssal circulation (Stommel 1958; Stommel, Arons, and Faller, 1958; Stommel and Arons, 1960). The theory differed so greatly from what was expected that Stommel and Arons devised laboratory experiments with rotating fluids to confirmed their theory. The theory for the deep circulation has been further discussed by Marotzke (2000) and Munk and Wunsch (1998).

The Stommel, Arons, Faller theory is based on three fundamental ideas:

- Cold, deep water is supplied by deep convection at a few high-latitude locations in the Atlantic, notably in the Irminger and Greenland Seas in the north and the Weddell Sea in the south.
- Uniform mixing in the ocean brings the cold, deep water back to the surface.
- The deep circulation is strictly geostrophic in the interior of the ocean, and therefore potential vorticity is conserved.

Notice that the deep circulation is driven by mixing, not by the sinking of cold water at high latitudes. Munk and Wunsch (1998) point out that deep convection by itself leads to a deep, stagnant, pool of cold water. In this case, the deep circulation is confined to the upper layers of the ocean. Mixing or upwelling is required to pump cold water upward through the thermocline and drive the deep circulation. Winds and tides are the primary source of energy driving the mixing.

Notice also that convection and sinking are not the same, and they do not occur in the same place (Marotzke and Scott, 1999). Convection occurs in small regions a few kilometers on a side. Sinking, driven by Ekman pumping and geostrophic currents, can occur over far larger areas. In this chapter, we are discussing mostly sinking of water.

To describe the simplest aspects of the flow, we begin with the Sverdrup equation applied to a bottom current of thickness H in an ocean of constant depth:

$$\beta v = f \frac{\partial w}{\partial z} \quad (13.1)$$

where $f = 2\,\Omega \sin \phi$, $\beta = (2\Omega \cos \phi)/R$, Ω is earth's rotation rate, R earth's radius, and ϕ is latitude. Integrating (13.1) from the bottom of the ocean to the top of the abyssal circulation gives:

$$V = \int_0^H v\,dz = \int_0^H \frac{f}{\beta}\frac{\partial w}{\partial z}\,dz \quad (13.2)$$
$$V = R \tan\varphi\, W_0$$

where V is the vertical integral of the northward velocity, and $W0$ is the velocity at the base of the thermocline. $W0$ must be positive (upward) almost everywhere to balance the downward mixing of heat. Then V must be everywhere toward the poles. This is the abyssal flow in the interior of the ocean sketched by Stommel in figure 13.4. The U component of the flow is calculated from V and w using the continuity equation.

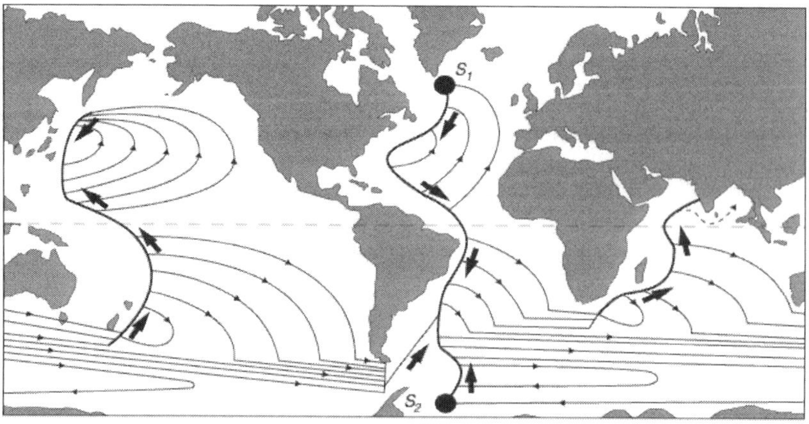

Figure 13.4 Idealized sketch of the deep circulation due to deep convection in the Atlantic (dark circles) and upwelling through the thermocline elsewhere. The real circulation is much different than the circulation shown in this sketch. After Stommel (1958).

To connect the streamlines of the flow in the west, Stommel added a deep western boundary current. The strength of the western boundary current depends on the volume of water *S* produced at the source regions.

Stommel and Arons calculated the flow for a simplified ocean bounded by the Equator and two meridians (a pie shaped ocean). First they placed the source *S0* near the pole to approximate the flow in the north Atlantic. If the volume of water sinking at the source equals the volume of water upwelled in the basin, and if the upwelled velocity is constant everywhere, then the transport *Tw* in the western boundary current is:

$$T_w = -2\,S_0 \sin\varphi \qquad (13.3)$$

The transport in the western boundary current at the poles is twice the volumeof the source, and the transport diminishes to zero at the Equator (Stommel and Arons, 1960a: eq, 7.3.15; see also Pedlosky, 1996: §7.3). The flow driven by the upwelling water adds a recirculation equal to the source. If *S0* exceeds the volume of water upwelled in the basin, then the western boundary current carries water across the Equator. This gives the western boundary current sketched in the north Atlantic in figure 13.4.

Next, Stommel and Arons calculated the transport in a western boundary current in a basin with no source. The transport is:

$$T_w = S\,[1 - 2\,\sin\varphi] \qquad (13.4)$$

where *S* is the transport across the Equator from the other hemisphere. In this basin Stommel notes:

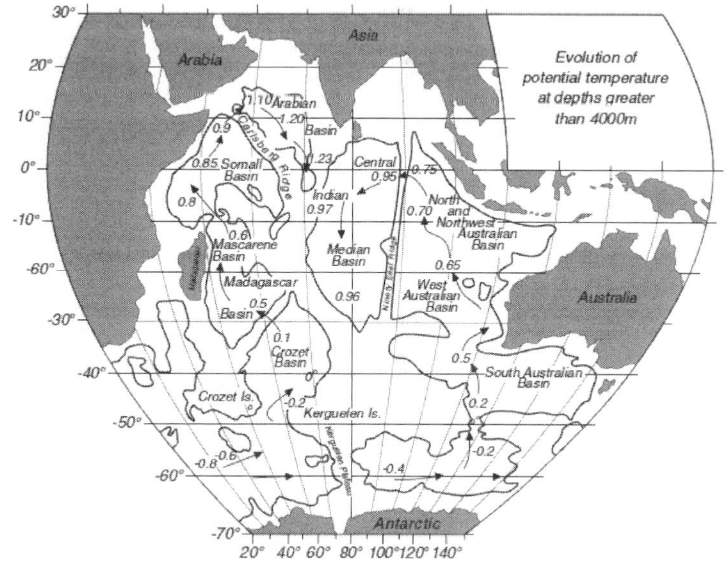

Figure 13.5 Deep flow in the Indian Ocean inferred from the temperature, given in °C. Note that the flow is constrained by the deep mid-ocean ridge system. After Tchernia (1980).

A current of re-circulated water equal to the source strength starts at the pole and flows toward the source . . . [and] gradually diminishes to zero at $\phi = 30^{\circ}$ north latitude. A northward current of equal strength starts at the equatorial source and also diminishes to zero at 30° north latitude.

This gives the western boundary current as sketched in the north Pacific in figure 13.4.

Note that the Stommel-Arons theory assumes a flat bottom. The mid-ocean ridge system divides the deep ocean into a series of basins connected by sills through which the water flows from one basin to the next. As a result, the flow in the deep ocean is not as simple as that sketched by Stommel. Boundary current flow along the edges of the basins, and flow in the eastern basins in the Atlantic comes through the mid-Atlantic ridge from the western basics. Figure 13.5 shows how ridges control the flow in the Indian Ocean.

Finally, Stommel-Arons theory gives some values for time required for water to move from the source regions to the base of the thermocline in various basins. The time varies from a few hundred years for basins near the sources to nearly a thousand years for the north Pacific, which is farther from the sources.

Comments on the Theory for the Deep Circulation: Our under- standing of the deep circulation is still evolving.

- Marotzke and Scott (1999) points out that deep convection and mixing are very different processes. Convection reduces the potential energy of the water column, and it is self powered. Mixing in a stratified fluid increases the potential energy, and it must be driven by an external process.
- Numerical models show that the deep circulation is very sensitive to the assumed value of vertical eddy diffusivity in the thermocline (Gargett and Holloway, 1992).
- Numerical calculations by Marotzke and Scott (1999) indicate that the mass transport is not limited by the rate of deep convection, but it is sensitive to the assumed value of vertical eddy diffusivity, especially near side boundaries.
- Cold water is mixed upward at the ocean's boundaries, above seamounts and mid-ocean ridges, along strong currents such as the Gulf Stream, and in the Antarctic Circumpolar Current (Toggweiler and Russell, 2008; Garabato et al, 2004, 2007). Because mixing is strong over mid-ocean ridges and small in nearby areas, flow is zonal in the ocean basins and poleward along the ridges (Hogg et al. 2001). A map of the circulation will not look like figure 13.4. Numerical models and measurements of deep flow by floats show the flow is indeed zonal.

- Because the transport of mass, heat, and salt are not closely related the transport of heat into the north Atlantic may not be as sensitive to surface salinity as described above.

13.4 OBSERVATIONS OF THE DEEP CIRCULATION

The abyssal circulation is less well known than the upper-ocean circulation. Direct observations from moored current meters or deep-drifting floats were difficult to make until recently, and there are few long-term direct measurements of current. In addition, the measurements do not produce a stable mean value for the deep currents. For example, if the deep circulation takes roughly 1,000 years to transport water from the north Atlantic to the Antarctic Circumpolar Current and then to the north Pacific, the mean flow is about 1 mm/s. Observing this small mean flow in the presence of typical deep currents having variable velocities of up to 10 cm/s or greater, is very difficult.

Most of our knowledge of the deep circulation is inferred from measured distribution of water masses with their distinctive temperature and salinity and their concentrations of oxygen, silicate, tritium, fluorocarbons and other tracers. These measurements are much more stable than direct current measurements, and observations made decades apart can be used to trace the circulation. Tomczak (1999) carefully describes how the techniques can be made quantitative and how they can be applied in practice.

Water Masses: The concept of water masses originates in meteorology. Vil helm Bjerknes, a Norwegian meteorologist, first described the cold air masses that form in the polar regions. He showed how they move southward, where they collide with warm air masses at places he called fronts, just as masses of troops collide at fronts in war (Friedman, 1989). In a similar way, water massesare formed in different regions of the ocean, and the water masses are often separated by fronts. Note, however, that strong winds are associated with fronts in the atmosphere because of the large difference in density and temperature on either side of the front. Fronts in the ocean sometimes have little contrast in density, and these fronts have only weak currents.

Tomczak (1999) defines a *water mass* as abody of water with a common formation history, having its origin in a physical region of the ocean. Just as air masses in the atmosphere, water masses are physical entities with a measurable volume and therefore occupy a finite volume in the ocean. In their formation region they have exclusive occupation of a particular part of the ocean. Elsewhere they share the ocean with other water masses with which they mix. The total volume of a water mass is given by the sum of all its elements regardless of their location.

Plots of salinity as a function of temperature, called *T-S* plots, are used to delineate water masses and their geographical distribution, to describe mixing among water masses, and to infer motion of water in the deep ocean. Here'swhy the plots are so useful: water properties, such as temperature and salinity, are formed only when the water is at the surface or in the mixed layer. Heating, cooling, rain, and evaporation all contribute. Once the water sinks below the mixed layer temperature and salinity can change only by mixing with adjacent water masses. Thus water from a particular region has a particular temperature associated with a particular salinity, and the relationship changes little as the water moves through the deep ocean.

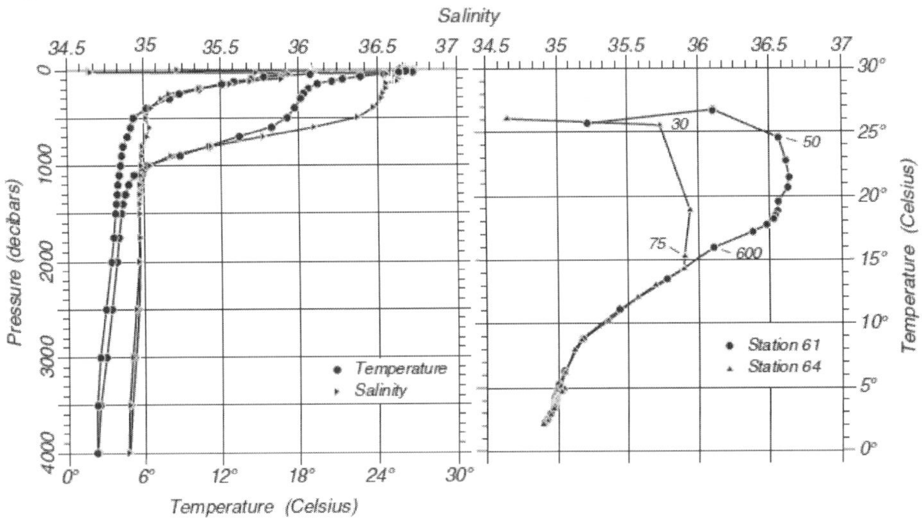

Figure 13.6 Temperature and salinity measured at hydrographic stations on either side of the Gulf Stream. Data are from tables 10.2 and 10.4. Left: Temperature and salinity plotted as a function of depth. Right: The same data, but salinity is plotted as a function of temperature in a *T-S* plot. Notice that temperature and salinity are uniquely related below the mixed layer. A few depths are noted next to data points.

Thus temperature and salinity are not independent variables. For example, the temperature and salinity of the water at different depths below the Gulf Stream are uniquely related (figure 13.6, right), indicating they came from the same source region, even though they do not appear related if temperature and salinity are plotted independently as a function of depth (figure 13.6, left).

Temperature and salinity are *conservative properties* because there are no sources or sinks of heat and salt in the interior of the ocean. Other properties,such as oxygen are non-conservative. For example, oxygen content may change slowly due to oxidation of organic material and respiration by animals.

Each point in the *T-S* plot is a *water type*. This is a mathematical ideal. Some water masses may be very homogeneous and they are almost points on the plot. Other water masses are less homogeneous, and they occupy regionson the plot.

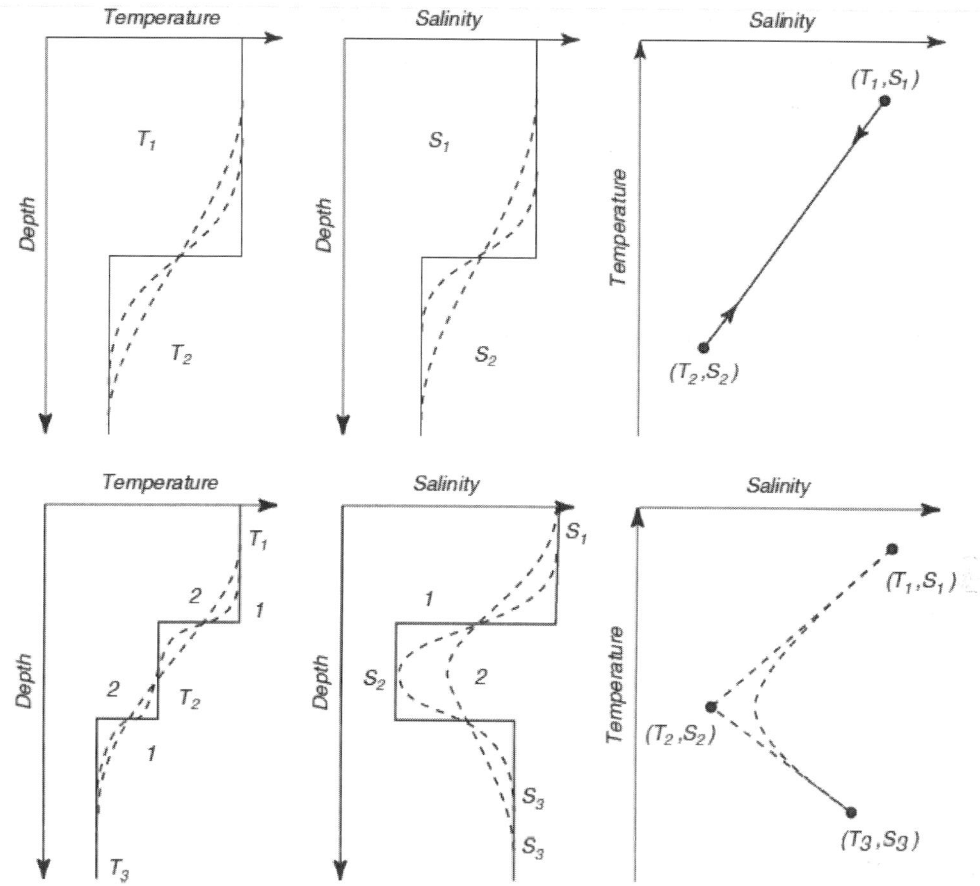

Figure 13.7 Upper: Mixing of two water masses produces a line on a *T-S* plot. Lower: Mixing among three water masses produces intersecting lines on a *T-S* plot, and the apex at the intersection is rounded by further mixing. After Defant (1961: 205).

Mixing two water types leads to a straight line on a *T-S* diagram (figure 13.7). Because the lines of constant density on a *T-S* plot are curved, mixing increases the density of the water. This is called *densification* (figure 13.8).

Water Masses and the Deep Circulation: Let's use these ideas of water masses and mixing to study the deep circulation. We start in the south Atlantic because it has very clearly defined water masses. A *T-S* plot calculated from hydrographic data collected in the south Atlantic (figure 13.9) shows three important water masses listed in order of decreasing depth (table 13.1): Antarctic Bottom Water AAB, North Atlantic Deep Water NADW, and Antarctic Intermediate Water AIW. All are deeper than one kilometer. The mixing among three water masses shows the characteristic rounded apexes shown in the idealized case shown in figure 13.7.

The plot indicates that the same water masses can be found throughout the western basins in the south Atlantic. Now let's use a cross section of salinity to trace the movement of the water masses using the core method.

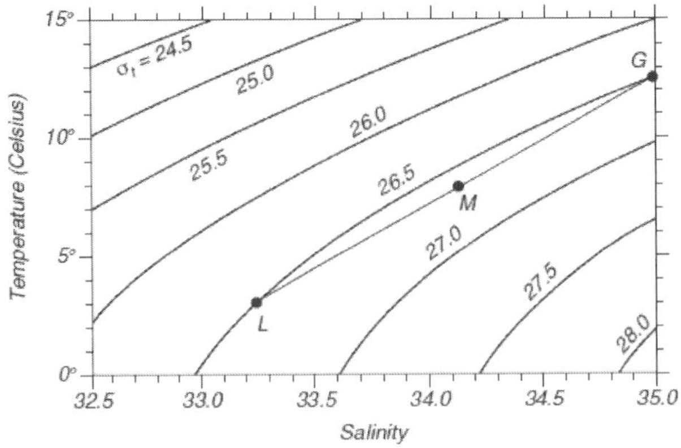

Figure 13.8 Mixing of two water types of the same density(L and G) produces water that is denser (M) than either water type. After Tolmazin (1985: 137).

Core Method: The slow variation from place to place in the ocean of a tracer such as salinity can be used to determine the source of the waters masses such as those in figure 13.9. This is called the *core method.* The method may also be used to track the slow movement of the water mass. Note, however, that as low drift of the water and horizontal mixing both produce the same observed properties in the plot, and they cannot be separated by the core method.

A *core* is a layer of water with extreme value (in the mathematical sense) of salinity or other property as a function of depth. An extreme value is a local maximum or minimum of the quantity as a function of depth. The method assumes that the flow is along the core. Water in the core mixes with the water masses above and below the core and it gradually loses its identity. Furthermore, the flow tends to be along surfaces of constant potential density.

Table 13.1 Water Masses of the south Atlantic between 33°S and 11°N

			Temp. (°C)	Salinity
Antarctic water	Antarctic Intermediate Water	AIW	3.3	34.15
	Antarctic Bottom Water	ABW	0.4	34.67
North Atlantic water	North Atlantic Deep Water	NADW	4.0	35.00
	North Atlantic Bottom Water	NABW	2.5	34.90
Thermocline water	Subtropical Lower Water	U	18.0	35.94

From Defant (1961: table 82)

Let's apply the method to the data from the south Atlantic to find the source of the water masses. As you might expect, this will explain their names.

We start with a north-south cross section of salinity in the western basins of the Atlantic (figure 13.10). It we locate the maxima and minima of salinity as a function of depth at different latitudes, we can see two clearly defined cores.

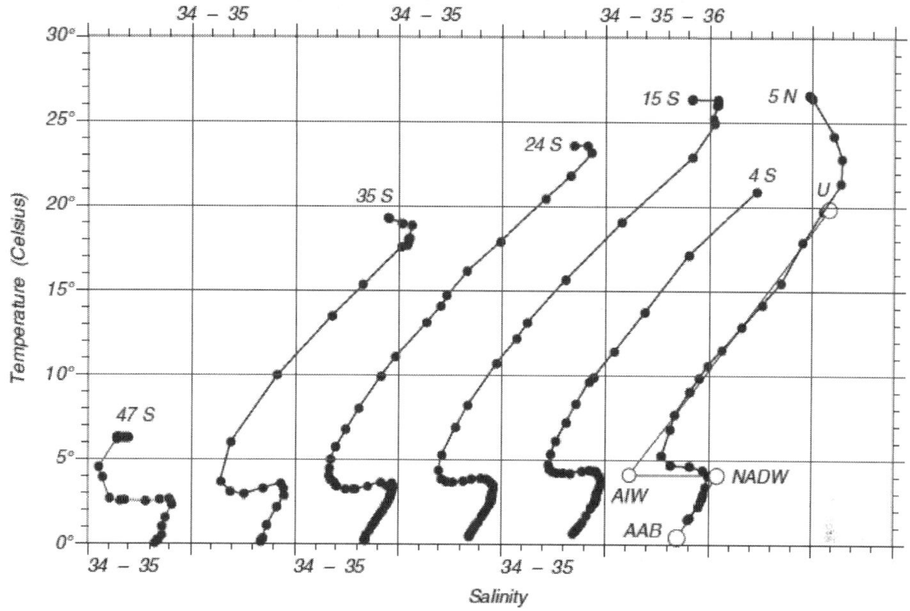

Figure 13.9 ***T-S*** **plot of data collected at various latitudes in the western basins of the south Atlantic. Lines drawn through data from 5°N, showing possible mixing between water masses: NADW–North Atlantic Deep Water, AIW–Antarctic Intermediate Water, AAB Antarctic Bottom Water, u Subtropical Lower Water.**

The upper low-salinity core starts near 55°S and it extends northward at depths near 1000 m. This water originates at the Antarctic Polar Front zone. This is the Antarctic Intermediate Water. Below this water mass is a core of salty water originating in the far north Atlantic. This is the North Atlantic Deep Water. Below this is the most dense water, the Antarctic Bottom Water. It originates in winter when cold, dense, saline water forms in the Weddell Sea and other shallow seas around Antarctica. The water sinks along the continental slope and mixes with Circumpolar Deep Water. It then fills the deep basins of the south Pacific, Atlantic, and Indian ocean.

The Circumpolar Deep Water is mostly North Atlantic Deep Water that has been carried around Antarctica. As it is carried along, it mixes with deep waters of the Indian and Pacific Ocean to form the circumpolar water.

The flow is probably not along the arrows shown in figure 13.10. The distribution of properties in the abyss can be explained by a combination of slow flow in the direction of the arrows plus horizontal mixing along surfaces of constant potential density with some weak vertical mixing. The vertical mixing probably occurs at the places where the density surface reaches the sea bottom at a lateral boundary such as seamounts, mid-ocean ridges, and along the western boundary. Flow in a plane perpendicular to that of the figure may be at least as strong as the flow in the plane of the figure shown by the arrows.

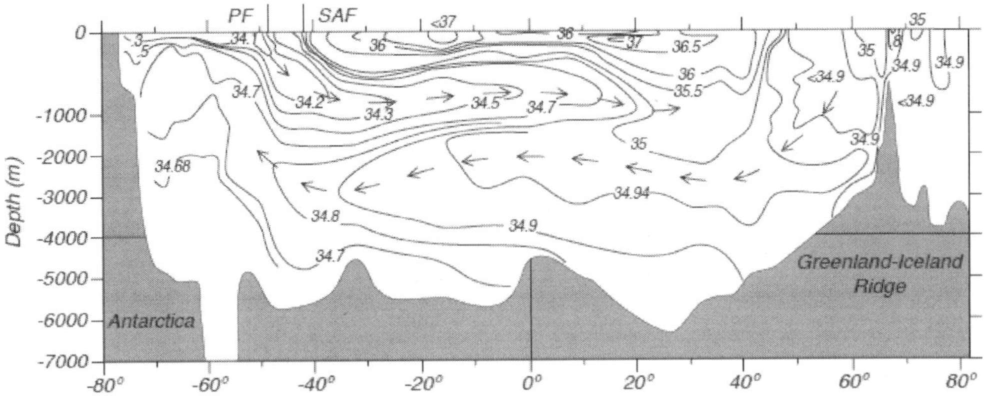

Figure 13.10 Contour plot of salinity as a function of depth in the western basins of the Atlantic from the Arctic Ocean to Antarctica. The plot clearly shows extensive cores, oneat depths near 1000 m extending from 50°S to 20°N, the other at is at depths near 2000 mextending from 20°N to 50°S. The upper is the Antarctic Intermediate Water, the lower isthe North Atlantic Deep Water. The arrows mark the assumed direction of the flow in the cores. The Antarctic Bottom Water fills the deepest levels from 50°S to 30°N. pf is the polarfront, SAF is the subantarctic front. See also figures 10.15 and 6.10. After Lynn and Reid (1968).

The core method can be applied only to a tracer that does not influence density. Hence temperature is usually a poor choice. If the tracer controls density, then flow will be around the core according to ideas of geostrophy, not along core as assumed by the core method.

The core method works especially well in the south Atlantic with its clearly defined water masses. In other ocean basins, the *T-S* relationship is more complicated. The abyssal waters in the other basins are a complex mixture of waters coming from different areas in the ocean (figure 13.11). For example, warm, salty water from the Mediterranean Sea enters the north Atlantic and spreads out at intermediate depths displacing intermediate water from Antarctica in the north Atlantic, adding additional complexity to the flow as seen in the lower right part of the figure.

Other Tracers: I have illustrated the core method using salinity as a tracer, but many other tracers are used. An ideal tracer is easy to measure even when its concentration is very small; it is conserved, which means that only mixing changes its concentration; it does not influence the density of the water; it exists in the water mass we wish to trace, but not in other adjacent water masses; and it does not influence marine organisms (we don't want to release toxic tracers).

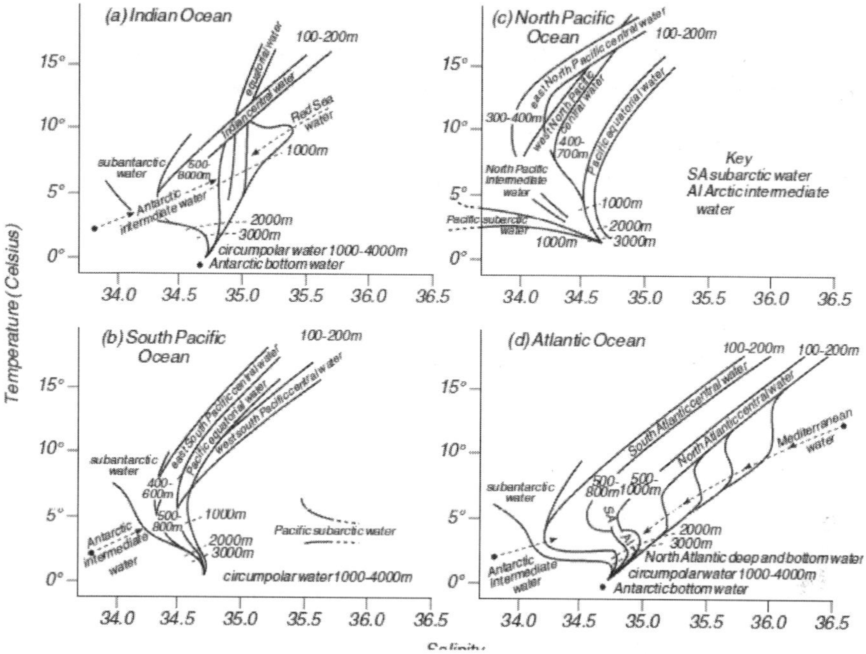

Figure 13.11 ***T-S*** **plots of water in the various ocean basins. After Tolmazin (1985: 138).**

Various tracers meet these criteria to a greater or lesser extent, and they are used to follow the deep and intermediate water in the ocean. Here are some of the most widely used tracers.

- Salinity is conserved, and it influences density much less than temperature.
- Oxygen is only partly conserved. Its concentration is reduced by the respiration by marine plants and animals and by oxidation of organic carbon.
- Silicates are used by some marine organisms. They are conserved at depths below the sunlit zone.
- Phosphates are used by all organisms, but they can provide additional information.
- 3He is conserved, but there are few sources, mostly at deep-sea volcanic areas and hot springs.
- 3H (tritium) was produced by atomic bomb tests in the atmosphere in the 1950s. It enters the ocean through the mixed layer, and it is useful for tracing the formation of deep water. It decays with a half life of 12.3 y and it is slowly disappearing from the ocean. Figure 10.16 shows the slow advection or perhaps mixing of the tracer into the deep north Atlantic. Note that after 25 years little tritium is found south of 30° N. This impliesa mean velocity of less than a mm/s.

- Fluorocarbons (Freon used in air conditioning) have been recently injected into atmosphere. They can be measured with very great sensitivity, and they are being used for tracing the sources of deep water.
- Sulphur hexafluoride SF6 can be injected into sea water, and the concentration can be measured with great sensitivity for many months.

Each tracer has its usefulness, and each provides additional information about the flow.

North Atlantic Meridional Overturning Circulation: The great importance of the meridional overturning circulation for European climate has led to programs to monitor the circulation. The Rapid Climate Change/Meridional Overturning Circulation and Heat Flux Array rapid/mocha deployed an array of instruments that measured bottom pressure plus temperature and salinity throughout the water column at 15 locations along 24°N near the western and eastern boundaries and on either side of the mid-Atlantic ridge beginning in 2004 (Church, 2007). At the same time, flow of the Gulf Stream was measured through the Strait of Florida, and wind stress, which gives the Ekman transports, was measured along 24°N by satellite instruments. The measurements show that transport across 24°N was zero, within the accuracy of the measurements, as expected. The one-year average of the Meridional Overturning Circulation was 18.7 ± 5.6 Sv, with variability ranging from 4.4 to 35.3 Sv. Accuracy of the measurement was ± 1.5 Sv.

13.5 ANTARCTIC CIRCUMPOLAR CURRENT

The Antarctic Circumpolar Current is an important feature of the ocean's deep circulation because it transports deep and intermediate water between the Atlantic, Indian, and Pacific Ocean, and because Ekman pumping driven by westerly winds is a major driver of the deep circulation. Because it is so important for understanding the deep circulation in all ocean, let's look at what is known about this current.

A plot of density across a line of constant longitude in the Drake Passage (figure 13.12) shows three fronts. They are, from north to south: 1) the Subantarctic Front, 2) the Polar Front, and 3) the Southern ACC Front. Each front is continuous around Antarctica (figure 13.13). The plot also shows that the constant-density surfaces slope at all depths, which indicates that the currents extend to the bottom.

Typical current speeds are around 10 cm/s with speeds of up to 50 cm/s near some fronts. Although the currents are slow, they transport much more water than western boundary currents because the flow is deep and wide. Whitworth and Peterson (1985) calculated transport through the Drake Passage using several years of data from an array of 91 current meters on 24 moorings spaced approximately 50 km apart along a line spanning the passage. They also used measurements of bottom pressure measured by gauges on either side of the passage. They found that the average transport through the Drake Passage was 125 ± 11 Sv, and that the transport varied from 95 Sv to 158 Sv. The maximum transport tended to occur in late winter and early spring (figure 13.14).

Because the Antarctic currents reach the bottom, they are influenced by topographic steering. As the current crosses ridges such as the Kerguelen Plateau, the Pacific-Antarctic Ridge, and the Drake Passage, it is deflected by the ridges.

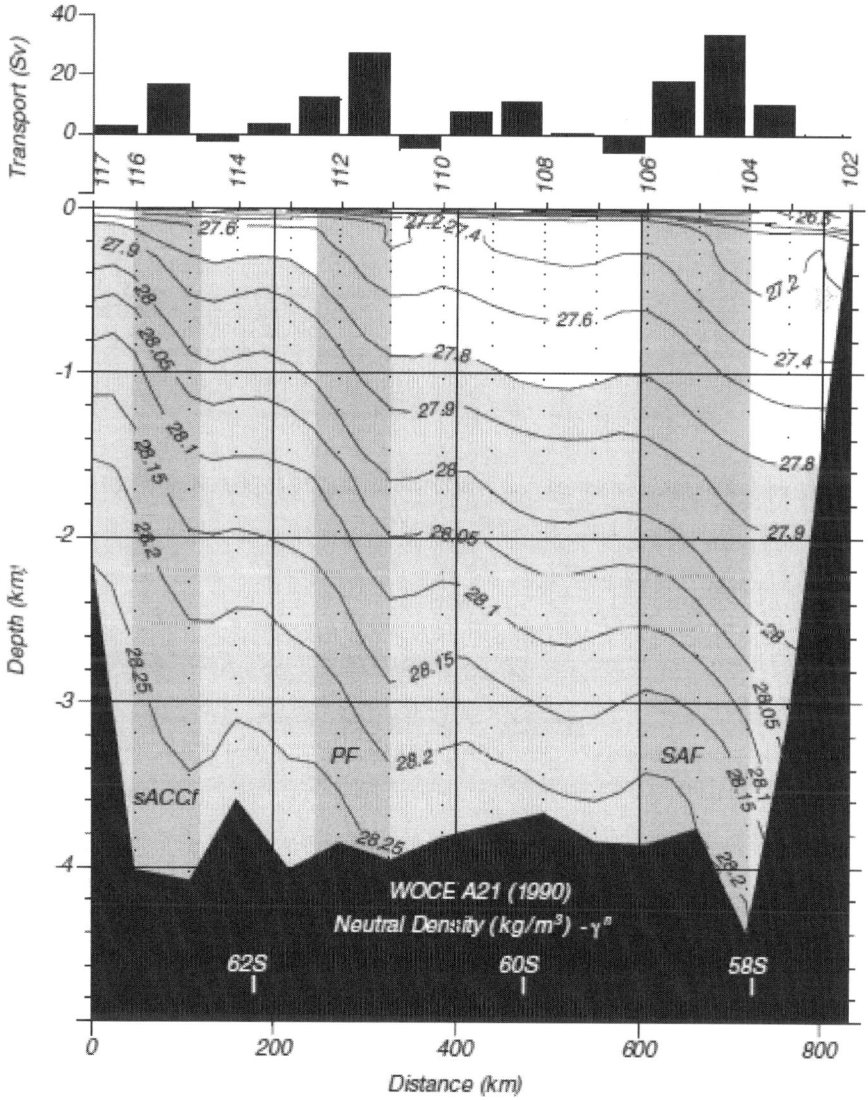

Figure 13.12 Cross section of neutral density across the Antarctic Circumpolar Current in the Drake Passage from the World Ocean Circulation Experiment section A21 in 1990. The current has three streams associated with the three fronts (dark shading): sf = Southern ACC Front, pf = Polar Front, and SAF = Subantarctic Front. Hydrographic station numbers are given at the top, and transports are relative to 3,000 dbar. Circumpolar deep water is indicated by light shading. Data from Alex Orsi, Texas A&M University.

The core of the current is composed of Circumpolar Deep Water, a mixture of deep water from all ocean. The upper branch of the current contains oxygen-poor water from all ocean. The lower (deeper) branch contains a core of high-salinity water from the Atlantic, including contributions from the north Atlantic deep water mixed with salty Mediterranean Sea water. As the different water masses circulate around Antarctica they mix with other water masses with similar density. In a sense, the current is a giant 'mix-master' taking deep water from each ocean, mixing it with deep water from other ocean, and then redistributing it back to each ocean (Garabato et al, 2007).

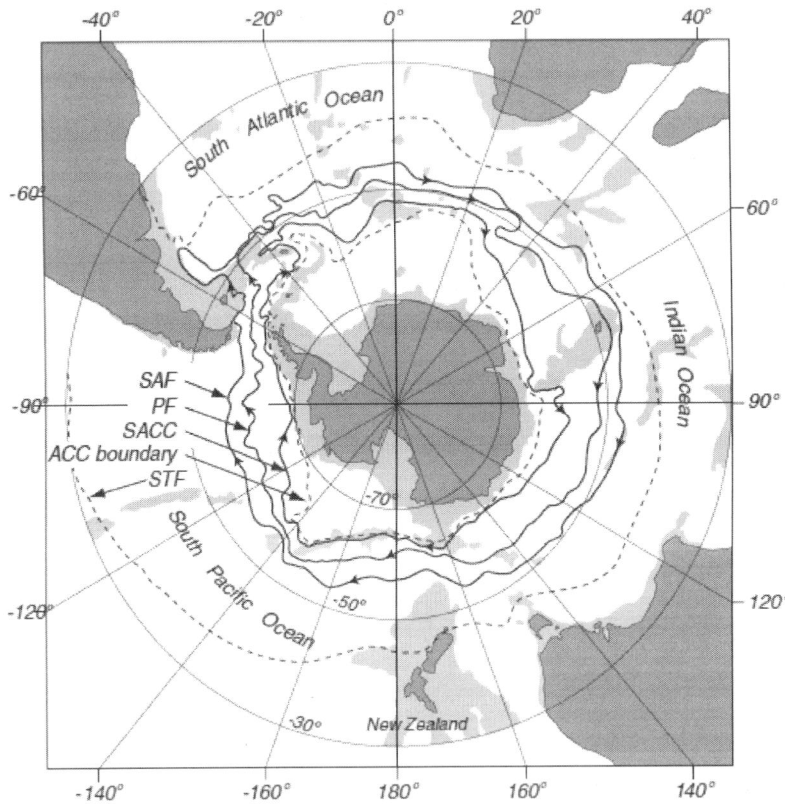

Figure 13.13 Distribution of fronts around Antarctica: STF: Subtrobical Front; SAF: Subantarctic Front; PF: Polar Front; SACC: Southern Antarctic Circumpolar Front. Shaded areas are shallower than 3 km. From Orsi (1995).

The coldest, saltiest water in the ocean is produced on the continental shelf around Antarctica in winter, mostly from the shallow Weddell and Ross seas. The cold salty water drains from the shelves, entrains some deep water, and spreads out along the sea floor. Eventually, 8–10 Sv of bottom water are formed (Orsi, Johnson, and Bullister, 1999). This dense water then seeps into all the ocean basins. By definition, this water is too dense to cross through the Drake Passage, so it is not circumpolar water.

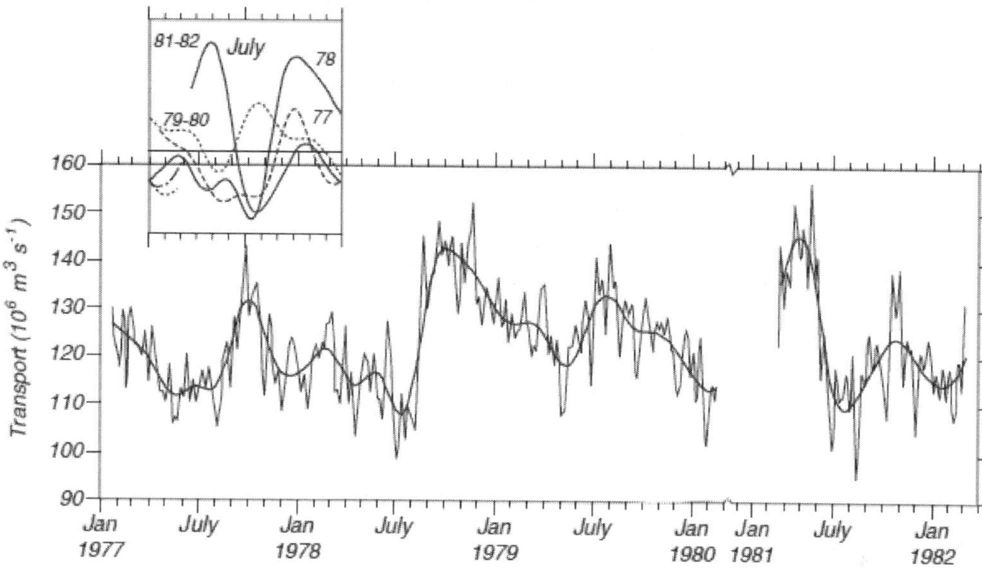

Figure 13.14 Variability of the transport in the Antarctic Circumpolar Current as measured by an array of current meters deployed across the Drake Passage. The heavier line is smoothed, time-averaged transport. From Whitworth (1988).

The Antarctic currents are wind driven. Strong west winds with maximum speed near 50°S drive the currents (see figure 4.2), and the north-south gradient of wind speed produces convergence and divergence of Ekman transports. Divergence south of the zone of maximum wind speed, south of 50°S leadsto upwelling of the Circumpolar Deep Water. Convergence north of the zone of maximum winds leads to downwelling of the Antarctic intermediate water. The surface water is relatively fresh but cold, and when they sink they define characteristics of the Antarctic intermediate water.

The position of the circumpolar current relative to the maximum of the westerly winds influences the meridional overturning circulation and climate. North of the maximum, Ekman transports converge, pushing water downward into the Antarctic Intermediate Water north of the Polar Front. South of the maximum winds, Ekman transports diverge, pulling Circumpolar Atlantic Deep Water to the surface south of the Polar Front, which helps drive the deep circulation (figure 13.10). When the maximum winds are further from the pole, less deep water is pulled upward, and the deep circulation is weak, as it was during the last ice age. As the earth warmed after the ice age, the maximum winds shifted south. The winds were more aligned with the Circumpolar Current, and they pulled more deep water to the surface. Since 1960, the winds have strengthened and shifted southward, further strengthening Circumpolar Current and the deep circulation Toggweiler and Russell, 2008).

Because wind constantly transfers momentum to the Antarctic Circumpolar Current, causing it to accelerate, the acceleration must be balanced by drag, and we are led to ask: What keeps the flow from accelerating to very high speeds? Munk and Palmen (1951), suggest form drag dominates. Form drag is due to the current crossing sub-sea ridges, especially at the Drake Passage. Form drag is also the drag of the wind on a fast moving car. In both cases, the flow is diverted, by the ridge or by your car, creating a low pressure zone downstream of the ridge or down wind of the car. The low pressure zone transfers momentum into the solid earth, slowing down the current.

13.6 CONCISE SUMMARY

- The deep circulation of the ocean is very important because it determines the vertical stratification of the ocean and because it modulates climate.
- The ocean absorbs CO2 from the atmosphere reducing atmospheric CO2 concentrations. The deep circulation carries the CO2 deep into the ocean temporarily keeping it from returning to the atmosphere. Eventually, however, most of the CO2 must be released back to the atmosphere. But, some remains in the ocean. Phytoplankton convert CO2 into organic carbon, some of which sinks to the sea floor and is buried in sediments. Some CO2 is used to make sea shells, and it too remains in the ocean.
- The production of deep bottom waters in the north Atlantic draws a petawatt of heat into the northern hemisphere which helps warm Europe.
- Variability of deep water formation in the north Atlantic has been tied to large fluctuations of northern hemisphere temperature during the last ice ages.
- Deep convection which produces bottom water occurs only in the far north Atlantic and at a few locations around Antarctica.
- The deep circulation is driven by vertical mixing, which is largest above mid-ocean ridges, near seamounts, and in strong boundary currents.
- The deep circulation is too weak to measure directly. It is inferred from observations of water masses defined by their temperature and salinity and from observation of tracers.
- The Antarctic Circumpolar Current mixes deep water from the Atlantic, Pacific, and Indian Ocean and redistributes it back to each ocean. The current is deep and slow with a transport of 125 Sv.

14 CONCEPTS OF EQUATORIAL PROCESSES

Equatorial processes are at the center of our understanding the influence of the ocean on the atmosphere, and they dominate the inter annual fluctuations in global weather patterns. The sun warms the vast expanses of the tropical Pacific and Indian ocean, evaporating water. When the water condenses as rain it releases so much heat that these areas are the primary engine driving the atmospheric circulation (figure 14.1). Rainfall over extensive areas exceeds three meters per year (figure 5.5), and some oceanic regions receive more than five meters of rain per year. To put the numbers in perspective, five meters of rain per year releases on average 400 W/m^2 of heat to the atmosphere. Equatorial currents modulate the air-sea interactions, especially through the phenomenon known as El Niño, with global consequences. I describe here first the basic equatorial processes, then the year-to-year variability of the processes and the influence of the variability on weather patterns.

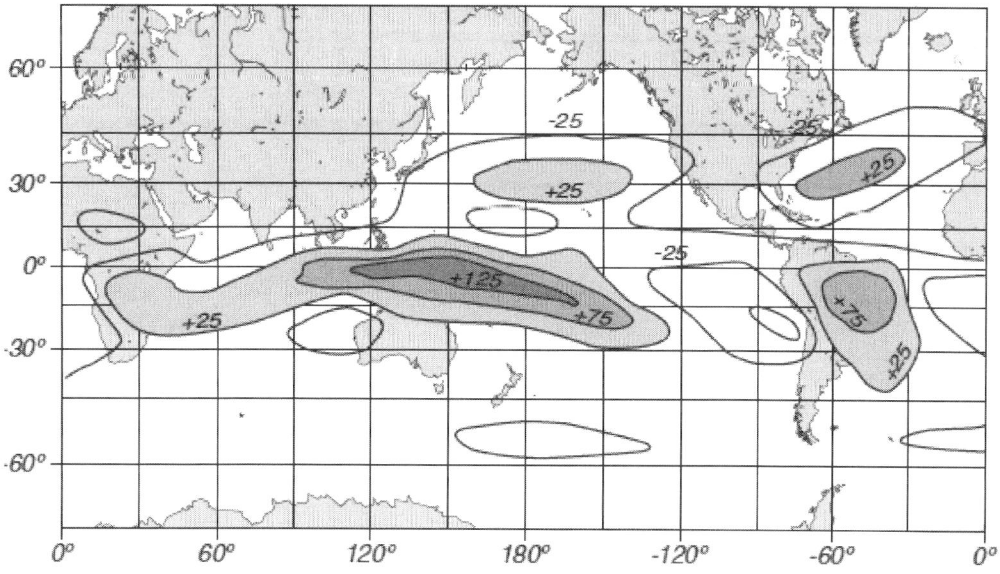

Figure 14.1 Average diabatic heating between 700 and 50 mb in the atmosphere during December, January and February calculated from ECMWF data for 1983–1989. Most of the heating is due to the release of latent heat by rain. After Webster et al. (1992).

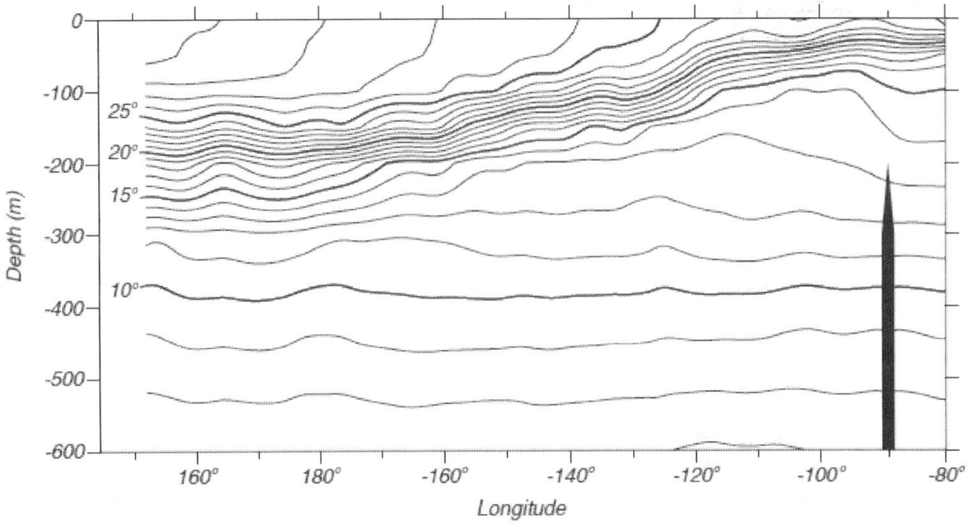

Figure 14.2 The mean, upper-ocean, thermal structure along the equator in the Pacific from north of New Guinea to Ecuador calculated from data in Levitus (1982).

14.1 INTRODUCTION TO EQUATORIAL PROCESSES

The tropical ocean is characterized by a thin, permanent, shallow layer of warm water over deeper, colder water. In this respect, the vertical stratification is similar to the summer stratification at higher latitudes. Surface waters are hottest in the west (figure 6.3) in the great Pacific warm pool. The mixed layer is deep in the west and very shallow in the east (Figure 14.2).

The shallow thermocline has important consequences. The southeast trade winds blow along the equator (figure 4.2) although they tend to be strongest in the east. North of the equator, Ekman transport is northward. South of the equator it is southward. The divergence of the Ekman flow causes upwelling on the equator. In the west, the upwelled water is warm. But in the east the upwelled water is cold because the thermocline is so shallow. This leads to a cold tongue of water at the sea surface extending from South America to near the dateline (figure 6.3).

Surface temperature in the east is a balance among four processes:

- The strength of the upwelling, which is determined by the westward component of the wind.
- The speed of westward currents which carry cold water from the coast of Peru and Ecuador.
- North-south mixing with warmer waters on either side of the equator.
- Heat fluxes through the sea surface along the equator.

The east-west temperature gradient on the equator drives a zonal circulation in the atmosphere, the Walker circulation. Thunderstorms over the warm pool carry air upward, and sinking air in the east feeds the return flow at the surface. Variations in the temperature gradient influences the Walker circulation, which, in turn, influences the gradient. The feedback can lead to an instability, the El Niño-Southern Oscillation (ENSO) discussed in the next section.

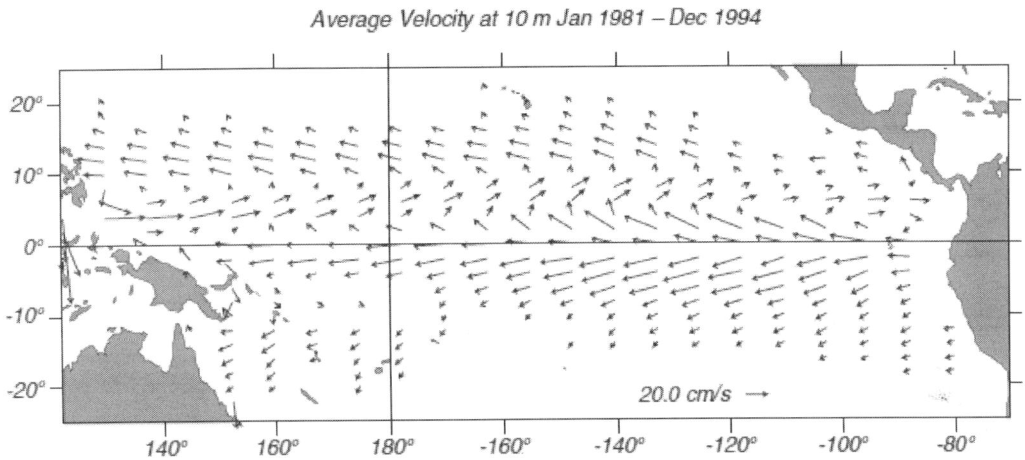

Figure 14.3 Average currents at 10 m calculated from the Modular Ocean Model driven by observed winds and mean heat fluxes from 1981 to 1994. The model, operated by the NOAA National Centers for Environmental Prediction, assimilates observed surface and subsurface temperatures. After Behringer, Ji, and Leetmaa (1998).

Surface Currents: The strong stratification confines the wind-driven circulation to the mixed layer and upper thermocline. Sverdrup's theory and Munk's extension, described in §11.1 and §11.3, explain the surface currents in the tropical Atlantic, Pacific, and Indian ocean. The currents include (figure 14.3):

- The North Equatorial Counter current between 3°N and 10°N, which flowseastward with a typical surface speed of 50 cm/s. The current is centered on the band of weak winds, the *doldrums*, around 5–10°N where the north and south trade winds converge, the *tropical convergence zone.*
- The North and South Equatorial Currents which flow westward in the zonal band on either side of the countercurrent. The currents are shallow, less than 200 m deep. The northern current is weak, with a speed lessthan roughly 20 cm/s. The southern current has a maximum speed of around 100 cm/s, in the band between 3°N and the equator.

The currents in the Atlantic are similar to those in the Pacific because the trade winds in that ocean also converge near 5°–10°N. The South Equatorial Current in the Atlantic continues northwest along the coast of Brazil, where it is known as the North Brazil Current. In the Indian Ocean, the doldrums occur in the southern hemisphere and only during the northern-hemisphere winter. In the northern hemisphere, the currents reverse with the monsoon winds.

There is, however, much more to the story of equatorial currents.

Equatorial Undercurrent: Observations: Just a few meters below the surface on the equator is a strong eastward flowing current, the Equatorial Under- current, the last major oceanic current to be discovered. Here's the story:

In September 1951, aboard the U.S. Fish and Wildlife Service research vessel long-line fishing on the equator south of Hawaii, it was noticed that the subsurface gear drifted steadily to the east. The next year Cromwell, in company with Montgomery and Stroup, led an expedition to investigate the vertical distribution of horizontal velocity at the equator. Using floating drogues at the surface and at various depths, they were able to establish the presence, near the equator in the central Pacific, of a strong, narrow eastward current in the lower part of the surface layer and the upper part of the thermocline (Cromwell, *et. al.*, 1954). A few years later the Scripps *Eastropac* Expedition, under Cromwell's leadership, found the current extended toward the east nearly to the Galapagos Islands but was not present between those islands and the South American continent.

The current is remarkable in that, even though comparable in trans- port to the Florida Current, its presence was unsuspected ten years ago. Even now, neither the source nor the ultimate fate of its waters has been established. No theory of oceanic circulation predicted its existence, and only now are such theories being modified to account for the important features of its flow.—Warren S. Wooster (1960).

The Equatorial Undercurrent in the Atlantic was first discovered by Buchanan in 1886, and in the Pacific by the Japanese Navy in the 1920s and 1930s (McPhaden, 1986).

However, no attention was paid to these observations. Other earlier hints regarding this undercurrent were mentioned by Mattha¨us (1969). Thus the old experience becomes even more obvious which says that discoveries not attracting the attention of contemporaries simply do not exist.— Dietrich et al. (1980).

Bob Arthur (1960) summarized the major aspects of the flow:

- Surface flow may be directed westward at speeds of 25–75 cm/s;
- Current reverses at a depth of from 20 to 40 m;
- Eastward undercurrent extends to a depth of 400 meters with a transport of as much as 30 Sv = 30×10^6 m^3/s;
- Core of maximum eastward velocity (0.50–1.50 m/s) rises from a depth of 100 m at 140°W to 40 m at 98°W, then dips down;
- Undercurrent appears to be symmetrical about the equator and becomes much thinner and weaker at 2°N and 2°S.

In essence, the Pacific Equatorial Undercurrent is a ribbon with dimensions of 0.2 km × 300 km × 13, 000 km (figure 14.4).

Equatorial Undercurrent: Theory: Although we do not yet have a complete theory for the undercurrent, we do have a clear understanding of some of the more important processes at work in the equatorial regions. Pedlosky(1996), in his excellent chapter on Equatorial Dynamics of the Thermocline: The Equatorial Undercurrent, points out that the basic dynamical balances we have used in mid latitudes break down near or on the equator.

Near the equator:

1. The Coriolis parameter becomes very small, going to zero at the equator:

$$f = 2\Omega \sin \varphi = \beta y \approx 2\Omega\, \varphi \qquad (14.1)$$

where ϕ is latitude, $\beta - \partial f/\partial y \approx 2\Omega/R$ near the equator, and $y = R\,\phi$.

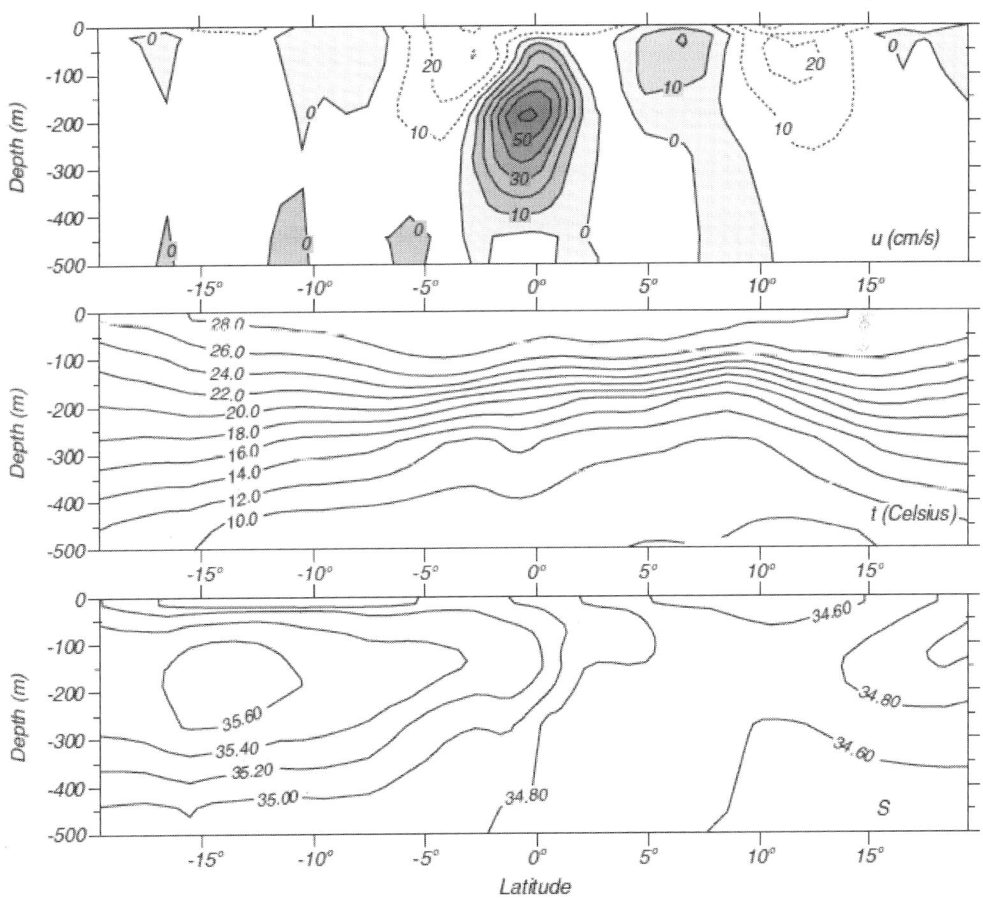

Figure 14.4 Cross section of the Equatorial Undercurrent in the Pacific calculated from Modular Ocean Model with assimilated surface data (See §14.5). The section is an average from 160°E to 170°E from January 1965 to December 1999. Stippled areas are westward flowing. From Nevin S. Fučkar.

2. Planetary vorticity f is also small, and the advection of relative vorticity cannot be neglected. Thus the Sverdrup balance (11.7) must be modified.
3. The geostrophic and vorticity balances fail when the meridional distance L to the equator is $O\ U/\beta$, where $\beta = \partial f/\partial y$. If $U = 1$ m/s, then$L = 200$ km or 2° of latitude. Lagerloeff et al (1999), using measured currents, show that currents near the equator can be described by the geostrophic balance for $|\phi| > 2.2^\circ$. They also show that flow closer to the equator can be described using a β-plane approximation $f = \beta y$.
4. The geostrophic balance for *zonal* currents works so well near the equator because f and $\partial\zeta/\partial y \rightarrow 0$ as $\phi \rightarrow 0$, where ζ is sea surface topography.

Upwelled water along the equator produced by Ekman pumping is not part of a two-dimensional flow in a north-south, meridional plane. The flow is three-dimensional. Water tends to flow along the contours of constant density (isopycnal surfaces), close to the lines of constant temperature in figure 14.2. Cold water enters the undercurrent in the far west Pacific, and it moves eastward and upward along the equator. For example, the 25°isotherm enters the under- current at a depth near 125 m in the western Pacific at 170°E and eventually reaches the surface at 125°W in the eastern Pacific.

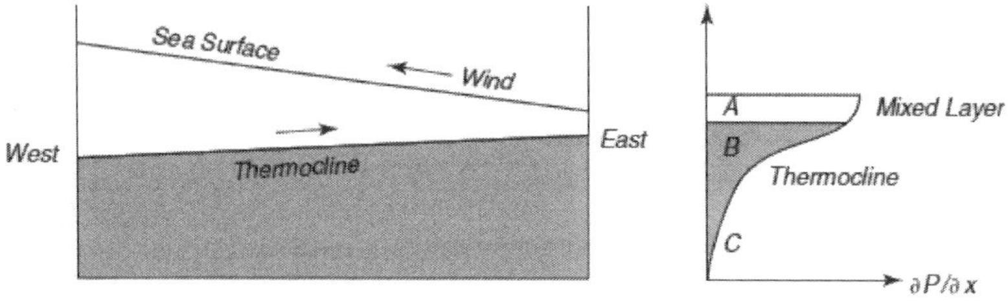

Figure 14.5 Left: Cross-sectional sketch of the thermocline and sea-surface topography along the equator. Right: Eastward pressure gradient in the central Pacific caused by the density structure at left.

The meridional geostrophic balance near the equator gives the speed of the zonal currents, but it does not explain what drives the undercurrent. A very simplified theory for the undercurrent is based on a balance of zonal pressure gradients along the equator. Wind stress pushes water westward, producing the deep thermocline and warm pool in the west. The deepening of the thermocline causes the sea-surface topography ζ to be higher in the west, assuming that flow below the thermocline is weak. Thus there is an eastward pressure gradient along the equator in the surface layers to a depth of a few hundred meters. The eastward pressure gradient at the surface (layer A in figure 14.5) is balanced by the wind stress Tx, and $Tx/H = -\partial p/\partial x$, where H is the mixed-layer depth

Below a few tens of meters in layer B, the influence of the wind stress is small, and the pressure gradient is unbalanced, leading to an accelerated flow toward the east, the equatorial undercurrent. Within this layer, the flow accelerates until the pressure gradient is balanced by frictional forces which tend to slowthe current. At depths below a few hundred meters in layer C, the eastward pressure gradient is too weak to produce a current, $\partial p/\partial x \approx 0$.

Coriolis forces keep the equatorial undercurrent centered on the equator. If the flow strays northward, the Coriolis force deflects the current southward. The opposite occurs if the flow strays southward.

14.2 VARIABLE EQUATORIAL CIRCULATION: EL NIÑO/LA NINA

The trades are remarkably steady, but they do vary from month to month and year to year, especially in the western Pacific. One important source of variability are Madden-Julian waves in the atmosphere (McPhaden, 1999). If the trades in the west weaken or reverse, the air-sea system in the equatorial regions can be thrown into another state called El Niño. This disruption of the equatorial system in the Pacific is the most important cause of changing weather patterns around the globe.

Although the modern meaning of the term El Niño denotes a disruption of the entire equatorial system in the Pacific, the term has been used in the past to describe several very different processes. This causes a lot of confusion. To reduce the confusion, let's learn a little history.

A Little History: In the 19th century, the term was applied to conditions off the coast of Peru. The following quote comes from the introduction to Philander's (1990) excellent book *El Niño, La Nin˜a, and the Southern Oscillation*:

- In the year 1891, Sen˜or Dr. Luis Carranza of the Lima Geographical Society, contributed a small article to the Bulletin of that Society, calling attention to the fact that a counter-current flowing from north to south had been observed between the ports of Paita and Pacasmayo.
- The Paita sailors, who frequently navigate along the coast in small craft, either to the north or the south of that port, name this counter- current the current of "El Niño" (the Child Jesus) because it has been observed to appear immediately after Christmas.

As this counter-current has been noticed on different occasions, and its appearance along the Peruvian coast has been concurrent with rains in latitudes where it seldom if ever rains to any great extent, I wish, on the present occasion, to call the attention of the distinguished geographers here assembled to this phenomenon, which exercises, undoubtedly, a very great influence over the climatic conditions of that part of the world.— Sen˜or Frederico Alfonso Pezet's address to the Sixth International Geo- graphical Congress in Lima, Peru 1895.

The Peruvians noticed that in some years the El Niño current was stronger than normal, it penetrated further south, and it is associated with heavy rains in Peru. This occurred in 1891 when (again quoting from Philander's book)

. . . it was then seen that, whereas nearly every summer here and there there is a trace of the current along the coast, in that year it was so visible, and its effects were so palpable by the fact that large dead alligators and trunks of trees were borne down to Pacasmayo from the north, and that the whole temperature of that portion of Peru suffered such a change owing to the hot current that bathed the coast. . . . —Sen˜or Frederico Alfonso Pezet.

. . . the sea is full of wonders, the land even more so. First of all the desert becomes a garden . . . . The soil is soaked by the heavy downpour, and within a few weeks the whole country is covered by abundant pasture. The natural increase of flocks is practically doubled and cotton can be grown in places where in other years vegetation seems impossible.—From Mr. S.M. Scott & Mr. H. Twiddle quoted from a paper by Murphy, 1926.

The El Niño of 1957 was even more exceptional. So much so that it attracted the attention of meteorologists and oceanographers throughout the Pacific basin.

By the fall of 1957, the coral ring of Canton Island, in the memory of man ever bleak and dry, was lush with the seedlings of countless tropical trees and vines.

One is inclined to select the events of this isolated atoll as epitomizing the year, for even here, on the remote edges of the Pacific, vast concerted shifts in the ocean and atmosphere had wrought dramatic change.

Elsewhere about the Pacific it also was common knowledge that the year had been one of extraordinary climatic events. Hawaii had its first recorded typhoon; the seabird-killing *El Niño* visited the Peruvian coast;the ice went out of Point Barrow at the earliest time in history; and on the Pacific's western rim, the tropical rainy season lingered six weeks beyond its appointed term—Sette and Isaacs (1960).

Just months after the event, in 1958, a distinguished group of oceanographers and meteorologists assembled in Rancho Santa Fe, California to try to under- stand the *Changing Pacific Ocean in 1957 and 1958* (Sette and Isaacs (1960). There, for perhaps the first time, they began the synthesis of atmospheric and oceanic events leading to our present understanding of El Niño.

While oceanographers had been mostly concerned with the eastern equatorial Pacific and El Niño, meteorologists had been mostly concerned with the western tropical Pacific, the tropical Indian Ocean, and the Southern Oscillation. Hildebrandsson, the Lockyers, and Sir Gilbert Walker noticed in the early decades of the 20th century that pressure fluctuations throughout that region are highly correlated with pressure fluctuations in many other regions of the world (figure

14.6). Because variations in pressure are associated with winds and rainfall, they wanted to find out if pressure in one region could be used to forecast weather in other regions using the correlations.

The early studies found that the two strongest centers of the variability are near Darwin, Australia and Tahiti. The fluctuations at Darwin are opposite those at Tahiti, and resemble an oscillation. Furthermore, the two centers had strong correlations with pressure in areas far from the Pacific. Walker named the fluctuations the *Southern Oscillation.*

The *Southern Oscillation Index* is sea-level pressure at Tahiti minus sea-level pressure at Darwin (figure 14.7) normalized by the standard deviation of the difference. The index is related to the trade winds. When the index is high, the pressure gradient between east and west in the tropical Pacific is large, and the trade winds are strong. When the index is negative trades, are weak.

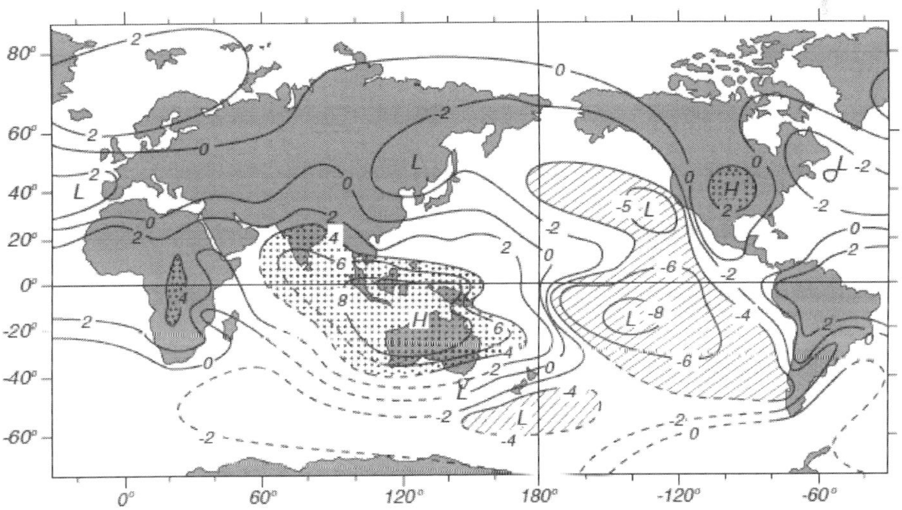

Figure 14.6 Correlation coefficient of annual-mean sea-level pressure with pressure at Darwin. - - - - Coefficient < -0.4. After Trenberth and Shea (1987).

The connection between the Southern Oscillation and El Niño was made soon after the Rancho Santa Fe meeting. Ichiye and Petersen (1963) and Bjerknes (1966) noticed the relationship between equatorial temperatures in the Pacific during the 1957 El Niño and fluctuations in the trade winds associated with the Southern Oscillation. The theory was further developed by Wyrtki (1975).

Because El Niño and the Southern Oscillation are so closely related, the phenomenon is often referred to as the *El Niño–Southern Oscillation* or enso. More recently, the oscillation is referred to as El Niño/La Nin˜a, where La Nin˜arefers to the positive phase of the oscillation when trade winds are strong, and water temperature in the eastern equatorial region is very cold.

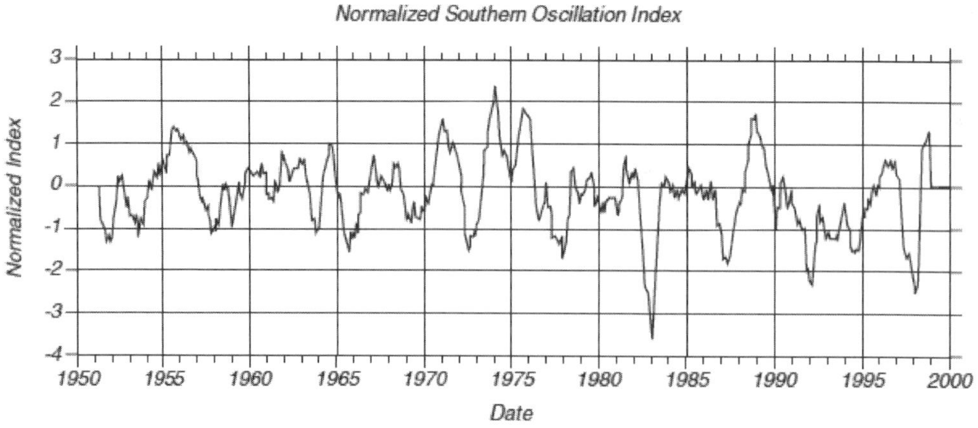

Figure 14.7 Normalized Southern Oscillation Index from 1951 to 1999. The normalized index is sea-level pressure anomaly at Tahiti divided by its standard deviation minus sea-level pressure anomaly at Darwin divided by its standard deviation then the difference is divided by the standard deviation of the difference. The means are calculated from 1951 to 1980. Monthly values of the index have been smoothed with a 5-month running mean. Strong El Niño events occurred in 1957–58, 1965–66, 1972–73, 1982–83, 1997–98. Data from NOAA.

Definition of El Niño: Philander (1990) points out that each El Niño is unique, with different temperature, pressure, and rainfall patterns. Some are strong, some are weak. So, exactly what events deserve to be called El Niño? The icoads data show that the best indicator of El Niño is sea-level pressure anomaly in the eastern equatorial Pacific from 4°S to 4°N and from 108°W to98°W (Harrison and Larkin, 1996). It correlates better with sea-surface temperature in the central Pacific than with the Southern-Oscillation Index. Thus the importance of the El Niño is not exactly proportional to the Southern Oscillation Index—the strong El Niño of 1957–58, has a weaker signal in figure 14.7 than the weaker El Niño of 1965–66.

Trenberth (1997) recommends that those disruptions of the equatorial system in the Pacific shall be called an El Niño only when the 5-month running mean of sea-surface temperature anomalies in the region 5°N–5°S, 120°W–170°W exceeds 0.4°C for six months or longer.

So El Niño, which started life as a change in currents off Peru each Christmas, has grown into a giant. It now means a disruption of the ocean-atmosphere system over the whole equatorial Pacific.

Theory of El Niño: Wyrtki (1975) gives a clear description of El Niño.

During the two years preceding El Niño, excessively strong southeast trades are present in the central Pacific. These strong southeast trades intensify the subtropical gyre of the South Pacific, strengthen the South Equatorial Current, and increase the east-west slope of sea level by build- ing up water in the western equatorial Pacific. As soon as the wind stress in the central Pacific relaxes, the accumulated water flows eastward, probably in the form of an equatorial Kelvin wave. This wave leads to the ac- cumulation of warm water off Ecuador and Peru and to a depression of the usually shallow thermocline. In total, El Niño is the result of the response of the equatorial Pacific to atmospheric forcing by the trade winds.

Sometimes the trades in the western equatorial Pacific not only weaken, they actually reverse direction for a few weeks to a month, producing *westerly wind bursts* that quickly deepen the thermocline there. The deepening of the thermocline launches an eastward propagating Kelvin wave and a westward propagating Rossby wave. (If you are asking, What are Kelvin and Rossby waves? I will answer that in a minute. So please be patient.)

The Kelvin wave deepens the thermocline as it moves eastward, and it carries warm water eastward. Both processes cause a deepening of the mixed layer in the eastern equatorial Pacific a few months after the wave is launched in the western Pacific. The deeper thermocline in the east leads to upwelling of warm water, and the surface temperatures offshore of Ecuador and Peru warms by 2–4°. The warm water reduces the temperature contrast between east and west,further reducing the trades. The strong positive feedback between sea-surface temperature and the trade winds causes rapid development of El Niño.

With time, the warm pool spreads east, eventually extending as far as 140°W(figure 14.8). Plus, water warms in the east along the equator due to upwelling of warm water, and to reduced advection of cold water from the east due to weaker trade winds.

The warm waters along the equator in the east cause the areas of heavy rain to move eastward from Melanesia and Fiji to the central Pacific. Essentially, a major source of heat for the atmospheric circulation moves from the west to the central Pacific, and the whole atmosphere responds to the change. Bjerknes (1972), describing the interaction between the ocean and the atmosphere over the eastern equatorial Pacific concluded:

In the cold ocean case (1964) the atmosphere has a pronounced stable layer between 900 and 800 mb, preventing convection and rainfall, and in the warm case (1965) the heat supply from the ocean eliminates the atmospheric stability and activates rainfall. . . . A side effect of the widespread warming of the tropical belt of the atmosphere shows up in the increase of exchange of angular momentum with the neighboring subtropical belt, whereby the subtropical westerly jet strengthens . . . The variability of the heat and moisture supply to the global atmospheric thermal engine from the equatorial Pacific can be shown to have far-reaching large-scale effects.

Klaus Wyrtki (1985), drawing on extensive observations of El Niño, writes:

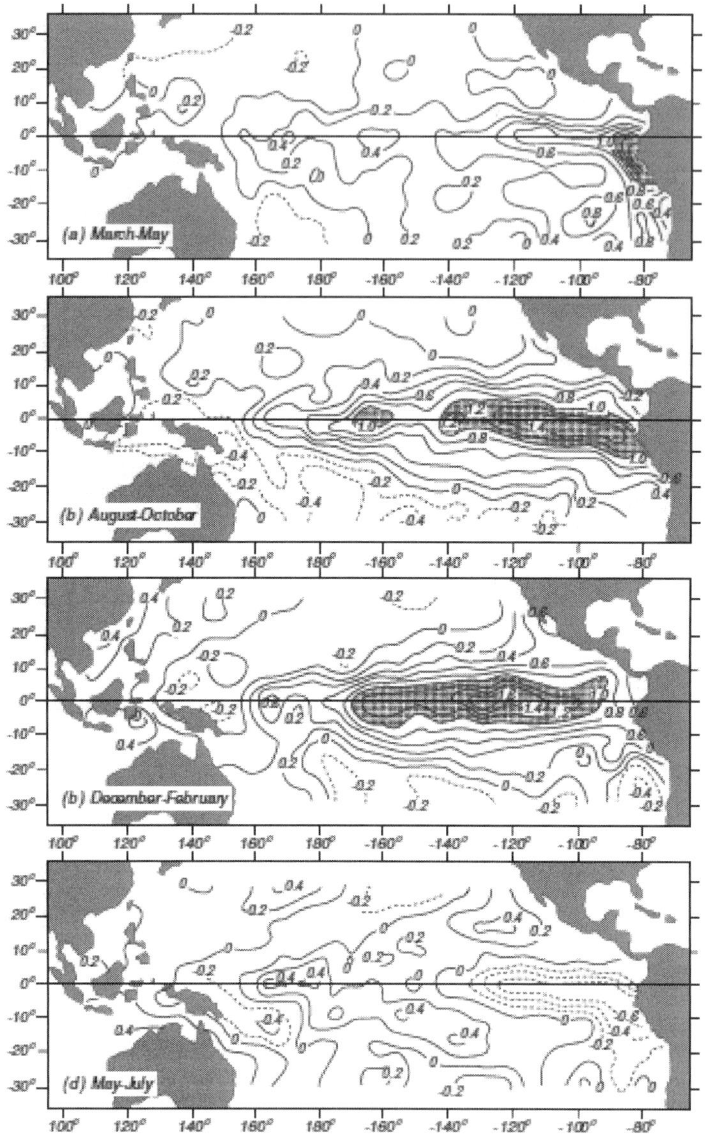

Figure 14.8 Anomalies of sea-surface temperature (in °C) during a typical El Niño obtained

by averaging data from El Niños between 1950 and 1973. Months are after the onset of the event. After Rasmusson and Carpenter (1982).

A complete El Niño cycle results in a net heat discharge from the tropical Pacific toward higher latitudes. At the end of the cycle the tropical Pacific is depleted of heat, which can only be restored by the slow accumulation of warm water in the western Pacific by the normal trade winds. Conse- quently, the time scale of the

Southern Oscillation is given by the time required for the accumulation of warm water in the western Pacific.

It is these far reaching events that make El Niño so important. Few people care about warm water off Peru around Christmas, many care about global changes the weather. El Niño is important because of its atmospheric influence. When the Kelvin wave reaches the coast of Ecuador, part is reflected as an westward propagating Rossby wave, and part propagates north and south as a coastal trapped Kelvin wave carrying warm water to higher latitudes. For example, during the 1957 El Niño, the northward propagating Kelvin wave produced unusually warm water off shore of California, and it eventually reached Alaska. This warming of the west coast of North America further influences climate in North America, especially in California.

As the Kelvin wave moves along the coast, it forces Rossby waves which move west across the Pacific at a velocity that depends on the latitude (14.4). The velocity is very slow at high latitudes and fastest on the equator, where the reflected wave moves back as a deepening of the thermocline, reaching the central equatorial Pacific a year later. Similarly, the westward propagating Rossby wave launched at the start of the El Niño in the west, reflects off Asia and returns to the central equatorial Pacific as a Kelvin wave, again about a year later.

El Niño ends when the Rossby waves reflected from Asia and Ecuador meet in the central Pacific about a year after the onset of El Niño (Picaut, Masia, and du Penhoat, 1997). The waves push the warm pool at the surface toward the west. At the same time, the Rossby wave reflected from the western boundary causes the thermocline in the central Pacific to become shallower when the waves reaches the central Pacific. Then any strengthening of the trades causes upwelling of cold water in the east, which increases the east-west temperature gradient, which increases the trades, which increases the upwelling (Takayabu et al 1999). The system is then thrown into the La Nin˜a state with strong trades, and a very cold tongue along the equator in the east.

La Nin˜a tends to last longer than El Niño, and the cycle from La Nin˜a to El Niño and back takes about three years. The cycle isn't exact. El Niño comes back at intervals from 2-7 years, with an average near four years (figure 14.7).

Equatorial Kelvin and Rossby Waves: Kelvin and Rossby waves are the ocean's way of adjusting to changes in forcing such as westerly wind bursts. The adjustment occurs as waves of current and sea level that are influenced by gravity, Coriolis force f, and the north-south variation of Coriolis force $\partial f/\partial y = \beta$. There are many kinds of these waves with different frequencies, wavelengths, and velocities. If gravity and f are the restoring forces, the waves are called Kelvin and

PoiNCAR´e waves. If β is the restoring force, the waves are called planetary waves. One important type of planetary wave is the Rossby wave.

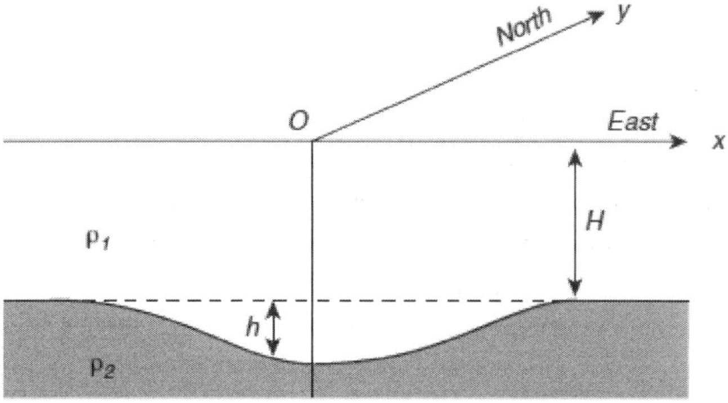

Figure 14.9 Sketch of the two-layer model of the equatorial ocean used to calculate planetary waves in those regions. After Philander (1990: 107).

Two types of waves are especially important for El Niño: internal Kelvinwaves and Rossby waves. Both waves can have modes that are confined to a narrow, north-south region centered on the equator. These are *equatorially trapped waves.* Both exist in slightly different forms at higher latitudes.

Kelvin and Rossby wave theory is beyond the scope of this book, so I will just tell you what they are without deriving the properties of the waves. If you are curious, you can find the details in Philander (1990): Chapter 3; Pedlosky (1987): Chapter 3; and Apel (1987): §6.10–6.12. If you know little about waves, their wavelength, frequency, group and phase velocities, skip to Chapter 16 and read §16.1.

The theory for equatorial waves is based on a two-layer model of the ocean (figure 14.9). Because the tropical ocean have a thin, warm, surface layer above a sharp thermocline, such a model is a good approximation for those regions.

$$c_{Kg} = c \equiv \sqrt{g'H}; \qquad \text{where} \qquad g' = \frac{\rho_2 - \rho_1}{\rho_1} g \qquad (14.2)$$

g' is *reduced gravity* , ρ1, ρ2 are the densities above and below the thermocline, and *g* is gravity. Trapped Kelvin waves propagate only to the east. Note, that *c*is the phase and group velocity of a shallow-water, internal, gravity wave. It is the maximum velocity at which disturbances can travel along the thermocline. Typical values of the quantities in (14.2) are:

$$\frac{\rho_2 - \rho_1}{\rho_1} = 0.003; \qquad H = 150 \text{ m}; \qquad c = 2.1 \text{ m/s}$$

At the equator, Kelvin waves propagate eastward at speeds of up to 3 m/s, and they cross the Pacific in a few months. Currents associated with the wave are everywhere eastward with north-south component (figure 14.10).

Kelvin waves can also propagate poleward as a trapped wave along an east coast of an ocean basin. Their group velocity is also given by (14.3), and they are confined to a coastal zone with width $x = c/(\beta y)$

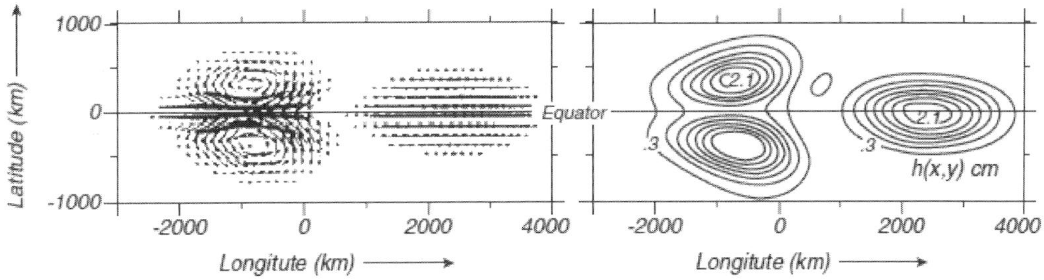

Figure 14.10 Left: Horizontal currents associated with equatorially trapped waves generated by a bell-shaped displacement of the thermocline. Right: Displacement of the thermocline due to the waves. The figures shows that after 20 days, the initial disturbance has separated into an westward propagating Rossby wave (left) and an eastward propagating Kelvin wave (right). After Philander et al. (1984: 120).

The important Rossby waves on the equator have frequencies much less than the Coriolis frequency. They can travel only to the west. The group velocity is:

$$c_{Rg} = -\frac{c}{(2n+1)}; \qquad n = 1, 2, 3, \ldots \qquad (14.3)$$

The fastest wave travels westward at a velocity near 0.8 m/s. The currents associated with the wave are almost in geostrophic balance in two counter- rotating eddies centered on the equator (figure 14.10).

Away from the equator, low-frequency, long-wavelength Rossby waves also travel only to the west, and the currents associated with the waves are again almost in geostrophic balance. Group velocity depends strongly on latitude:

$$c_{Rg} = -\frac{\beta g' H}{f^2} \qquad (14.4)$$

Wave dynamics in the equatorial regions differ markedly from wave dynamics at mid-latitudes. Baroclinic waves are much faster, and the response of the ocean to changes in wind forcing is much faster than at mid-latitudes. For planetary waves near the equator, we can speak of an *equatorial wave guide*. Now, let's return to El Niño and its "far-reaching large-scale effects."

14.3 EL NIÑO TELECONNECTIONS

Teleconnections are statistically significant correlations between weather events that occur at different places on the earth. Figure 14.11 shows the dominant global tele connections associated with the El Niño/Southern Oscillation.

The influence of enso is through its influence on convection and associated latent heat release in the equatorial Pacific. As the area of heavy rain moves east, the source of atmospheric heating moves with the rain, leading to widespread changes in atmospheric circulation and weather patterns outside the tropical Pacific (McPhaden, Zebiak and Glantz, 2006), including perturbations in atmospheric pressure (figure 14.12). This sequence of events leads to some predictability of weather patterns a season in advance over North America, Brazil, Australia, South Africa and other regions.

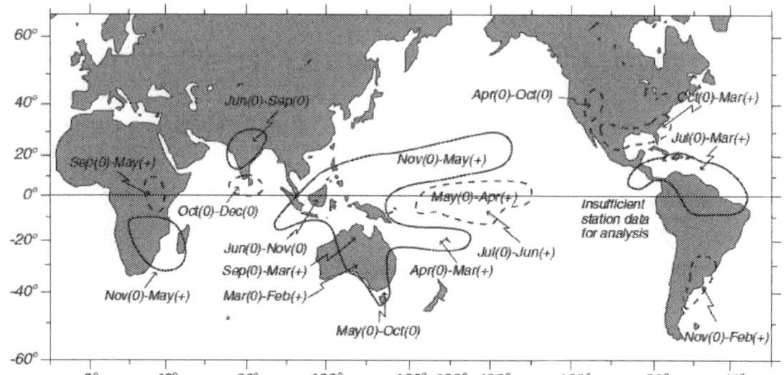

Figure 14.11 Sketch of regions receiving enhanced rain (dashed lines) or drought (solid lines) during an El Niño event. (0) indicates that rain changed during the year in which El Niño began, (+)indicates that rain changed during the year after El Niño began. After Ropelewski and Halpert (1987).

The ENSO perturbations to mid-latitude and tropical weather systems leads to dramatic changes in rainfall in some regions (figure 14.11). As the convective regions migrate east along the equator, they bring rain to the normally arid, central-Pacific islands. The lack of rain the western Pacific leads to drought in Indonesia and Australia.

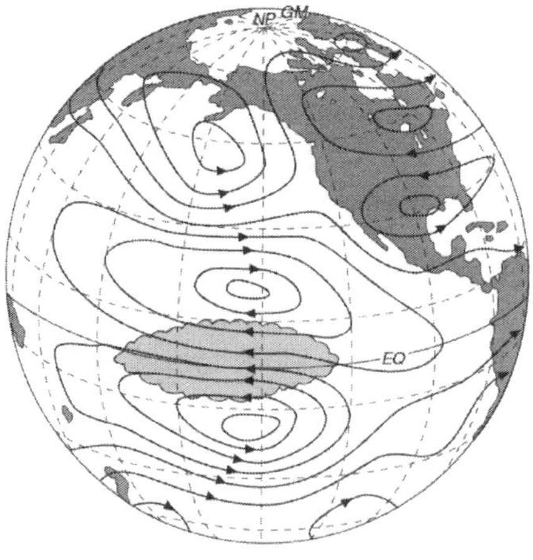

Figure 14.12 Changing patterns of convection in the equatorial Pacific during an El Niño, set up a pattern of pressure anomalies in the atmosphere (solid lines) which influence the extratropical atmosphere. After Rasmusson and Wallace (1983).

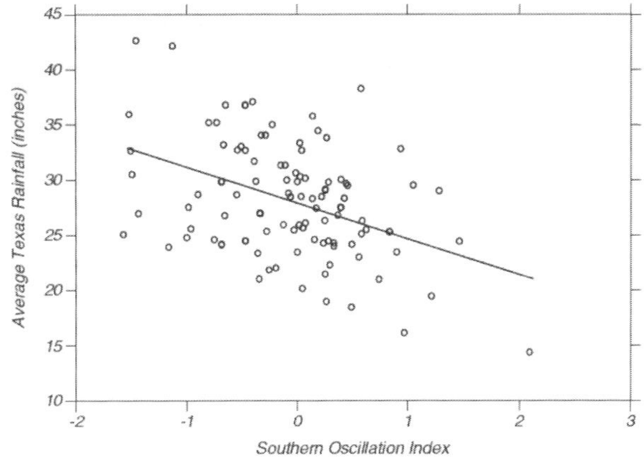

Figure 14.13 Correlation of yearly averaged rainfall averaged over Texas plotted as a function of the Southern Oscillation Index averaged for the year. From Stewart (1995).

An Example: Variability of Texas Rainfall: Figure 14.11 shows a global view of teleconnections. Let's zoom in to one region, Texas, that I chose only because I live there. The global figure shows that the region should have higher than normal rainfall in the winter season after El Niño begins. I therefore corre- lated yearly averaged rainfall for the state of Texas to the Southern Oscillation Index (figure 14.13). Wet years correspond to El Niño years in the equatorial Pacific. During El Niño, convection normally found in the western equatorial Pacific moved east into the central equatorial Pacific. The subtropical jet also moves east, carrying tropical moisture across Mexico to Texas and the Mississippi Valley. Cold fronts in winter interact with the upper level moisture to produce abundant winter rains from Texas eastward.

14.4 OBSERVING EL NIÑO

The tropical and equatorial Pacific is a vast, remote area seldom visited by ships. To observe the region NOAA's Pacific Marine Environmental Laboratory in Seattle deployed an array of buoys to measure oceanographic and meteorological variables (figure 14.14). The first buoy was successfully deployed in 1976 by David Halpern. Since then, new moorings have been added to the array, new instruments have been added to the moorings, and the moorings have been im- proved. The program has now evolved into the Tropical Atmosphere Ocean TAOarray of approximately 70 deep-ocean moorings spanning the equatorial Pacific Ocean between 8°N and 8°S from 95°W to 137°E (McPhaden et al, 1998).

The array began full operation in December 1994, and it continues to evolve. The work necessary to design and calibrate instruments, deploy moorings, and process data is coordinated through the TAO Project. It is a multi-national effort involving the United States, Japan, Korea, Taiwan, and France with a project office at the Pacific Marine Environmental Laboratory.

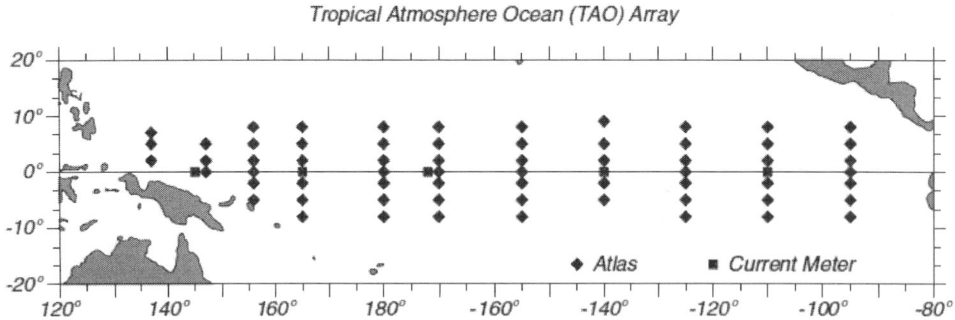

Figure 14.14 Tropical Atmosphere Ocean TAO array of moored buoys operated by the NOAA Pacific Marine Environmental Laboratory with help from Japan, Korea, Taiwan, and France. Figure from NOAA Pacific Marine Environmental Laboratory.

The TAO moorings measure air temperature, relative humidity, surface wind velocity, sea-surface temperatures, and subsurface temperatures from 10 meters down to 500 meters. Five moorings located on the equator at 110°W, 140°W,170°W, 165°E, and 147°E also carry upward-looking Acoustic Doppler Current Profilers ADCP to measure upper-ocean currents between 10 m and 250 m. Data are sent back through the Argos system, and data are processed and made available in near real time. The moorings are recovered and replaced yearly. All sensors are calibrated prior to deployment and after recovery.

Data from TAO are merged with altimeter data from Jasin, and ers-2 to obtain a more comprehensive measurement of El Niño. Jasin and Topex/Poseidon data have been especially useful because they could be used to produce accurate maps of sea level every ten days. The maps provided detailed views of the development of the 1997–1998 El Niño in near real time that were widely reproduced throughout the world. The observations (figure 10.6) show high sea level propagating from west to east, peaking in the eastern equatorial Pacific in November 1997. In addition, satellite data extended beyond the TAO data region to include the entire tropical Pacific. This allowed oceanographers to look for extra-tropical influences on El Niño.

Rain rates are measured by NASA's Tropical Rainfall Measuring Mission which was specially designed for this purpose. It was launched on 27 November 1997, and it carries five instruments: the first space borne precipitation radar, a five-frequency microwave radiometer, a visible and infrared scanner, a cloud and earth radiant energy system, and a lightning imaging sensor. Working together, the instruments provide data necessary to produce monthly maps of tropical rainfall averaged over 500 km by 500 km areas with 15% accuracy.

The grid is global between ±35° latitude. In addition, the satellite data are used to measure latent heat released to the atmosphere by rain, thus providing continuous monitoring of heating of the atmosphere in the tropics.

14.5 FORECASTING EL NIÑO

The importance of El Niño to global weather patterns has led to many schemes for forecasting events in the equatorial Pacific. Several generations of models have been produced, but the skill of the forecasts has not always increased. Models worked well for a few years, then failed. Failure was followed by improved models, and the cycle continued. Thus, the best models in 1991 failed to predict weak El Niños in 1993 and 1994 (Ji, Leetmaa, and Kousky, 1996). The best model of the mid 1990s failed to predict the onset of the strong El Niño of 1997-1998 although a new model developed by the National Centers for Environmental Prediction made the best forecast of the development of the event. In general, the more sophisticated the model, the better the forecasts (Kerr, 1998).

The following recounts some of the more recent work to improve the forecasts. For simplicity, I describe the technique used by the National Centers for Environmental Prediction (Ji, Behringer, and Leetmaa, 1998). But Chen et al. (1995), Latif et al. (1993), and Barnett et al. (1993), among others, have all developed useful prediction models.

Atmospheric Models : How well can we model atmospheric processes over the Pacific? To help answer the question, the World Climate Research Program's Atmospheric Model Intercomparison Project (Gates, 1992) compared output from 30 different atmospheric numerical models for 1979 to 1988. *The Variability in the Tropics: Synoptic to Intraseasonal Timescales* subproject is especially important because it documents the ability of 15 atmospheric general-circulation models to simulate the observed variability in the tropical atmosphere (Slingo et al. 1995). The models included several operated by government weather forecasting centers, including the model used for day-to-day forecasts by the European Center for Medium-Range Weather Forecasts.

The first results indicate that none of the models were able to duplicate all important inter seasonal variability of the tropical atmosphere on time-scales of 2 to 80 days. Models with weak intra seasonal activity tended to have a weak annual cycle. Most models seemed to simulate some important aspects of the inter annual variability including El Niño. The length of the time series was, however, too short to provide conclusive results on inter annual variability.

The results of the substudy imply that numerical models of the atmospheric general circulation need to be improved if they are to be used to study tropical variability and the response of the atmosphere to changes in the tropical ocean.

Oceanic Models: Our ability to understand El Niño also depends on our ability to model the oceanic circulation in the equatorial Pacific. Because the models provide the initial conditions used for the forecasts, they must be able to assimilate up-to-date measurements of the Pacific along with heat fluxes and surface winds calculated from the atmospheric models. The measurements include sea-surface winds from scatterometers and moored buoys, surface temperature from the optimal-interpolation data set (see §6.6), subsurface temperatures frombuoys and XBTs, and sea level from altimetry and tide-gauges on islands.

Ji, Behringer, and Leetmaa (1998) at the National Centers for Environ- mental Prediction have modified the Geophysical Fluid Dynamics Laboratory's Modular Ocean Model for use in the tropical Pacific (see §15.3 for more information about this model). It's domain is the Pacific between 45°S and 55°N and between 120°E and 70°W. The zonal resolution is 1.5°. The meridional resolution is 1/3° within 10° of the equator, increasing smoothly to 1° poleward of 20° latitude. It has 28 vertical levels, with 18 in the upper 400 m toresolve the mixed layer and thermocline. The model is driven by mean winds from Hellerman and Rosenstein (1983), anomalies in the wind field from Florida State University, and mean heat fluxes from Oberhuber (1988). It assimilates subsurface temperature from the TAO array and XBTs, and surface temperatures from the monthly optimal-interpolation data set (Reynolds and Smith, 1994).

The output of the model is an ocean analysis, the density and current field that best fits the data used in the analysis (figures 14.3 and 14.4). This is used to drive a coupled ocean-atmosphere model to produce forecasts.

Coupled Models: Coupled models are separate atmospheric and oceanic models that pass information through their common boundary at the sea surface, thus coupling the two calculations. The coupling can be one way, from the atmosphere, or two way, into and out of the ocean. In the scheme used by the NOAA National Centers for Environmental Prediction the ocean model is the same Modular Ocean Model described above. It is coupled to a low-resolution version of the global, medium-range forecast model operated by the National Centers (Kumar, Leetmaa, and Ji, 1994). Anomalies of wind stress, heat, and fresh-water fluxes calculated from the atmospheric model are added to the mean annual values of the fluxes, and the sums are used to drive the ocean model. Sea-surface temperature calculated from the ocean model is used to drive the atmospheric model from 15°N to 15°S.

As computer power decreases in cost, models are becoming ever more complex. The trend is to global coupled models able to include other coupled ocean atmosphere systems in addition to ENSO. I return to the problem in §15.6 where I describe global coupled models.

Statistical Models: Statistical models are based on an analysis of weather patterns in the Pacific using data going back many decades. The basic idea is that if weather patterns today are similar to patterns at some time in the past, then todays patterns will evolve as they did at that past time. For example, if winds and temperatures in the tropical Pacific today are similar to wind and temperatures just before the 1976 El Niño, then we might expect a similar El Niño to start in the near future.

Forecasts: In general, the coupled ocean-atmosphere models produce forecasts that are no better than the statistical forecasts (Jan van Oldenborgh, 2005). The forecasts include not only events in the Pacific but also the global consequences of El Niño. The forecasts are judged two ways:

- Using the correlation between the area-averaged sea-surface-temperature anomalies from the model and the observed temperature anomalies in the eastern equatorial Pacific. The area is usually from 170°W to 120° ᶻW between 5°S and 5°N. Useful forecasts have correlations exceeding 0.6.
- Using the root-mean-square difference between the observed and predicted sea-surface temperature in the same area.

The forecasts of the very strong 1997 El Niño have been carefully studied. Jan van Oldenborg et al (2005) and Barnston et al (1999) found no models successfully forecast the earliest onset of the El Niño in late 1996 and early1997. The first formal announcements of the El Niño were made in May 1997. Nor did any model forecast the large temperature anomalies observed in the eastern equatorial Pacific until the area had already warmed. There was no clear distinction between the accuracy of the dynamical or statistical forecasts.

14.6 CONCISE SUMMARY

- Equatorial processes are important because heat released by rain in the equatorial region helps drives much of the atmospheric circulation.
- Solar energy absorbed by the Pacific is the most important driver of atmospheric circulation. Solar energy is lost from the ocean mainly by evaporation. The heat warms the atmosphere and drives the circulation when the latent heat of evaporation is released in rainy areas, primarily in the western tropical Pacific and the Intertropical Convergence Zone.
- The interannual variability of currents and temperatures in the equatorial Pacific modulates the oceanic forcing of the atmosphere. This inter annual variability is associated with El Niño/La Nin˜a.
- Changes in equatorial dynamics cause changes in atmospheric circulation by changing the location of rain in tropical Pacific and therefore the location of the major heat source driving the atmospheric circulation.
- El Niño causes the biggest changes in equatorial dynamics. During El Niño, trade-winds weaken in the western Pacific, the thermocline becomes less deep in the west. This drives a Kelvin wave eastward along the equator,

which deepens the thermocline in the eastern Pacific. The warm pool in the west moves eastward toward the central Pacific, and the intense tropical rain areas move with the warm pool.

- El Ni˜[n]o is the largest source of year-to-year fluctuations in global weather patterns.
- As a result of El Niño, drought occurs in the Indonesian area and northern Australia, and floods occur in western, tropical South America. Variations in the atmospheric circulation influence more distant areas through interconnections.
- Forecasts of El Niño are made using coupled ocean-atmospheric numerical models. Forecasts appear to have useful accuracy for 3–6 months in advance, mostly after the onset of El Niño.

15 NUMERICAL MODELS

We saw earlier that analytic solutions of the equations of motion are impossible to obtain for typical oceanic flows. The problem is due to non-linear terms in the equations of motion, turbulence, and the need for realistic shapes for the sea floor and coastlines. We have also seen how difficult it is to describe the ocean from measurements. Satellites can observe some processes almost everywhere every few days. But they observe only some processes, and only near or at the surface. Ships and floats can measure more variables, and deeper into the water, but the measurements are sparse. Hence, numerical models provide the only useful, global view of ocean currents. Let's look at the accuracy and validity of the models, keeping in mind that although they are only models, they provide a remarkably detailed and realistic view of the ocean.

15.1 INTRODUCTION: AN OVERVIEW

Numerical models of ocean currents have many advantages. They simulate flows in realistic ocean basins with a realistic sea floor. They include the in- fluence of viscosity and non-linear dynamics. And they can calculate possible future flows in the ocean. Perhaps, most important, they interpolate between sparse observations of the ocean produced by ships, drifters, and satellites.

Numerical models are not without problems. "There is a world of difference between the character of the fundamental laws, on the one hand, and the nature of the computations required to breathe life into them, on the other"—Berlinski (1996). The models can never give complete descriptions of the oceanic flows even if the equations are integrated accurately. The problems arise from several sources.

Discrete equations are not the same as continuous equations. In Chapter 7 we wrote down the differential equations describing the motion of a continuous fluid. Numerical models use algebraic approximations to the differential equations. We assume that the ocean basins are filled with a grid of points, and time moves forward in tiny steps. The value of the current, pressure, temperature, and salinity are calculated from their values at nearby points and previous times. Ian Stewart (1992), a noted mathematician, points out that

Discretization is essential for computer implementation and cannot be dispensed with. The essence of the difficulty is that the dynamics of discrete systems is only loosely related to that of continuous systems—indeed the dynamics of discrete systems is far richer than that of their continuous counterparts—and the approximations involved can create spurious solutions.

Calculations of turbulence are difficult. Numerical models provide information only at grid points of the model. They provide no information about the flow between the points. Yet, the ocean is turbulent, and any oceanic model capable of resolving the turbulence needs grid points spaced millimeters apart, with time steps of milliseconds.

Practical ocean models have grid points spaced tens to hundreds of kilometers apart in the horizontal, and tens to hundreds of meters apart in the vertical. This means that turbulence cannot be calculated directly, and the influence of turbulence must be parameterized. Holloway (1994) states the problem succinctly:

Ocean models retain fewer degrees of freedom than the actual ocean (by about 20 orders of magnitude). We compensate by applying 'eddy- viscous goo' to squash motion at all but the smallest retained scales. (We also use non-conservative numerics.) This is analogous to placing a partition in a box to prevent gas molecules from invading another region of the box. Our oceanic models cannot invade most of the real oceanic degrees of freedom simply because the models do not include them.

Given that we cannot do things 'right', is it better to do nothing? That is not an option. 'Nothing' means applying viscous goo and wishing for the ever bigger computer. Can we do better? For example, can we guess a higher entropy configuration toward which the eddies tend to drive the ocean (that tendency to compete with the imposed forcing and dissipation)?

By "degrees of freedom" Holloway means all possible motions from the smallest waves and turbulence to the largest currents. Let's do a calculation. We know that the ocean is turbulent with eddies as small as a few millimeters. Tocompletely describe the ocean we need a model with grid points spaced 1 mm apart and time steps of about 1 ms. The model must therefore have $360^\circ \times 180^\circ \times (111 \text{ km/degree})^2 \times 10^{12} (\text{mm/km})^2 \times 3 \text{ km} \times 10^6 (\text{mm/km}) = 2.4 \times 10^{27}$ data points for a 3 km deep ocean covering the globe. The global Parallel Ocean Program Model described in the next section has 2.2×10^7 points. So we need 10^{20} times more points to describe the real ocean. These are the missing 10^{20} degrees of freedom.

Practical models must be simpler than the real ocean. Models of the ocean must run on available computers. This means oceanographers further simplify their models. We use the hydrostatic and Boussinesq approximations, and we often use equations integrated in the vertical, the shallow-water equations (Haid- vogel and Beckmann, 1999: 37). We do this because we cannot yet run the most detailed models of oceanic circulation for thousands of years to understand the role of the ocean in climate.

Numerical code has errors. Do you know of any software without bugs? Numerical models use many subroutines each with many lines of code which are converted into instructions understood by processors using other software called a compiler. Eliminating all software errors is impossible. With careful testing, the output may be correct, but the accuracy cannot be guaranteed. Plus, numerical calculations cannot be more accurate than the accuracy of the floating-point numbers and integers used by the computer. Round-off errors cannot be ignored. Lawrence et al (1999), examining the output of an atmospheric numerical model found an error in the code produced by the fortran-90 compiler used on the cray Research supercomputer used to run the code. They also found round-off errors in the concentration of tracers calculated from the model. Both errors produced important errors in the output of the model.

Most models are not well verified or validated (Post & Votta, 2005). Yet, without adequate verification and validation, output from numerical models is not credible.

Summary: Despite these many sources of error, most are small in practice. Numerical models of the ocean are giving the most detailed and complete views of the circulation available to oceanographers. Some of the simulations contain unprecedented details of the flow. I included the words of warning not to lead you to believe the models are wrong, but to lead you to accept the output with a grain of salt.

15.2 APPLICATION OF NUMERICAL MODELS IN OCEANOGRAPHY

Numerical models are very widely used for many purposes in oceanography. For our purpose we can divide models into two classes:

- Mechanistic models are simplified models used for studying processes. Because the models are simplified, the output is easier to interpret than output from more complex models. Many different types of simplified models have been developed, including models for describing planetary waves, the interaction of the flow with sea-floor features, or the response of the upper ocean to the wind. These are perhaps the most useful of all models because they provide insight into the physical mechanisms influencing the ocean. The development and use of mechanistic models is, unfortunately, beyond the scope of this book.
- Simulation models are used for calculating realistic circulation of oceanic regions. The models are often very complex because all important processes are included, and the output is difficult to interpret.

- The first simulation model was developed by Kirk Bryan and Michael Cox (Bryan, 1969) at the Geophysical Fluid Dynamics laboratory in Princeton. They calculated the 3-dimensional flow in the ocean using the continuity and momentum equation with the hydrostatic and Boussinesq approximations and a simple equation of state. Such models are called *primitive equation* models because they use the basic, or primitive form of the equations of motion. The equation of state allows the model to calculate changes in density due to fluxes of heat and water through the surface, so the model includes thermodynamic processes.
- The Bryan-Cox model used large horizontal and vertical viscosity and diffusion to eliminate turbulent eddies having diameters smaller about 500 km, which is a few grid points in the model. It had complex coastlines, smoothed sea-floor features, and a rigid lid. The rigid lid was needed to eliminate ocean-surface waves, such as tides and tsunamis, that move far too fast for the coarse time steps used by all simulation models. The rigid lid had, however, disadvantages. Islands substantially slowed the computation, and the sea-floor features were smoothed to eliminate steep gradients.
- The first simulation model was regional. It was quickly followed by a global model (Cox, 1975) with a horizontal resolution of 2° and with 12 levels in the vertical. The model ran far too slowly even on the fastest computers of the day, but it laid the foundation for more recent models. The coarse spatial resolution required that the model have large values for viscosity, and even regional models were too viscous to have realistic western boundary currents or mesoscale eddies. Since those times, the goal has been to produce models with ever finer resolution, more realistic modeling of physical processes, and better numerical schemes. Computer technology is changing rapidly, and models are evolving rapidly. The output from the most recent models of the north Atlantic, which have resolution of 0.03° look very much like the real ocean. Models of other areas show previously unknown currents near Australia and in the south Atlantic.
- Ocean and Atmosphere Models use very different spacing of grid points. As a result, ocean modeling lags about a decade behind atmosphere modelling. Dominant ocean eddies are 1/30 the size of dominant atmosphere eddies(storms). But, ocean features evolve at a rate that is 1/30 the rate in the atmosphere. Thus ocean models running for say one year have (30×30) morehorizontal grid points than the atmosphere, but they have 1/30 the number of time steps. Both have about the same number of grid points in the vertical. As a result, ocean models run 30 times slower than atmosphere models of the same complexity.

15.3 TYPES OF GLOBAL OCEAN MODELS

Several types of global models are widely used in oceanography. Most have grid points about one tenth of a degree apart, which is sufficient to resolve mesoscale eddies, such as those seen in figures 11.10, 11.11, and 15.2, that have a diameter larger than two to three times the distance between grid points. Vertical resolution is typically around 30 vertical levels. Models include: i) realistic coasts and bottom features; ii) heat and water fluxes though the surface; iii)eddy dynamics; and iv) the meridional-overturning circulation. Many assimilate satellite and float data using techniques described in §15.5. The models range in complexity from those that can run on desktop workstations to those that require the world's fastest computers.

All models must be run to calculate one to two decades of variability before they can be used to simulate the ocean. This is called *spin-up*. Spin-up is needed because initial conditions for density, fluxes of momentum and heat through the sea-surface, and the equations of motion are not all consistent.

Models are started from rest with values of density from the Levitus (1982) atlas and integrated for a decade using mean-annual wind stress, heat fluxes, and water flux. The model may be integrated for several more years using monthly wind stress, heat fluxes, and water fluxes.

The Bryan-Cox models evolved into several widely used models which are providing impressive views of the global ocean circulation.

Geophysical Fluid Dynamics Laboratory Modular Ocean Model MOM: consists of a large set of modules that can be configured to run on many different computers to model many different aspects of the circulation. The source code is open and free, and it is in the public domain. The model is widely use for climate studies and for studying the ocean's circulation over a wide range of space and time scales (Pacanowski and Griffies, 1999).

Because mom is used to investigate processes which cover a wide range of time and space scales, the code and manual are lengthy. However, it is far from necessary for the typical ocean modeler to become acquainted with all of its aspects. Indeed, mom can be likened to a growing city with many different neighborhoods. Some of the neighborhoods communicate with one another, some are mutually incompatible, and others are basically independent. This diversity is quite a challenge to coordinate and support. Indeed, over the years certain "neighborhoods" have been jettisoned or greatly renovated for various reasons.—Pacanowski and Griffies.

The model uses the momentum equations, equation of state, and the hydrostatic and Boussinesq approximations. Subgrid-scale motions are reduced by use of eddy viscosity. Version 4 of the model has improved numerical schemes, a free surface, realistic bottom features, and many types of mixing including horizontal mixing along surfaces of constant density. Plus, it can be coupled to atmospheric models.

Parallel Ocean Program Model: produced by Smith and colleagues at Los Alamos National Laboratory (Maltrud et al, 1998) is another widely used model growing out of the original Bryan-Cox code. The model includes improved numerical algorithms, realistic coasts, islands, and unsmoothed bottom features.

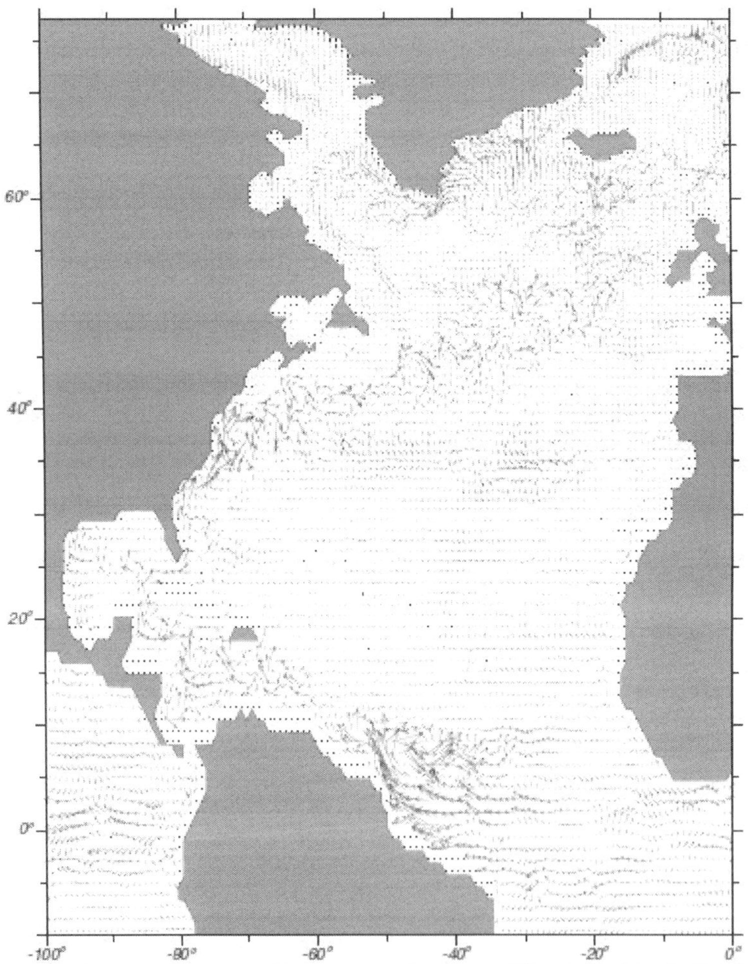

Figure 15.1 Near-surface geostrophic currents on October 1, 1995 calculated by the Parallel Ocean Program numerical model developed at the Los Alamos National Laboratory. The length of the vector is the mean speed in the upper 50 m of the ocean. The direction is the mean direction of the current. From Richard Smith, Los Alamos National Laboratory.

It has model has 1280×896 equally spaced grid points on a Mercator projection extending from 77°Sto 77°N, and 20 levels in the vertical. Thus it has 2.2×10^7 points giving a resolution of$0.28^\circ \times 0.28^\circ$ cos°, which varies from 0.28° (31.25 km) at the equator to 0.06° (6.5 km) at the highest latitudes. The average resolution is about

0.2°. The model was is forced by ECMWF wind stress and surface heat and water fluxes (Barnier et al, 1995).

Hybrid Coordinate Ocean Model: HYCOM All the models just described use *x*, *y*, *z* coordinates. Such a coordinate system has both advantages and disadvantages. It can have high resolution in the surface mixed layer and in shallower regions. But it is less useful in the interior of the ocean. Below the mixed layer, mixing in the ocean is easy along surfaces of constant density, and difficult across such surfaces. A more natural coordinate system in the interior of the ocean uses *x*, *y*, ρ, where ρ is density. Such a model is called an *isopycnalmodel.* Essentially, ρ(*z*) is replaced with *z*(ρ). Because isopycnal surfaces are surfaces of constant density, horizontal mixing is always on constant-density surfaces in this model.

The Hybrid Coordinate Ocean Model HYCOM model uses different vertical coordinates in different regions of the ocean, combining the best aspects of *z*-coordinate model and isopycnal-coordinate model (Bleck, 2002). The hybrid model has evolved from the Miami Isopycnic-Coordinate Ocean Model (figure 15.2). It is a primitive-equation model driven by wind stress and heat fluxes. It has realistic mixed layer and improved horizontal and vertical mixing schemes that include the influences of internal waves, shear instability, and double-diffusion (see §8.5). The model results from collaborative work amonginvestigators at many oceanographic laboratories.

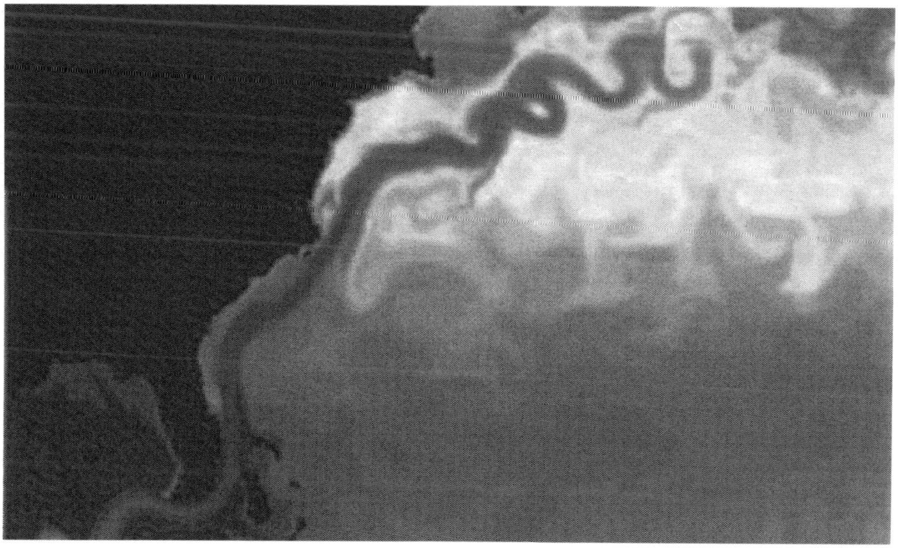

Figure 15.2 Output of Bleck's Miami Isopycnal Coordinate Ocean Model micom. It is a high-resolution model of the Atlantic showing the Gulf Stream, its variability, and the circulation of the north Atlantic. From Bleck.

Regional Oceanic Modeling System: roms is a regional model that can be imbedded in models of much larger regions. It is widely used for studying coastal current systems closely tied to flow further offshore, for example, the California

Current. roms is a hydrostatic, primitive equation, terrain-following model using stretched vertical coordinates, driven by surface fluxes of momentum, heat, and water. It has improved surface and bottom boundary layers (Shchepetkin and McWilliams, 2004).

Climate models are used for studies of large-scale hydrographic structure, climate dynamics, and water-mass formation. These models are the same as the eddy-admitting, primitive equation models I have just described except the horizontal resolution is much coarser because they must simulate ocean processes for decades or centuries. As a result, they must have high dissipation for numericalstability, and they cannot simulate mesoscale eddies. Typicalhorizontal resolutions are 2° to 4°. The models tend, however, to have highvertical resolution necessary for describing the deep circulation important for climate.

15.4 COASTAL MODELS

The great economic importance of the coastal zone has led to the development of many different numerical models for describing coastal currents, tides, and storm surges. The models extend from the beach to the continental slope, and they can include a free surface, realistic coasts and bottom features, river runoff, and atmospheric forcing. Because the models don't extend very far into deep water, they need additional information about deep-water currents or conditions at the shelf break.

The many different coastal models have many different goals, and many different implementations. Several of the models described above, including mom and rom, have been used to model coastal processes. But many other specialized models have also been developed. Heaps (1987), Lynch et al (1996), and Haidvogel and Beckman (1998) provide good overviews of the subject. Rather than look at a menu of models, let's look at two typical models.

Princeton Ocean Model developed by Blumberg and Mellor (1987, and Mellor, 1998) and is widely used for describing coastal currents. It includes thermodynamic processes, turbulent mixing, and the Boussinesq and hydrostatic approximations. The Coriolis parameter is allowed to vary using a beta-plane approximation. Because the model must include a wide range of depths, Blum- berg and Mellor used a vertical coordinate σ scaled by the depth of the water:

$$\sigma = \frac{z - \eta}{H + \eta} \qquad (15.1)$$

where $z = \eta(x, y, t)$ is the sea surface, and $z = -H(x, y)$ is the bottom.

Sub-grid turbulence is parameterized using a closure scheme proposed byMellor and Yamada (1982) whereby eddy diffusion coefficients vary with the size of the eddies producing the mixing and the shear of the flow.

The model is driven by wind stress and heat and water fluxes from meteorological models. The model uses known geostrophic, tidal, and Ekman currents at the outer boundary.

The model has been used to calculate the three-dimensional distribution of velocity, salinity, sea level, temperature, and turbulence for up to 30 days over a region roughly 100–1000 km on a side with grid spacing of 1–50 km.

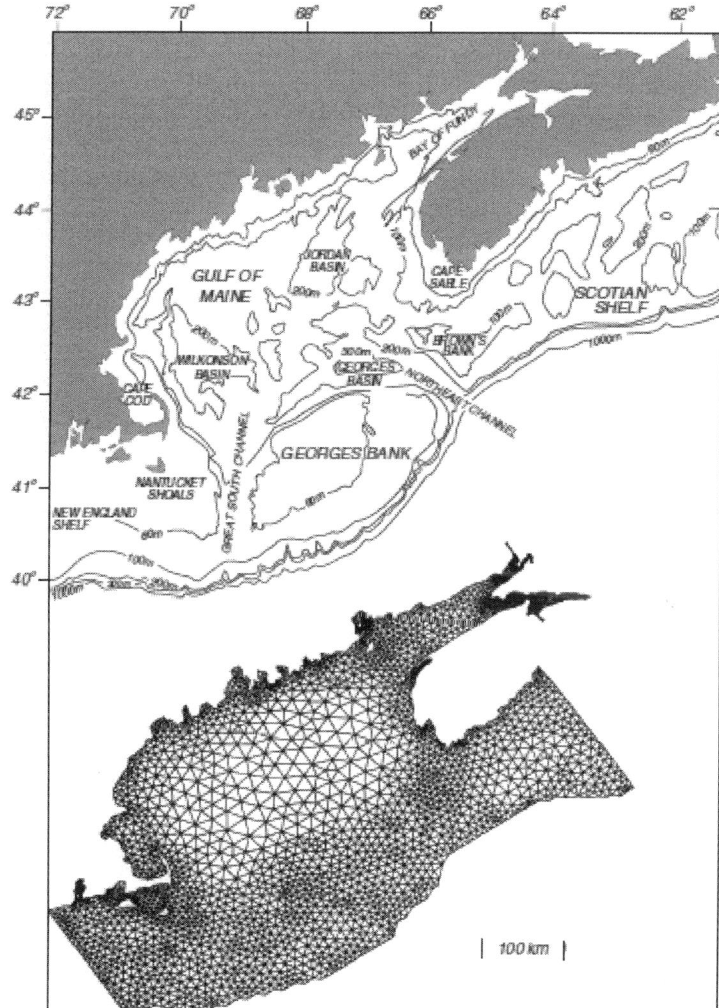

Figure 15.3 Top: Topographic map of the Gulf of Maine showing important features. Inset: Triangular, finite-element grid used to compute flow in the gulf. The size of the triangles varies with depth and rate of change of depth. After Lynch et al, (1996).

Dartmouth Gulf of Maine Model developed by Lynch et al (1996) is a 3-dimensional model of the circulation using a triangular, finite-element grid. The size of the triangles is proportional to both depth and the rate of change of depth. The triangles are small in regions where the bottom slopes are large and the depth is shallow, and they are large in deep water. The variable mesh is especially useful in coastal regions where the depth of water varies greatly. Thus the variable grid gives highest resolution where it is most needed.

The model uses roughly 13,000 triangles to cover the Gulf of Maine and nearby waters of the north Atlantic (figures 15.3). Minimum size of the elements is roughly one kilometer. The model has 10 to 40 horizontal layers. The vertical spacing of the layers is not uniform. Layers are closer together near the top and bottom and they are more widely spaced in the interior. Minimum spacing is roughly one meter in the bottom boundary layer.

The model integrates the three-dimensional, primitive equations in shallow-water form. The model has a simplified equation of state and a depth-averaged continuity equation, and it uses the hydrostatic and Boussinesq assumptions. Sub-grid mixing of momentum, heat and mass is parameterized using the Mellor and Yamada (1982) turbulence-closure scheme which gives vertical mixing coefficients that vary with stratification and velocity shear. Horizontal mixing coefficients were calculated from Smagorinski (1963). A carefully chosen, turbulent, eddy viscosity is used in the bottom boundary layer. The model is forced by wind, heating, and tidal forcing from the deep ocean.

The model is spun up from rest for a few days using a specified density field at all grid points, usually from a combination of CTD data plus historical data. This gives a velocity field consistent with the density field. The model is thenforced with local winds and heat fluxes to calculate the evolution of the density and velocity fields.

Comments on Coastal Models: Roed et al. (1995) examined the accuracy of coastal models by comparing the ability of five models, including Blumberg and Mellor's to describe the flow in typical cases. They found that the models produced very different results, but that after the models were adjusted, the differences were reduced. The differences were due to differences in vertical and horizontal mixing and spatial and temporal resolution.

Hackett et al. (1995) compared the ability of two of the five models to describe observed flow on the Norwegian shelf. They conclude that. . . both models are able to qualitatively generate many of the observed features of the flow, but neither is able to quantitatively reproduce de- tailed currents . . . [Differences] are primarily attributable to inadequate parameterizations of subgrid scale turbulent mixing, to lack of horizontal resolution and to imperfect initial and boundary conditions.

Storm-Surge Models: Storms coming ashore across wide, shallow, continental shelves drive large changes of sea level at the coast called storm surges (see§17.3 for a description of surges and processes influencing surges). The surgescan cause great damage to coasts and coastal structures. Intense storms in theBay of Bengal have killed hundreds of thousands in a few days in Bangladesh. Because surges are so important, government agencies in many countries have developed models to predict the changes of sea level and the extent of coastal flooding.

Calculating storm surges is not easy. Here are some reasons, in a rough order of importance.

- The distribution of wind over the ocean is not well known. Numerical weather models calculate wind speed at a constant pressure surface, storm-surge models need wind at a constant height of 10 m. Winds in bays and lagoons tend to be weaker than winds just offshore because nearby land distorts the airflow, and this is not included in the weather models.
- The shoreward extent of the model's domain changes with time. For example, if sea level rises, water will flood inland, and the boundary between water and sea moves inland with the water.
- The drag coefficient of wind on water is not well known for hurricane force winds.
- The drag coefficient of water on the seafloor is also not well known.
- The models must include waves and tides which influence sea level in shallow waters.
- Storm surge models must include the currents generated in a stratified, shallow sea by wind.

To reduce errors, models are tuned to give results that match conditions seen in past storms. Unfortunately, those past conditions are not well known. Changes in sea level and wind speed are rarely recorded accurately in storms except at a few, widely paced locations. Yet storm-surge heights can change by more than a meter over distances of tens of kilometers.

Despite these problems, models give very useful results. Let's look at the official NOAA model, and a new experimental model developed by the Corps of Engineers.

Sea, Lake, and Overland Surges Model slosh is used by NOAA for forecasting storm surges produced by hurricanes coming ashore along the Atlantic and Gulfcoasts of the United States (Jelesnianski, Chen, and Shaffer, 1992).

The model is the result of a lifetime of work by Chester Jelesnianski. In developing the model, Jelesnianski paid careful attention to the relative importance of errors in the model. He worked to reduce the largest errors, and ignored the smaller ones. For example, the distribution of winds in a hurricane is not well known, so it makes little sense to use a spatially varying drag coefficient for the wind. Thus, Jelesnianski used a constant drag coefficient in the air, and a constant eddy stress coefficient in the water.

slosh calculates water level from depth-integrated, quasi-linear, shallow-water equations. Thus it ignores stratification. It also ignores river inflow, rain, and tides. The latter may seem strange, but the model is designed for forecasting. The time of landfall cannot be forecast accurately, and hence the height of the tides is mostly unknown. Tides can be added to the calculated surge, but the nonlinear interaction of tides and surge is ignored.

The model is forced by idealized hurricane winds. It needs only atmospheric pressure at the center of the storm, the distance from the center to the area of maximum winds, the forecast storm track and speed along the track.

In preparation for hurricanes coming ashore near populated areas, the model has been adapted for 27 basins from Boston Harbor Massachusetts to Laguna Madre Texas. The model uses a fixed polar mesh. Mesh spacing begins with a fine mesh near the pole, which is located near the coastal city for which the model is adapted. The grid stretches continuously to a coarse mesh at distant boundaries of a large basin. Such a mesh gives high resolution in bays and near the coast where resolution is most needed. Using measured depths at sea and elevations on land, the model allows flooding of land, over topping of levees and dunes, and sub-grid flow through channels between offshore islands.

Sea level calculated from the model has been compared with heights measured by tide gauges for 13 storms, including Betsy: 1965, Camile: 1969, Donna: 1960, and Carla: 1961. The overall accuracy is ±20%.

Advanced Circulation Model ADCIRC is an experimental model for forecasting storm surges produced by hurricanes coming ashore along the Atlantic and Gulfcoasts of the United States (Graber et al, 2006). The model uses a finite-element grid, the Boussinesq approximation, quadratic bottom friction, and vertically integrated continuity and momentum equations for flow on a rotating earth. It can be run as either a two-dimensional, depth-integrated model, or as a three-dimensional model. Because waves contribute to storm surges, the modelincludes waves calculated from the wam third-geneation wave model (see §16.5).The model is forced by:

- High resolution winds and surface pressure obtained by combining weather forecasts from the NOAA National Weather Service and the National Hurricane Center along the official and alternate forecast storm tracks.
- Tides at the open-ocean boundaries of the model.
- Sea-surface height and currents at the open-ocean boundaries of the model.

The model successfully forecast the Hurricane Katrina storm surge, giving values in excess of 6.1 m near New Orleans.

15.5 ASSIMILATION MODELS

Many of the models I have described so far have output, such as currentvelocity or surface topography, constrained by oceanic observations of the variables they calculate. Such models are called *assimilation models.* In this section, I will consider how data can be assimilated into numerical models.

Let's begin with a primitive-equation, eddy-admitting numerical model used to calculate the position of the Gulf Stream. Let's assume that the model is driven with real-time surface winds from the ECMWF weather model. Using the model, we can calculate the position of the current and also the sea-surface topography associated with the current. We find that the position of the Gulf Stream wiggles offshore of Cape Hatteras due to instabilities, and the position calculated by the model is just one of many possible positions for the same wind forcing. Which position is correct, that is, what is the position of the current today? We know, from satellite altimetry, the position of the current at a few points a few days ago. Can we use this information to calculate the current's position today? How do we assimilate this information into the model?

Many different approaches are being explored (Malanotte-Rizzoli, 1996). Roger Daley (1991) gives a complete description of how data are used with atmospheric models. Andrew Bennet (1992) and Carl Wunsch (1996) describe oceanic applications.

The different approaches are necessary because assimilation of data into models is not easy.

- Data assimilation is an *inverse problem*: A finite number of observations are used to estimate a continuous field—a function, which has an infinite number of points. The calculated fields, the solution to the inverse problem, are completely under-determined. There are many fields that fit the observations and the model precisely, and the solutions are not unique. In our example, the position of the Gulf Stream is a function. We may not need an infinite number of values to specify the position of the stream ifwe assume the position is somewhat smooth in space. But we certainly need hundreds of values along the stream's axis. Yet, we have only a few satellite points to constrain the position of the Stream.To learn more about inverse problems and their solution, read Parker (1994) who gives a very good introduction based on geophysical examples.
- Ocean dynamics are non-linear, while most methods for calculating solutions to inverse problems depend on linear approximations. For example the position of the Gulf Stream is a very nonlinear function of the forcing by wind and heat fluxes over the north Atlantic.

- Both the model and the data are incomplete and both have errors. Forexample, we have altimeter measurements only along the tracks such as those shown in figure 2.6, and the measurements have errors of ±2 cm.
- Most data available for assimilation into data comes from the surface, such as AVHRR and altimeter data. Surface data obviously constrain the surface geostrophic velocity, and surface velocity is related to deeper velocities. The trick is to couple the surface observations to deeper currents.

While various techniques are used to constrain numerical models in oceanography, perhaps the most practical are techniques borrowed from meteorology.

Most major ocean currents have dynamics which are significantly nonlinear. This precludes the ready development of inverse methods. . . Accordingly, most attempts to combine ocean models and measurements have followed the practice in operational meteorology: measurements are used to prepare initial conditions for the model, which is then integrated forward in time until further measurements are available. The model is thereupon re-initialized. Such a strategy may be described as sequential.—Bennet (1992).

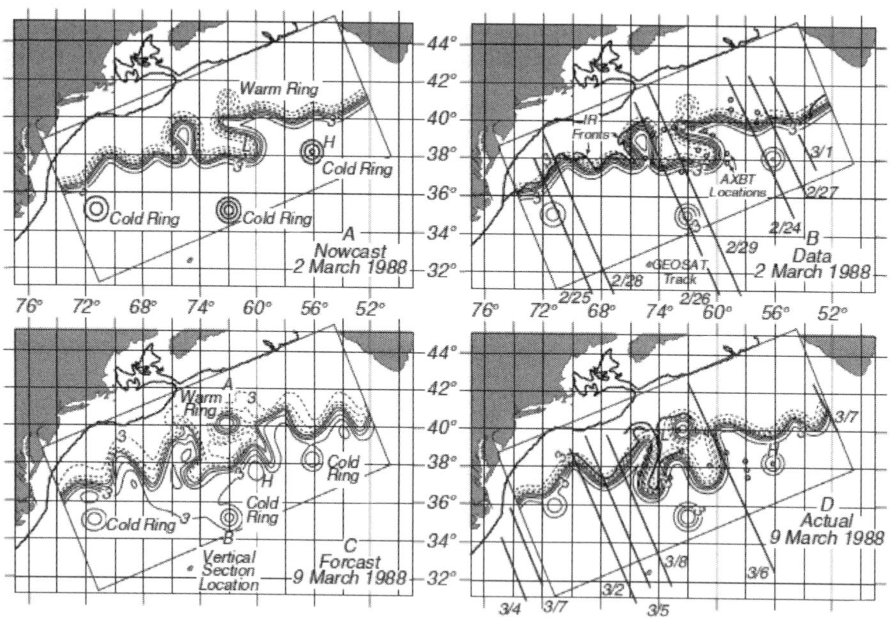

Figure 15.4 Output from the Harvard Open-Ocean Model: A the initial state of the model, the analysis, and B Data used to produce the analysis for 2 March 1988. C The forecast for 9 March 1988. D The analysis for 9 March. Although the Gulf Stream changed substantially in one week, the model forecasts the changes well. After Robinson et al. (1989).

Let's see how Professor Allan Robinson and colleagues at Harvard University used sequential estimation techniques to make the first forecasts of the Gulf Stream using a very simple model.

The Harvard Open-Ocean Model was an eddy-admitting, quasigeostropic model of the Gulf Stream east of Cape Hatteras (Robinson et al. 1989). It hadsix levels in the vertical, 15 km resolution, and one-hour time steps. It used a filter to smooth high-frequency variability and to damp grid-scale variability.

By *quasi-geostrophic* we mean that the flow field is close to geostrophic balance. The equations of motion include the acceleration terms D/Dt, whereD/Dt is the substantial derivative and t is time. The flow can be stratified, but there is no change in density due to heat fluxes or vertical mixing. Thus the quasi-geostrophic equations are simpler than the primitive equations, and they could be integrated much faster. Cushman-Roisin (1994: 204) gives a good description of the development of quasi-geostrophic equations of motion.

The model reproduces the important features of the Gulf Stream and it's extension, including meanders, cold- and warm-core rings, the interaction of rings with the stream, and baroclinic instability (figure 15.4). Because themodel was designed to forecast the dynamics of the Gulf Stream, it must be constrained by oceanic measurements:

- Data provide the initial conditions for the model. Satellite measurements of sea-surface temperature from the AVHRR and topography from an altimeter are used to determine the location of features in the region. Expendable bathythermograph, AXBT measurements of subsurface temperature, and historical measurements of internal density are also used. The features are represented by an analytic functions in the model.
- The data are introduced into the numerical model, which interpolates andsmoothes the data to produce the best estimate of the initial fields of density and velocity. The resulting fields are called an *analysis*.
- The model is integrated forward for one week, when new data are available, to produce a forecast.
- Finally, the new data are introduced into the model as in the first step above, and the processes is repeated.

The model made useful, one-week forecasts of the Gulf Stream region. Much more advanced models with much higher resolution are now being used to make global forecasts of ocean currents up to one month in advance in support of the Global Ocean Data Assimilation Experiment GODAE that started in 2003. The goal of GODAE is produce routine oceanic forecasts similar to todays weather forecasts.

An example of a GODAE model is the global US Navy Layered Ocean Model. It is a primitive equation model with $1/32^{\circ}$resolution in the horizontal andseven layers in the vertical. It assimilates altimeter data from Jason, Geosat Follow-on (GFO), and ers-2 satellites and sea-surface temperature from AVHRR on NOAA satellites.

The model is forced with winds and heat fluxes for up to five days in the future using output from the Navy Operational Global Atmospheric Prediction System. Beyond five days, seasonal mean winds and fluxes are used. The model is run daily (figure 15.5) and produces forecasts for up to one month in the future. The model has useful skill out to about 20 days.

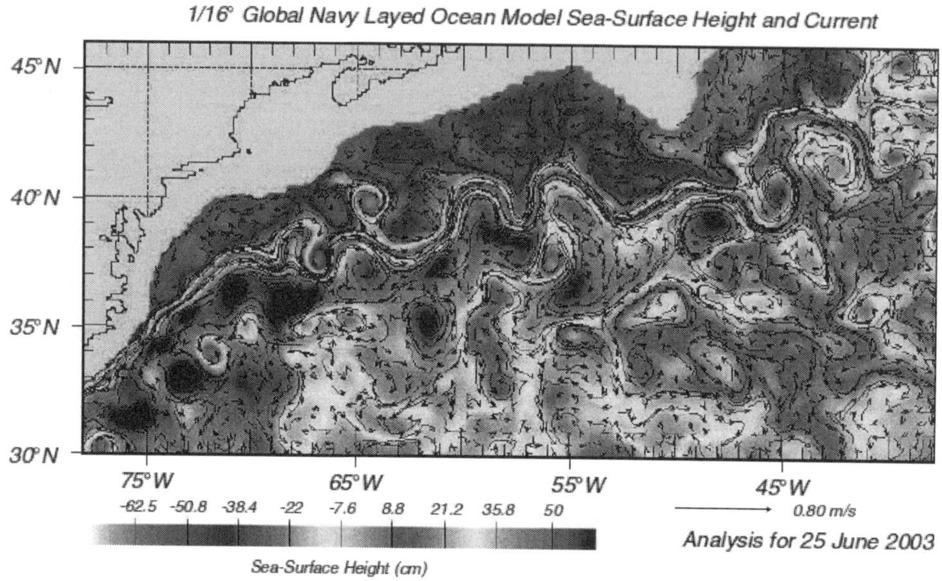

Figure 15.5 Analysis of the Gulf Stream region from the Navy Layered Ocean Model. From the U.S. Naval Oceanographic Office.

A group of French laboratories and agencies operates a similar operational forecasting system, Mercator, based on assimilation of altimeter measurements of sea-surface height, satellite measurements of sea-surface temperature, and internal density fields in the ocean, and currents at 1000 m from thousands ofArgo floats. Their model has 1/15° resolution in the Atlantic and 2° globally.

15.6 COUPLED NUMERICAL MODELS OF THE ATMOSPHERE AND OCEAN

Coupled numerical models of the atmosphere and ocean are used to study the climate, its variability, and its response to external forcing. The most important use of the models has been to study how earth's climate might respond to a doubling of CO2 in the atmosphere. Much of the literature on climate change is based on studies using such models. Other important uses of coupled models include studies of El Niño and the meridional overturning circulation. The former varies over a few years, the latter varies over a few centuries.

Development of the coupled models tends to be coordinated through the World Climate Research Program of the World Meteorological Organization WCRP/WMO, and recent progress is summarized in Chapter 8 of the *ClimateChange 2001: The Scientific Basis* report by the Intergovernmental Panel on Climate Change (ipcc, 2007).

Many coupled ocean and atmosphere models have been developed. Some include only physical processes in the ocean, atmosphere, and the ice-covered polar seas. Others add the influence of land and biological activity in the ocean. Let's look at the oceanic components of a few models.

Climate System Model: The Climate System Model developed by the National Center for Atmospheric Research NCAR includes physical and biogeochemical influence on the climate system (Boville and Gent, 1998). It has atmosphere, ocean, land-surface, and sea-ice components coupled by fluxes between components. The atmospheric component is the NCAR Community Climate Model, the oceanic component is a modified version of the Princeton Modular Ocean Model, using the Gent and McWilliams (1990) scheme for parameterizing mesoscale eddies. Resolution is approximately $2^{\circ} \times 2^{\circ}$ with 45 vertical levels in the ocean.

The model has been spun up and integrated for 300 years, the results arerealistic, and there is no need for a flux adjustment. (See the special issue of*Journal of Climate*, June 1998).

Princeton Coupled Model: The model consists of an atmospheric model with a horizontal resolution of 7.5° longitude by 4.5° latitude and 9 levels in the vertical, an ocean model with a horizontal resolution of 4° and 12 levels in the vertical, and a land-surface model. The ocean and atmosphere are coupled through heat, water, and momentum fluxes. Land and ocean are coupled through river runoff. And land and atmosphere are coupled through water and heat fluxes.

Hadley Center Model: This is an atmosphere-ocean-ice model that minimizes the need for flux adjustments (Johns et al, 1997). The ocean component is based on the Bryan-Cox primitive equation model, with realistic bottom features,vertical mixing coefficients from Pacanowski and Philander (1981). Both the ocean and the atmospheric component have a horizontal resolution of 96×73 grid points, the ocean has 20 levels in the vertical.

In contrast to most coupled models, this one is spun up as a coupled system with flux adjustments during spin up to keep sea surface temperature and salinity close to observed mean values. The coupled model was integrated from rest using Levitus values for temperature and salinity for September. The initial integration was from 1850 to 1940. The model was then integrated for another

1000 years. No flux adjustment was necessary after the initial 140-year integration because drift of global-averaged air temperature was ≤ 0.016 K/century.

Comments on Accuracy of Coupled Models: Models of the coupled, land- air-ice-ocean climate system must simulate hundreds to thousands of years. Yet,It will be very hard to establish an integration framework, particularly on a global scale, as present capabilities for modelling the earth system are rather limited. A dual approach is planned. On the one hand, the relatively conventional approach of improving coupled atmosphere-ocean- land-ice models will be pursued. Ingenuity aside, the computational demands are extreme, as is borne out by the Earth System Simulator — 640linkedsupercomputers providing 40 teraflops [10^{12} floating-point operations per second] and a cooling system from hell under one roof — to be built in Japan by 2003.— Newton, 1999.

Because models must be simplified to run on existing computers, the models must be simpler than models that simulate flow for a few years (WCRP, 1995). In addition, the coupled model must be integrated for many years for the ocean and atmosphere to approach equilibrium. As the integration proceeds, the coupled system tends to drift away from reality due to errors in calculating fluxes of heat and momentum between the ocean and atmosphere. For example, very small errors in precipitation over the Antarctic Circumpolar Current leads to small changes the salinity of the current, which leads to large changes in deep convection in the Weddell Sea, which greatly influences the volume of deep watermasses.

Some modelers allow the system to drift, others adjust sea-surface temperature and the calculated fluxes between the ocean and atmosphere. Returning to the example, the flux of fresh water in the circumpolar current could be adjusted to keep salinity close to the observed value in the current. There is no good scientific basis for the adjustments except the desire to produce a "good" coupled model. Hence, the adjustments are ad hoc and controversial. Such adjustments are called *flux adjustments* or *flux corrections.*

Fortunately, as models have improved, the need for adjustment or the magnitude of the adjustment has been reduced. For example, using the Gent- McWilliams scheme for mixing along constant-density surfaces in a coupled ocean-atmosphere model greatly reduced climate drift in a coupled ocean-atmosphere model because the mixing scheme reduced deep convection in the Antarctic Circumpolar Current and elsewhere (Hirst, O'Farrell, and Gordon, 2000).

Grassl (2000) lists four capabilities of a credible coupled general circulation model:

- "Adequate representation of the present climate.
- "Reproduction (within typical inter annual and decades time-scale climate variability) of the changes since the start of the instrumental record for a given history of external forcing;

- "Reproduction of a different climate episode in the past as derived from paleoclimate records for given estimates of the history of external forcing; and
- "Successful simulation of the gross features of an abrupt climate change event from the past."

McAvaney et al. (2001) compared the oceanic component of twenty-four coupled models, including models with and without flux adjustments. They found substantial differences among the models. For example, only five models calculated a meridional overturning circulation within 10% the observed value of 20 Sv. Some had values as low as 3 Sv, others had values as large as 36 Sv. Most models could not calculate a realistic transport for the Antarctic Circumpolar Current.

Grassl (2000) found that many coupled climate models, including models with and without flux adjustment, meet the first criterion. Some models meet the second criterion (Smith et al. 2002, Stott et al. 2000), but external solar forcing is still not well known and more work is needed. And a few models are starting to reproduce some aspects of the warm event of 6,000 years ago.

But how useful are these models in making projections of future climate? Opinion is polarized. At one extreme are those who take the model results as gospel. At the other are those who denigrate results simply because they distrust models, or on the grounds that the model performance is obviously wrong in some respects or that a process is not adequately included. The truth lies in between. All models are of course wrong because, by design, they give a simplified view of the system being modelled. Nevertheless, many—but not all—models are very useful.—Trenberth, 1997.

15.7 CONCISE SUMMARY

- Numerical models are used to simulate oceanic flows with realistic and useful results. The most recent models include heat fluxes through the surface, wind forcing, mesoscale eddies, realistic coasts and sea-floor features, and more than 20 levels in the vertical.
- Recent models are now so good, with resolution near 0.1°, that they show previously unknown aspects of the ocean circulation,
- Numerical models are not perfect. They solve discrete equations, which are not the same as the equations of motion described in earlier chapters. And,
- Numerical models cannot reproduce all turbulence of the ocean because the grid points are tens to hundreds of kilometers apart. The influence of turbulent motion over smaller distances must be calculated from theory, and this introduces errors.

- Numerical models can be forced by real-time oceanographic data from ships and satellites to produce forecasts of oceanic conditions, including El Niño in the Pacific, and the position of the Gulf Stream in the Atlantic.
- Coupled ocean-atmosphere models have much coarser spatial resolution so that they can be integrated for hundreds of years to simulate the natural variability of the climate system and its response to increased CO_2 in the atmosphere.

16 INTRODUCTION TO OCEAN WAVES

Looking out to sea from the shore, we can see waves on the sea surface. Looking carefully, we notice the waves are undulations of the sea surface with a height of around a meter, where height is the vertical distance between the bottom of a trough and the top of a nearby crest. The wavelength, which we might take to be the distance between prominent crests, is around 50-100 meters. Watching the waves for a few minutes, we notice that wave height and wave length are not constant. The heights vary randomly in time and space, and the statistical properties of the waves, such as the mean height averaged for a few hundred waves, change from day to day. These prominent offshore waves are generated by wind. Sometimes the local wind generates the waves, other times distant storms generate waves which ultimately reach the coast. For example, waves breaking on the Southern California coast on a summer day may come from vast storms offshore of Antarctica 10,000 km away.

If we watch closely for a long time, we notice that sea level changes from hour to hour. Over a period of a day, sea level increases and decreases relative to a point on the shore by about a meter. The slow rise and fall of sea level is due to the tides, another type of wave on the sea surface. Tides have wavelengths of thousands of kilometers, and they are generated by the slow, very small changes in gravity due to the motion of the sun and the moon relative to earth.

In this chapter you will learn how to describe ocean-surface waves quantitatively. In the next chapter I will describe tides and waves along coasts.

16.1. LINEAR THEORY OF OCEAN SURFACE WAVES

Surface waves are inherently nonlinear: The solution of the equations of motion depends on the surface boundary conditions, but the surface boundary conditions are the waves we wish to calculate. How can we proceed?

We begin by assuming that the amplitude of waves on the water surface is infinitely small so the surface is almost exactly a plane. To simplify the mathematics, we can also assume that the flow is 2-dimensional with waves traveling in the x-direction. We also assume that the Coriolis force and viscosity can be neglected. If we retain rotation, we get Kelvin waves discussed in §14.2.

With these assumptions, the sea-surface elevation ζ of a wave traveling in the x direction is:

$$\zeta = a \sin(k\,x - \omega\,t) \quad (16.1)$$

with

$$\omega = 2\pi f = \frac{2\pi}{T}; \qquad k = \frac{2\pi}{L} \quad (16.2)$$

where ω is wave frequency in radians per second, f is the wave frequency in Hertz (Hz), k is wave number, T is wave period, L is wave length, and where we assume, as stated above, that $k_a = O(0)$.

The *wave period* T is the time it takes two successive wave crests or troughs to pass a fixed point. The *wave length* L is the distance between two successive wave crests or troughs at a fixed time.

Dispersion Relation: Wave frequency ω is related to wave number k by the*dispersion relation* (Lamb, 1945 §228):

$$\omega^2 = g\,k \tanh(kd) \quad (16.3)$$

where d is the water depth and g is the acceleration of gravity.

Two approximations are especially useful.

1. *Deep-water approximation* is valid if the water depth d is much greater than the wave length L. In this case, $d \gg L$, $kd \gg 1$, and tanh $(kd) = 1$.
2. *Shallow-water approximation* is valid if the water depth is much less than a wavelength. In this case, $d \ll L$, $kd \ll 1$, and tanh $(kd) = kd$.

For these two limits of water depth compared with wavelength the dispersion relation reduces to:

$\omega^2 = g\,k$ Deep-water dispersion relation (16.4)
$d > L/4$

$\omega^2 = g\,k^2\,d$ Shallow-water dispersion relation (16.5)
$d < L/11$

The stated limits for d/L give a dispersion relation accurate within 10%. Because many wave properties can be measured with accuracies of 5–10%, the approximations are useful for calculating wave properties. Later we will learn to calculate wave properties as the waves propagate from deep to shallow water.

Phase Velocity: The phase velocity c is the speed at which a particular phase of the wave propagates, for example, the speed of propagation of the wave crest. In one wave period T the crest advances one wave length L and the phase speed is $c = L/T = \omega/k$. Thus, the definition of phase speed is:

$$c \equiv \frac{\omega}{k} \quad (16.6)$$

The direction of propagation is perpendicular to the wave crest and toward the positive *x* direction.

The deep- and shallow-water approximations for the dispersion relation give:

$$c = \sqrt{\frac{g}{k}} = \frac{g}{\omega} \text{ Deep-water phase velocity (16.7)}$$

$$c = \sqrt{g\,d} \text{ Shallow-water phase velocity (16.8)}$$

The approximations are accurate to about 5% for limits stated in (16.4, 16.5). In deep water, the phase speed depends on wave length or wave frequency. Longer waves travel faster. Thus, deep-water waves are said to be dispersive.

In shallow water, the phase speed is independent of the wave; it depends only on the depth of the water. Shallow-water waves are non-dispersive.

Group Velocity: The concept of group velocity *cg* is fundamental for understanding the propagation of linear and nonlinear waves. First, it is the velocity at which a group of waves travels across the ocean. More importantly, it is alsothe propagation velocity of wave energy. Whitham (1974, §1.3 and §11.6) gives a clear derivation of the concept and the fundamental equation (16.9).

The definition of group velocity in two dimensions is:

$$c_g \equiv \frac{\partial \omega}{\partial k} \quad (16.9)$$

Using the approximations for the dispersion relation:

$$c_g = \frac{g}{2\omega} = \frac{c}{2} \text{ Deep-water group velocity (16.10)}$$

$$c_g = \sqrt{g\,d} = c \text{ Shallow-water group velocity (16.11)}$$

For ocean-surface waves, the direction of propagation is perpendicular to the wave crests in the positive *x* direction. In the more general case of other types of waves, such as Kelvin and Rossby waves that we met in §14.2, the group velocity is not necessarily in the direction perpendicular to wave crests.

Notice that a group of deep-water waves moves at half the phase speed of the waves making up the group. How can this happen? If we could watch closely a group of waves crossing the sea, we would see waves crests appear at the back of the wave train, move through the train, and disappear at the leading edge of the group. Each wave crest moves at twice the speed of the group.

Do real ocean waves move in groups governed by the dispersion relation? Yes. Walter Munk and colleagues (1963) in a remarkable series of experiments in the

1960s showed that ocean waves propagating over great distances are dispersive, and that the dispersion could be used to track storms. They recorded waves for many days using an array of three pressure gauges just offshore of San Clemente Island, 60 miles due west of San Diego, California. Wave spectra were calculated for each day's data. (The concept of a spectra is discussed below.) From the spectra, the amplitudes and frequencies of the low-frequency waves and the propagation direction of the waves were calculated. Finally, they plotted contours of wave energy on a frequency-time diagram (figure 16.1).

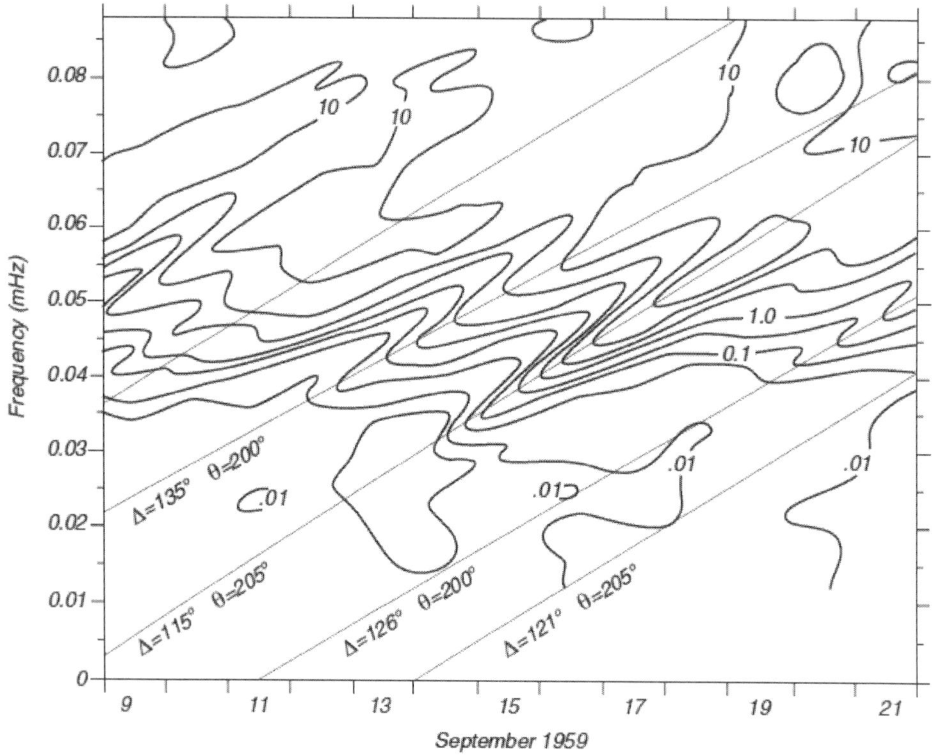

Figure 16.1 Contours of wave energy on a frequency-time plot calculated from spectra of waves measured by pressure gauges offshore of southern California. The ridges of high wave energy show the arrival of dispersed wave trains from distant storms. The slope of the ridge is inversely proportional to distance to the storm. Δ is distance in degrees, θ is direction of arrival of waves at California. After Munk et al. (1963).

To understand the figure, consider a distant storm that produces waves of many frequencies. The lowest-frequency waves (smallest ω) travel the fastest (16.11), and they arrive before other, higher-frequency waves. The further away the storm, the longer the delay between arrivals of waves of different frequencies. The ridges of high wave energy seen in the figure are produced by individual storms. The slope of the ridge gives the distance to the storm in degrees Δ along a great circle;

and the phase information from the array gives the angle to the storm θ. The two angles give the storm's location relative to San Clemente.

Thus waves arriving from 15 to 18 September produce a ridge indicating the storm was 115° away atan angle of 205° which is south of new Zealand near Antarctica.

The locations of the storms producing the waves recorded from June through October 1959 were compared with the location of storms plotted on weather maps and in most cases the two agreed well.

Wave Energy: Wave energy E in Joules per square meter is related to the variance of sea-surface displacement ζ by:

$$E = \rho_w g \left\langle \zeta^2 \right\rangle \quad (16.12)$$

where ρw is water density, g is gravity, and the brackets denote a time or space average.

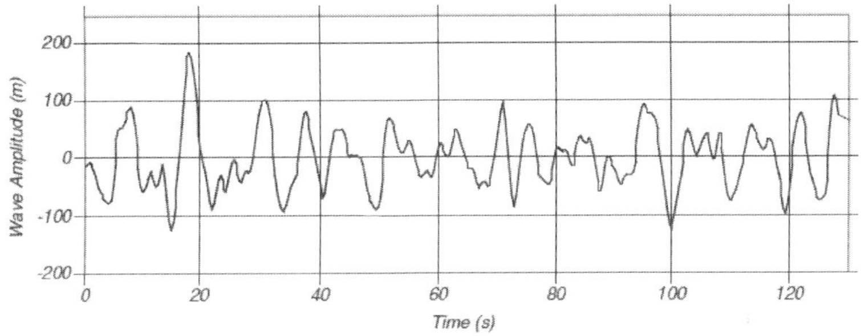

Figure 16.2 A short record of wave amplitude measured by a wavc buoy in the north Atlantic.

Significant Wave Height: What do we mean by wave height? If we look at a wind-driven sea, we see waves of various heights. Some are much larger than most, others are much smaller (figure 16.2). A practical definition that is often used is the height of the highest 1/3 of the waves, $H_1/3$. The height is computed as follows: measure wave height for a few minutes, pick out say 120 wave crests and record their heights. Pick the 40 largest waves and calculate the average height of the 40 values. This is $H_1/3$ for the wave record.

The concept of significant wave height was developed during the World WarII as part of a project to forecast ocean wave heights and periods. Wiegel (1964:p. 198) reports that work at the Scripps Institution of Oceanography showed. . . wave height estimated by observers corresponds to the average of the highest 20 to 40 per cent of waves. . . Originally, the term significant wave height was attached to the average of these observations, the highest 30 per cent of the waves, but has evolved to become the average of the highest one-third of the waves, (designated HS or $H1/3$)

More recently, significant wave height is calculated from measured wave displacement. If the sea contains a narrow range of wave frequencies, $H_1/3$ is related to the standard deviation of sea-surface displacement (NAS , 1963: 22; Hoffman and Karst, 1975)

$$H_{1/3} = 4\left\langle \zeta^2 \right\rangle^{1/2} \quad (16.13)$$

where $\zeta^2\ ^1/2$ is the standard deviation of surface displacement. This relationship is much more useful, and it is now the accepted way to calculate wave height from wave measurements.

16.2 NONLINEAR WAVES

We derived the properties of an ocean surface wave assuming the wave amplitude was infinitely small $ka = O(0)$. If $ka \ll 1$ but not infinitely small, the wave properties can be expanded in a power series of ka (Stokes, 1847). He calculated the properties of a wave of finite amplitude and found:

$$\zeta = a\cos(kx - \omega t) + \frac{1}{2}ka^2\cos 2(kx - \omega t) + \frac{3}{8}k^2a^3\cos 3(kx - \omega t) + \cdots \quad (16.14)$$

The phases of the components for the Fourier series expansion of ζ in (16.14) are such that non-linear waves have sharpened crests and flattened troughs. The maximum amplitude of the Stokes wave is $amax = 0.07L$ ($ka = 0.44$). Suchsteep waves in deep water are called Stokes waves (See also Lamb, 1945, §250).

Knowledge of non-linear waves came slowly until Hasselmann (1961, 1963a,1963b, 1966), using the tools of high-energy particle physics, worked out to 6th order the interactions of three or more waves on the sea surface. He, Phillips (1960), and Longuet-Higgins and Phillips (1962) showed that n free waves on the sea surface can interact to produce another free wave only if the frequencies and wave numbers of the interacting waves sum to zero:

$$\omega_1 \pm \omega_2 \pm \omega_3 \pm \cdots \omega_n = 0 \quad (16.15a)$$

$$\mathbf{k}_1 \pm \mathbf{k}_2 \pm \mathbf{k}_3 \pm \cdots \mathbf{k}_n = 0 \quad (16.15b)$$

$$\omega_i^2 = g\,k_i \quad (16.15c)$$

where we allow waves to travel in any direction, and k_i is the vector wave number giving wave length and direction. (16.15) are general requirements for any interacting waves. The fewest number of waves that meet the conditions of (16.15) are three waves which interact to produce a fourth. The interaction is weak; waves must interact for hundreds of wave lengths and periods to produce a fourth wave with amplitude comparable to the interacting waves. The Stokes wave does not meet the criteria of (16.15) and the wave components are not free waves; the higher harmonics are bound to the primary wave.

Wave Momentum: The concept of wave momentum has caused considerable confusion (McIntyre, 1981). In general, waves do not have momentum, a mass flux, but they do have a momentum flux. This is true for waves on the sea surface. Ursell (1950) showed that ocean swell on a rotating earth has no mass transport. His proof seems to contradict the usual textbook discussions of steep, non-linear waves such as Stokes waves. Water particles in a Stokes wave move along paths that are nearly circular, but the paths fail to close, and the particles move slowly in the direction of wave propagation. This is a mass transport, and the phenomena is called Stokes drift. But the forward transport near the surface is balanced by an equal transport in the opposite direction at depth, and there is no net mass flux.

16.3 WAVES AND THE CONCEPT OF A WAVE SPECTRUM

If we look out to sea, we notice that waves on the sea surface are not sinusoids. The surface appears to be composed of random waves of various lengths and periods. How can we describe this surface? The answer is, Not very easily. We can however, with some simplifications, come close to describing the surface. The simplifications lead to the concept of the spectrum of ocean waves. The spectrum gives the distribution of wave energy among different wave frequencies or wave lengths on the sea surface.

The concept of a spectrum is based on work by Joseph Fourier (1768–1830),who showed that almost any function $\zeta(t)$ (or $\zeta(x)$ if you like), can be represented over the interval $-T/2 \le t \le T/2$ as the sum of an infinite series of sineand cosine functions with harmonic wave frequencies:

$$\zeta(t) = \frac{a_0}{2} + \sum_{n=1}^{\infty}(a_n \cos 2\pi nft + b_n \sin 2\pi nft) \quad (16.16)$$

where

$$a_n = \frac{2}{T}\int_{-T/2}^{T/2} \zeta(t) \cos 2\pi nft \, dt, \qquad (n = 0, 1, 2, \ldots) \quad (16.17a)$$

$$b_n = \frac{2}{T}\int_{-T/2}^{T/2} \zeta(t) \sin 2\pi nft \, dt, \qquad (n = 0, 1, 2, \ldots) \quad (16.17b)$$

$f = 1/T$ is the fundamental frequency, and nf are harmonics of the fundamental frequency. This form of $\zeta(t)$ is called a *Fourier series* (Bracewell, 1986: 204; Whittaker and Watson, 1963: §9.1). Notice that a_0 is the mean value of $\zeta(t)$over the interval.

Equations (16.18 and 16.19) can be simplified using

$$\exp(2\pi inft) = \cos(2\pi nft) + i\sin(2\pi nft) \quad (16.18)$$

where $i = \sqrt{-1}$. Equations (16.18 and 16.19) then become:

$$\zeta(t) = \sum_{n=-\infty}^{\infty} Z_n \exp^{i2\pi nft} \quad (16.19)$$

where

$$Z_n = \frac{1}{T}\int_{-T/2}^{T/2} \zeta(t)\exp^{-i2\pi nft}\,dt, \qquad (n = 0, 1, 2, \ldots) \quad (16.20)$$

Zn is called the *Fourier transform* of ζ(*t*).

The spectrum *S*(*f*) of ζ(*t*) is:

$$S(nf) = Z_n Z_n^* \quad (16.21)$$

where Z* is the complex conjugate of Z. We will use these forms for the Fourier series and spectra when we describing the computation of ocean wave spectra. We can expand the idea of a Fourier series to include series that represent surfaces ζ(*x*, *y*) using similar techniques. Thus, any surface can be represented as an infinite series of sine and cosine functions oriented in all possible directions.

Now, let's apply these ideas to the sea surface. Suppose for a moment that the sea surface were frozen in time. Using the Fourier expansion, the frozen surface can be represented as an infinite series of sine and cosine functions of different wave numbers oriented in all possible directions. If we unfreeze the surface and let it move, we can represent the sea surface as an infinite series of sine and cosine functions of different wave lengths moving in all directions. Because wave lengths and wave frequencies are related through the dispersion relation, we can also represent the sea surface as an infinite sum of sine and cosine functions of different frequencies moving in all directions.

Note in our discussion of Fourier series that we assume the coefficients (*an, bn, Zn*) are constant. For times of perhaps an hour, and distances of per- haps tens of kilometers, the waves on the sea surface are sufficiently fixed that the assumption is true. Furthermore, non-linear interactions among waves are very weak. Therefore, we can represent a local sea surface by a linear super- position of real, sine waves having many different wave lengths or frequencies and different phases traveling in many different directions. The Fourier series in not just a convenient mathematical expression, it states that the sea surface is really, truly composed of sine waves, each one propagating according to the equations I wrote down in §16.1.

The concept of the sea surface being composed of independent waves can be carried further. Suppose I throw a rock into a calm ocean, making a big splash. According to Fourier, the splash can be represented as a superposition of cosine waves all of nearly zero phase so the waves add up to a big splash at the origin. Each individual Fourier wave begins to travel away from the splash. The longest waves travel fastest, and eventually, far from the splash, the sea consists of a dispersed train of waves with the longest waves further from the splash and the shortest waves closest. This is exactly what we see in figure 16.1. The storm makes the splash, and the waves disperse as seen in the figure.

Sampling the Sea Surface: Calculating the Fourier series that represents the sea surface is perhaps impossible. It requires that we measure the height of the sea surface $\zeta(x, y, t)$ everywhere in an area perhaps ten kilometers on a side for perhaps an hour. So, let's simplify. Suppose we install a wave staff somewhere in the ocean and record the height of the sea surface as a function of time $\zeta(t)$. We would obtain a record like that in figure 16.2. All waves on the sea surface will be measured, but we will know nothing about the direction of the waves. This is a much more practical measurement, and it will give the frequency spectrum of the waves on the sea surface.

Working with a trace of wave height on say a piece of paper is difficult, so let's digitize the output of the wave staff to obtain

$$\zeta_j \equiv \zeta(t_j), \qquad t_j \equiv j\Delta \qquad j = 0, 1, 2, \cdots, N-1 \qquad (16.22)$$

where Δ is the time interval between the samples, and N is the total number of samples. The length T of the record is $T = N\,\Delta$. Figure 16.3 shows the first 20 seconds of wave height from figure 16.2 digitized at intervals of $\Delta = 0.32$ s.

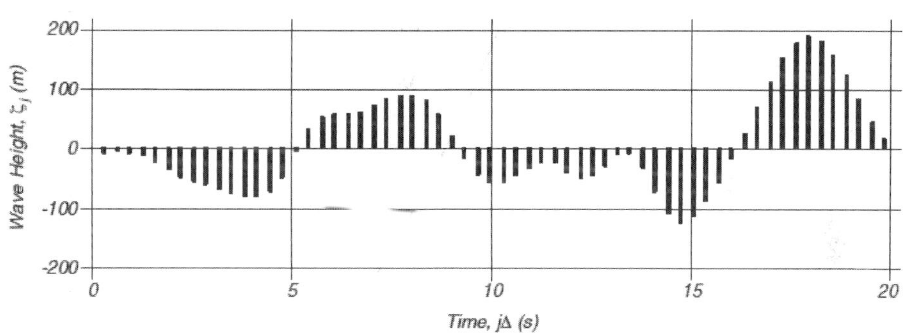

Figure 16.3 The first 20 seconds of digitized data from figure 16.2. Δ = 0.32 s.

Notice that j is not the same as (t). We have absolutely no information about the height of the sea surface between samples. Thus we have converted from an infinite set of numbers which describes (t) to a finite set of numbers which describe j. By converting from a continuous function to a digitized function, we have given up an infinite amount of information about the surface. The sampling interval Δ defines a *Nyquist critical frequency* (Press et al, 1992: 494)

$$Ny \equiv 1/(2\Delta) \quad (16.23)$$

The Nyquist critical frequency is important for two related, but dis- tinct, reasons. One is good news, the other is bad news. First the good news. It is the remarkable fact known as the *sampling theorem*: If a continuous function $\zeta(t)$, sampled at an interval Δ, happens to be *bandwidth limited* to frequencies smaller in magnitude than Ny, i.e., if $S(nf) = 0$ for all $|nf| \geq Ny$, then the function $\zeta(t)$ is *completely determined* by its samples ζj . . . This is a remarkable theorem for many reasons, among them that it shows that the "information content" of a bandwidth limited function is, in some sense, infinitely smaller than that of a general continuous function. . .

Now the bad news. The bad news concerns the effect of sampling a continuous function that is *not* bandwidth limited to less than the Nyquist critical frequency. In that case, it turns out that all of the power spectral density that lies outside the frequency range $-Ny \leq nf \leq Ny$ is spuriously moved into that range. This phenomenon is called *aliasing*. Any frequency component outside of the range ($-Ny$, Ny) is *aliased* (falsely translated) into that range by the very act of discrete sampling. . . There is little that you can do to remove aliased power once you have discretely sampled a signal. The way to overcome aliasing is to (i) know the natural bandwidth limit of the signal — or else enforce a known limit by analog filtering of the continuous signal, and then (ii) sample at a rate sufficiently rapid to give at least two points per cycle of the highest frequency present.

—Press et al 1992, but with notation changed to our notation.

Figure 16.4 illustrates the aliasing problem. Notice how a high frequency signal is aliased into a lower frequency if the higher frequency is above the critical frequency. Fortunately, we can can easily avoid the problem: (i) use instruments that do not respond to very short, high frequency waves if we are interested in the bigger waves; and (ii) chose Δt small enough that we lose little useful information. In the example shown in figure 16.3, there are no waves in the signal to be digitized with frequencies higher than $Ny = 1.5625$ Hz.

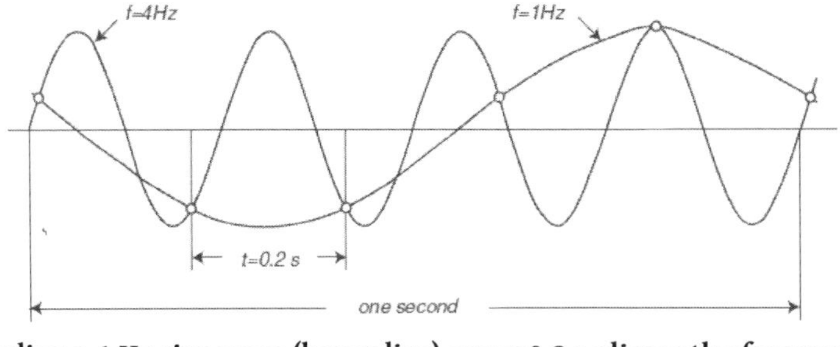

Figure 16.4 Sampling a 4 Hz sine wave (heavy line) every 0.2 s aliases the frequency to 1 Hz (light line). The critical frequency is 1/(2 × 0.2 s) = 2.5 Hz, which is less than 4 Hz.

Let's summarize. Digitized signals from a wave staff cannot be used to study waves with frequencies above the Nyquist critical frequency. Nor can the signal be used to study waves with frequencies less than the fundamental frequency determined by the duration T of the wave record. The digitized wave record contains information about waves in the frequency range:

$$\frac{1}{T} < f < \frac{1}{2\Delta} \quad (16.24)$$

where $T = N\Delta$ is the length of the time series, and f is the frequency in Hertz.

Calculating The Wave Spectrum: The digital Fourier transform Z_n of a wave record ζ_j equivalent to (16.21 and 16.22) is:

$$Z_n = \frac{1}{N}\sum_{j=0}^{N-1} \zeta_j \exp[-i2\pi jn/N] \qquad (16.25a)$$

$$\zeta_n = \sum_{n=0}^{N-1} Z_j \exp[i2\pi jn/N] \qquad (16.25b)$$

for $j = 0, 1, \cdots, N - 1$; $n = 0, 1, \cdots, N - 1$. These equations can be summed very quickly using the Fast Fourier Transform, especially if N is a power of 2 (Cooley, Lewis, and Welch, 1970; Press et al. 1992: 542).

This spectrum S_n of ζ, which is called the *periodogram*, is:

$$\begin{aligned} S_n &= \frac{1}{N^2}\left[|Z_n|^2 + |Z_{N-n}|^2\right]; \qquad n = 1, 2, \cdots, (N/2 - 1) \\ S_0 &= \frac{1}{N^2}|Z_0|^2 \\ S_{N/2} &= \frac{1}{N^2}|Z_{N/2}|^2 \end{aligned} \qquad (16.26)$$

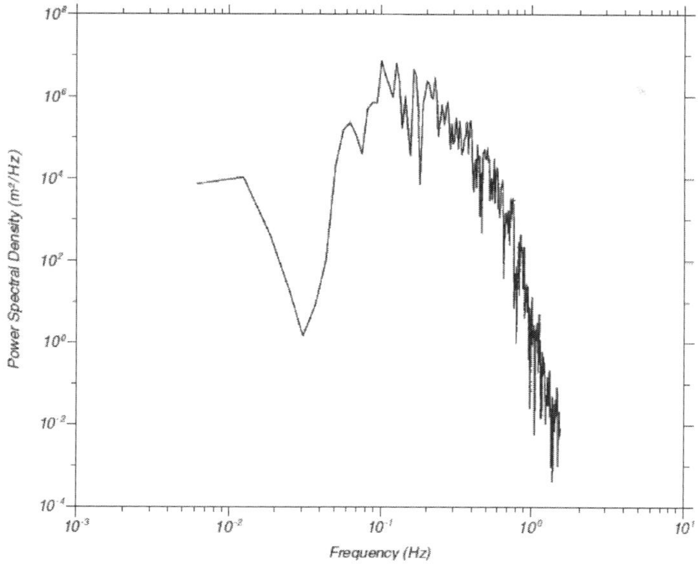

Figure 16.5 The periodogram calculated from the first 164 s of data from figure 16.2. The Nyquist frequency is 1.5625 Hz.

where *SN* is normalized such that:

$$\sum_{j=0}^{N-1} |\zeta_j|^2 = \sum_{n=0}^{N/2} S_n \qquad (16.27)$$

The variance of ζ_j is the sum of the $(N/2 + 1)$ terms in the periodogram. Note, the terms of Sn above the frequency $(N/2)$ are symmetric about that frequency. Figure 16.5 shows the periodogram of the time series shown in figure 16.2.

The periodogram is a very noisy function. The variance of each point is equal to the expected value at the point. By averaging together 10–30 periodograms we can reduce the uncertainty in the value at each frequency. The averaged periodogram is called the spectrum of the wave height (figure 16.6). It gives the distribution of the variance of sea-surface height at the wave staff as a function of frequency. Because wave energy is proportional to the variance (16.12) the spectrum is called the *energy spectrum* or the *wave-height spectrum*. Typically three hours of wave staff data are used to compute a spectrum of wave height.

Summary: We can summarize the calculation of a spectrum into the following steps:

1. Digitize a segment of wave-height data to obtain useful limits according to (16.26). For example, use 1024 samples from 8.53 minutes of data sampled at the rate of 2 samples/second.

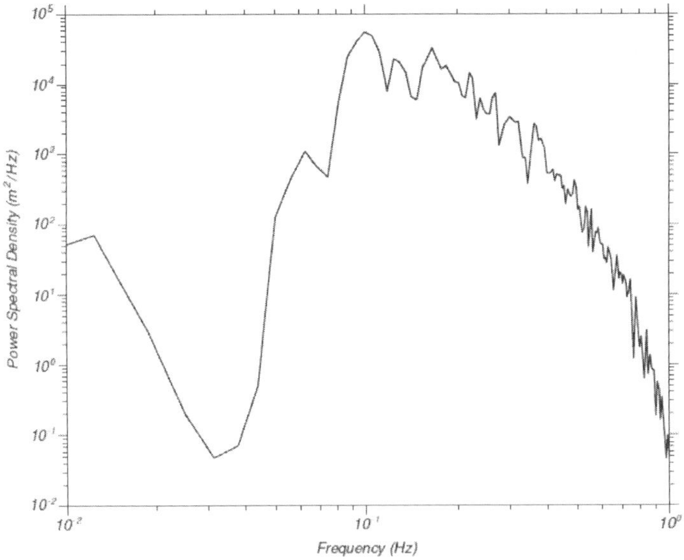

Figure 16.6 The spectrum of waves calculated from 11 minutes of data shown in figure 7.2 by averaging four periodograms to reduce uncertainty in the spectral values. Spectral values below 0.04 Hz are in error due to noise.

1. Calculate the digital, fast Fourier transform Zn of the time series.
2. Calculate the periodogram Sn from the sum of the squares of the real and imaginary parts of the Fourier transform.
3. Repeat to produce $M = 20$ periodograms.
4. Average the 20 periodograms to produce an averaged spectrum SM .

5. SM has values that are χ^2 distributed with $2M$ degrees of freedom.
6. This outline of the calculation of a spectrum ignores many details. For more complete information see, for example, Percival and Walden (1993), Press etal. (1992: §12), Oppenheimer and Schafer (1975), or other texts on digital signal processing.

16.4 OCEAN-WAVE SPECTRA

Ocean waves are produced by the wind. The faster the wind, the longer the wind blows, and the bigger the area over which the wind blows, the bigger the waves. In designing ships or offshore structures we wish to know the biggest waves produced by a given wind speed. Suppose the wind blows at 20 m/s for many days over a large area of the North Atlantic. What will be the spectrum of ocean waves at the downwind side of the area?

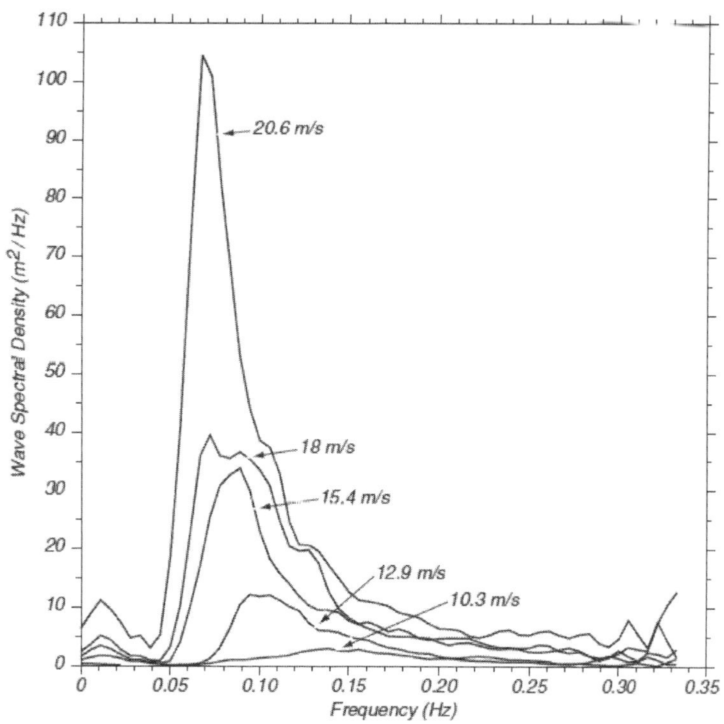

Figure 16.7 Wave spectra of a fully developed sea for different wind speeds according to Moskowitz (1964).

Pierson-Moskowitz Spectrum: Various idealized spectra are used to answer the question in oceanography and ocean engineering. Perhaps the simplest is that proposed by Pierson and Moskowitz (1964). They assumed that if the wind blew steadily for a long time over a large area, the waves would come into equilibrium with the wind. This is the concept of a *fully developed sea*. Here, a"long time" is roughly ten-thousand wave periods, and a "large area" is roughly five-thousand wave lengths on a side.

To obtain a spectrum of a fully developed sea, they used measurements of waves made by accelerometers on British weather ships in the north Atlantic. First, they selected wave data for times when the wind had blown steadily for long times over large areas of the north Atlantic. Then they calculated the wave spectra for various wind speeds (figure 16.7), and they found that the function

$$S(\omega) = \frac{\alpha g^2}{\omega^5} \exp\left[-\beta\left(\frac{\omega_0}{\omega}\right)^4\right] \quad (16.28)$$

was a good fit to the observed spectra, where = $2\pi f$, f is the wave frequency in Hertz, = 8.1×10^{-3}, β = 0.74, ω0 = $g/U19.5$ and $U19.5$ is the wind speed ata height of 19.5 m above the sea surface, the height of the anemometers on the weather ships used by Pierson and Moskowitz (1964).

For most airflow over the sea the atmospheric boundary layer has nearly neutral stability, and

$$U_{19.5} \approx 1.026\, U_{10} \quad (16.29)$$

assuming a drag coefficient of 1.3×10^{-3}.

The frequency of the peak of the Pierson-Moskowitz spectrum is calculated by solving dS/d= 0 for $\omega_{p,}$ to obtain

$$\omega_p = 0.877\, g/U_{19.5}. \quad (16.30)$$

The speed of waves at the peak is calculated from (16.7), which gives:

$$c_p = \frac{g}{\omega_p} = 1.14\, U_{19.5} \approx 1.17\, U_{10} \quad (16.31)$$

Hence waves with frequency ω_p travel 14% faster than the wind at a height of19.5 m or 17% faster than the wind at a height of 10 m. This poses a difficult problem: How can the wind produce waves traveling faster than the wind? I will return to the problem after I discuss the jonswap spectrum and the influence of nonlinear interactions among wind-generated waves.

By integrating $S(\omega)$ over all ω we get the variance of surface elevation:

$$\langle \zeta^2 \rangle = \int_0^\infty S(\omega)\, d\omega = 2.74 \times 10^{-3} \frac{(U_{19.5})^4}{g^2} \quad (16.32)$$

Remembering that $H_{1/3} = 4\langle \zeta^2 \rangle^{1/2}$, the significant wave height is:

$$H_{1/3} = 0.21 \frac{(U_{19.5})^2}{g} \approx 0.22 \frac{(U_{10})^2}{g} \quad (16.33)$$

Figure 16.8 gives significant wave heights and periods for the Pierson-Moskowitz spectrum.

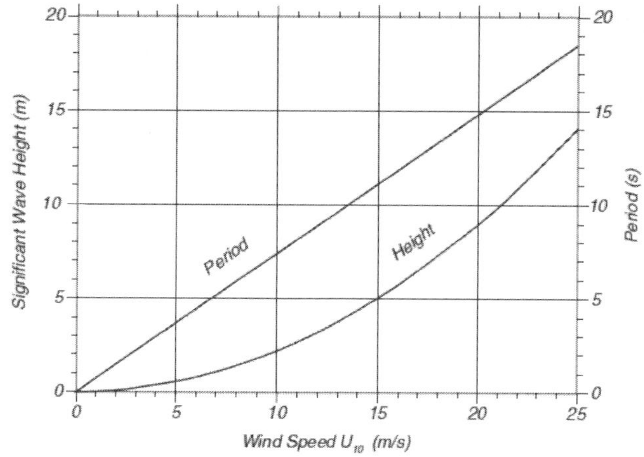

Figure 16.8 Significant wave height and period at the peak of the spectrum of a fully developed sea calculated from the Pierson-Moskowitz spectrum using (16.35 and 16.32).

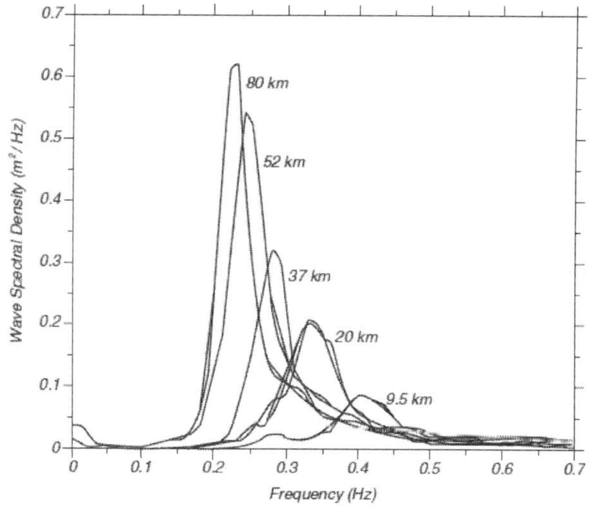

Figure 16.9 Wave spectra of a developing sea for different fetches measured at jonswap. After Hasselmann et al. (1973).

JONSWAP Spectrum: Hasselmann et al. (1973), after analyzing data collected during the Joint North Sea Wave Observation Project jonswap, found that the wave spectrum is never fully developed (figure 16.9). It continues to develop through non-linear, wave-wave interactions even for very long times and distances. They therefore proposed the spectrum:

$$S(\omega) = \frac{\alpha g^2}{\omega^5} \exp\left[-\frac{5}{4}\left(\frac{\omega_p}{\omega}\right)^4\right] \gamma^r \quad (16.34a)$$

$$r = \exp\left[-\frac{(\omega - \omega_p)^2}{2\,\sigma^2\,\omega_p^2}\right] \quad (16.34b)$$

Wave data collected during the jonswap experiment were used to determine the values for the constants in (16.36):

$$\alpha = 0.076 \left(\frac{U_{10}^2}{F g} \right)^{0.22} \quad (16.35a)$$

$$\omega_p = 22 \left(\frac{g^2}{U_{10} F} \right)^{1/3} \quad (16.35b)$$

$$\gamma = 3.3 \quad (16.35c)$$

$$\sigma = \begin{cases} 0.07 & \omega \leq \omega_p \\ 0.09 & \omega > \omega_p \end{cases} \quad (16.35d)$$

where F is the distance from a lee shore, called the *fetch*, or the distance over which the wind blows with constant velocity.

The energy of the waves increases with fetch:

$$\langle \zeta^2 \rangle = 1.67 \times 10^{-7} \frac{(U_{10})^2}{g} F \quad (16.36)$$

where F is fetch.

The jonswap spectrum is similar to the Pierson-Moskowitz spectrum except that waves continues to grow with distance (or time) as specified by the α term, and the peak in the spectrum is more pronounced, as specified by the γ term. The latter turns out to be particularly important because it leads to enhanced non-linear interactions and a spectrum that changes in time according to the theory of Hasselmann (1966).

Generation of Waves by Wind: We have seen in the last few paragraphs that waves are related to the wind. We have, however, put off until now just how they are generated by the wind. Suppose we begin with a mirror-smooth sea (Beaufort Number 0). What happens if the wind suddenly begins to blow steadily at say 8 m/s? Three different physical processes begin:

- The turbulence in the wind produces random pressure fluctuations at the sea surface, which produces small waves with wavelengths of a few centimeters (Phillips, 1957).
- Next, the wind acts on the small waves, causing them to become larger. Wind blowing over the wave produces pressure differences along the wave profile causing the wave to grow. The process is unstable because, as the wave gets bigger, the pressure differences get bigger, and the wave grows faster. The instability causes the wave to grow exponentially (Miles, 1957).
- Finally, the waves begin to interact among themselves to produce longer waves (Hasselmann et al. 1973). The interaction transfers wave energy from

short waves generated by Miles' mechanism to waves with frequencies slightly lower than the frequency of waves at the peak of the spectrum. Eventually, this leads to waves going faster than the wind, as noted by Pierson and Moskowitz.

16.5 UNDERSTANDING OF WAVE FORECASTING

Our understanding of ocean waves, their spectra, their generation by the wind, and their interactions are now sufficiently well understood that the wave spectrum can be forecast using winds calculated from numerical weather models.

If we observe some small ocean area, or some area just offshore, we can see waves generated by the local wind, the *wind sea*, plus waves that were generated in other areas at other times and that have propagated into the area we are observing, the *swell*. Forecasts of local wave conditions must include both sea and swell, hence wave forecasting is not a local problem. We saw, for example, that waves off California can be generated by storms more than 10,000 km away. Various techniques have been used to forecast waves. The earliest attempts were based on empirical relationships between wave height and wave length and wind speed, duration, and fetch. The development of the wave spectrum allowed evolution of individual wave components with frequency f travelling in direction θ of the directional wave spectrum $\psi(f, \theta)$ using where $\psi 0 = \psi 0(f, \theta; \mathbf{x}, t)$ varies in space ($\mathbf{x}$) and time t, Si is input from the wind given by the Phillips (1957) and Miles (1957) mechanisms, Snl is the transfer among wave components due to nonlinear interactions, and Sd is dissipation.

$$\frac{\partial \psi_0}{\partial t} + \mathbf{c_g} \cdot \nabla \psi_0 = S_i + S_{nl} + S_d \quad (16.37)$$

The third-generation wave-forecasting models now used by meteorological agencies throughout the world are based on integrations of (16.39) using many individual wave components (The swamp Group 1985; The wamdi Group, 1988; Komen et al, 1996). The forecasts follow individual components of the wave spectrum in space and time, allowing each component to grow or decay depending on local winds, and allowing wave components to interact according to Hasselmann's theory. Typically the sea is represented by 300 components: 25wavelengths going in 12 directions (30°). To reduce computing time, the models use a nested grid of points: the grid has a high density of points in storms and near coasts and a low density in other regions. Typically, grid points in the open ocean are 3° apart.

Some recent experimental models take the wave-forecasting process one step further by assimilating altimeter and scattero meter observations of wind speed and wave height into the model. Forecasts of waves using assimilated satellite data are available from the European Centre for Medium-Range Weather Forecasts.

NOAA's Ocean Modeling Branch at the National Centers for Environmental Predictions also produces regional and global forecasts of waves. The Branch use a third-generation model based on the Cycle-4 wam model. It accommodates ever-changing ice edge, and it assimilates buoy and satellite altimeter wave data. The model calculates directional frequency spectra in 12 directions and 25 frequencies at 3-hour intervals up to 72 hours in advance. The lowest frequency is 0.04177 Hz and higher frequencies are spaced logarithmically at increments of 0.1 times the lowest frequency. Wave spectral data are available on a 2.5°× 2.5° grid for deep-water points between 67.5°S and 77.5°N. The model is driven using 10-meter winds calculated from the lowest winds from the Centers weather model adjusted to a height of 10 meter by using a logarithmic profile (8.20). The Branch is testing an improved forecast with 1°× 1.25° resolution (figure 16.10).

16.6 MEASUREMENT OF WAVES

Because waves influence so many processes and operations at sea, many techniques have been invented for measuring waves. Here are a few of the more commonly used. Stewart (1980) gives a more complete description of wave measurement techniques, including methods for measuring the directional distribution of waves.

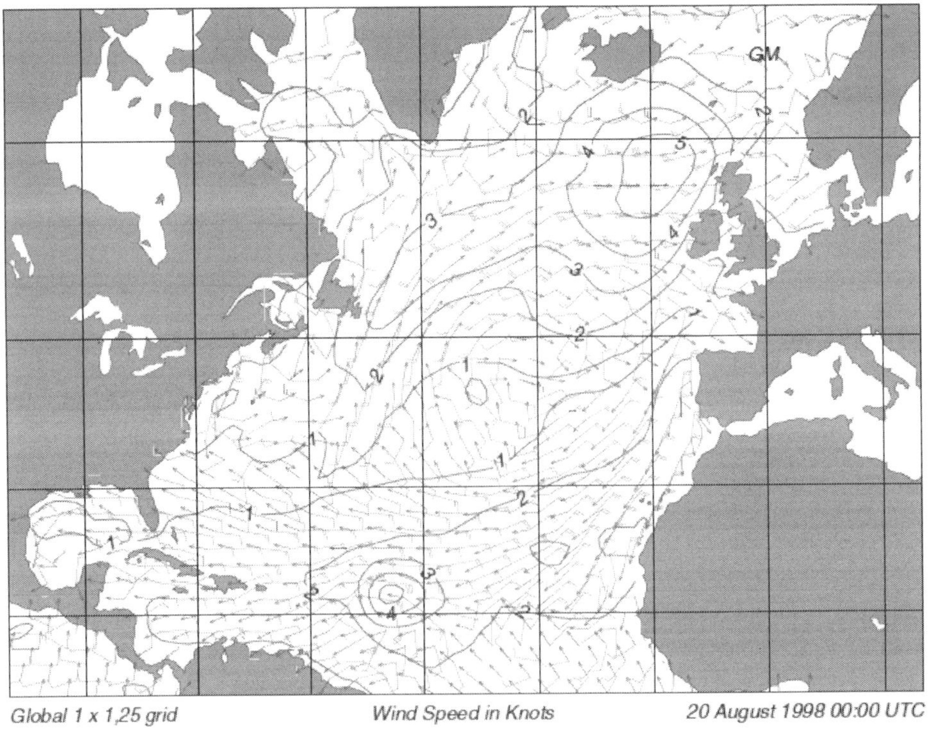

Figure 16.10 Output of a third-generation wave forecast model produced by NOAA's Ocean Modeling Branch for 20 August 1998. Contours are significant wave height in meters, arrows give direction of waves at the peak of the wave spectrum, and barbs give wind speed in m/s and direction. From NOAA Ocean Modeling Branch.

Sea State Estimated by Observers at Sea: This is perhaps the most common observation included in early tabulations of wave heights. These are the significant wave heights summarized in the U.S. Navy's *Marine Climatological Atlas* and other such reports printed before the age of satellites.

Satellite Altimeters: The satellite altimeters used to measure surface geostrophic currents also measure wave height. Altimeters were flown on Seasat in 1978, Geosat from 1985 to 1988, ers–1 &2 from 1991, Topex/Poseidon from 1992, and Jason from 2001. Altimeter data have been used to produce monthly mean maps of wave heights and the variability of wave energy density in time and space. The next step, just begun, is to use altimeter observation with wave forecasting programs, to increase the accuracy of wave forecasts.

The altimeter technique works as follows. Radio pulse from a satellite altimeter reflect first from the wave crests, later from the wave troughs. The reflection stretches the altimeter pulse in time, and the stretching is measured and used to calculate wave height (figure 16.11). Accuracy is ±10%.

Accelerometer Mounted on Meteorological or Other Buoy: This is a less common measurement, although it is often used for measuring waves during short experiments at sea. For example, accelerometers on weather ships measured wave height used by Pierson & Moskowitz and the waves shown in figure 16.2. The most accurate measurements are made using an accelerometer stabilized by a gyro so the axis of the accelerometer is always vertical.

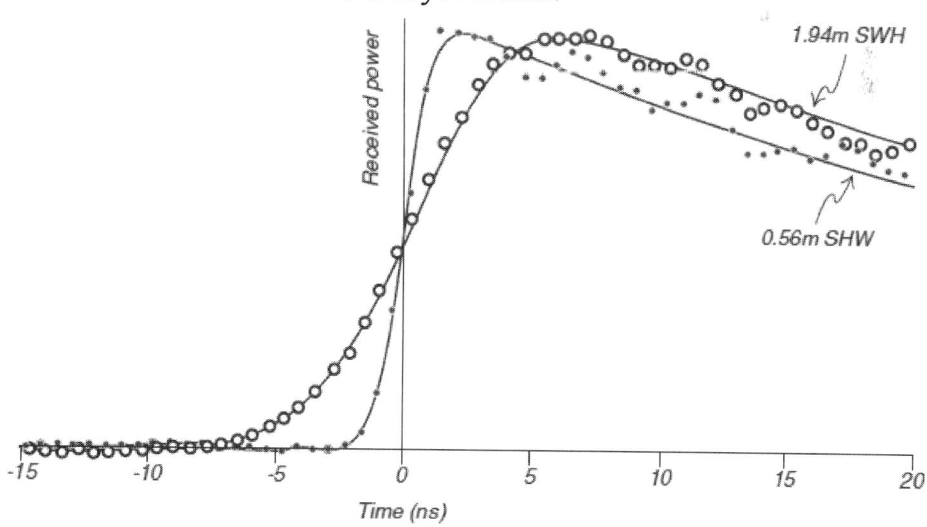

Figure 16.11 Shape of radio pulse received by the Seasat altimeter, showing the influence of ocean waves. The shape of the pulse is used to calculate significant wave height. After Stewart (1985: 264).

Double integration of vertical acceleration gives displacement. The double integration, however, amplifies low-frequency noise, leading to the low frequency signals seen in figures 16.5 and 16.6. In addition, the buoy's heave is not sensitive to wavelengths less than the buoy's diameter, and buoys measure only waves having wavelengths greater than the diameter of the buoy. Overall, careful measurements are accurate to ± 10% or better.

Wave Gages: Gauges may be mounted on platforms or on the sea floor in shallow water. Many different types of sensors are used to measure the height of the wave or subsurface pressure which is related to wave height. Sound, infrared beams, and radio waves can be used to determine the distance from the sensor to the sea surface provided the sensor can be mounted on a stable platform that does not interfere with the waves. Pressure gauges described in§6.8 can be used to measure the depth from the sea surface to the gauge. Arrays of bottom-mounted pressure gauges are useful for determining wave directions.

Thus arrays are widely used just offshore of the surf zone to determine offshore wave directions.

Pressure gauge must be located within a quarter of a wavelength of the surface because wave-induced pressure fluctuations decrease exponentially with depth. Thus, both gauges and pressure sensors are restricted to shallow water orto large platforms on the continental shelf. Again, accuracy is ± 10% or better.

Synthetic Aperture Radars on Satellites: These radars map the radar reflectivity of the sea surface with spatial resolution of 6–25 m. Maps of reflectivity often show wave-like features related to the real waves on the sea surface. I say 'wave-like' because there is not an exact one-to-one relationship between wave height and image density. Some waves are clearly mapped, others less so. The maps, however, can be used to obtain additional information about waves, especially the spatial distribution of wave directions in shallow water (Vesecky andStewart, 1982). Because the directional information can be calculated directly from the radar data without the need to calculate an image (Hasselmann, 1991), data from radars and altimeters on ers–1 & 2 are being used to determine if the radar and altimeter observations can be used directly in wave forecast programs.

16.7 CONCISE SUMMARY

- Wavelength and frequency of waves are related through the dispersion relation.
- The velocity of a wave phase can differ from the velocity at which wave energy propagates.
- Waves in deep water are dispersive, longer wavelengths travel faster than shorter wavelengths. Waves in shallow water are not dispersive.
- The dispersion of ocean waves has been accurately measured, and observations of dispersed waves can be used to track distant storms.

- The shape of the sea surface results from a linear superposition of waves of all possible wavelengths or frequencies travelling in all possible directions.
- The spectrum gives the contributions by wavelength or frequency to the variance of surface displacement.
- Wave energy is proportional to variance of surface displacement.
- Digital spectra are band limited, and they contain no information about waves with frequencies higher than the Nyquist frequency.
- Waves are generated by wind. Strong winds of long duration generate the largest waves.
- Various idealized forms of the wave spectrum generated by steady, homogeneous winds have been proposed. Two important ones are the Pierson- Moskowitz and jonswap spectra.
- Observations by mariners on ships and by satellite altimeters have been used to make global maps of wave height. Wave gauges are used on platforms in shallow water and on the continental shelf to measure waves. Bottom-mounted pressure gauges are used to measure waves just offshore of beaches. And synthetic-aperture radars are used to obtain information about wave directions.

17 COASTAL PROCESSES AND TIDES

The interplay between tidal phenomena and coastal processes represents a dynamic and intricate relationship within the Earth's aquatic realms. Tides, governed by celestial forces, exert a profound influence on coastal areas, shaping their morphology and influencing various geological and ecological processes. This intricate interaction between tidal forces and coastal dynamics is a subject of significant scientific interest, delving into the complexities of how these natural forces collaborate to shape the coastal environment. This exploration encompasses a range of phenomena, from tidal currents and sediment transport to the evolution of coastlines, offering insights into the delicate balance that exists between the ebb and flow of tides and the ever-changing coastal landscapes.

In the last chapter I described waves on the sea surface. Now we can consider several special and important cases: the transformation of waves as they come ashore and break; the currents and edge waves generated by the interaction of waves with coasts; tsunamis; storm surges; and tides, especially tides along coasts and in the deep ocean.

17.1 WAVES IN SHOALING AND COASTAL PROCESSES

Wave phase and group velocities are a function of depth when the depth is less than about one-quarter wavelength in deep water. Wave period and frequency are invariant (don't change as the wave comes ashore); and this is used to compute the properties of shoaling waves. Wave height increases as wave group velocity slows. Wave length decreases. Waves change direction due to refraction. Finally, waves break if the water is sufficiently shallow; and broken waves pour water into the surf zone, creating long-shore and rip currents.

Shoaling Waves: The dispersion relation (16.3) is used to calculate wave properties as the waves propagate from deep water offshore to shallow water just outside the surf zone. Because ω is constant, (16.3) leads to:

$$\frac{L}{L_0} = \frac{c}{c_0} = \frac{\sin \alpha}{\sin \alpha_0} = \tanh \frac{2\pi d}{L} \quad (17.1)$$

where

$$L_0 = \frac{gT^2}{2\pi}, \qquad c_0 = \frac{gT}{2\pi} \quad (17.2)$$

and L is wave length, c is phase velocity, α is the angle of the crest relative to contours of constant depth, and d is water depth. The subscript 0 indicates values in deep water.

The quantity d/L is calculated from the solution of using an iterative technique, or from figure 17.1, or from Wiegel (1964: A1).

$$\frac{d}{L_0} = \frac{d}{L}\tanh\frac{2\pi d}{L} \quad (17.3)$$

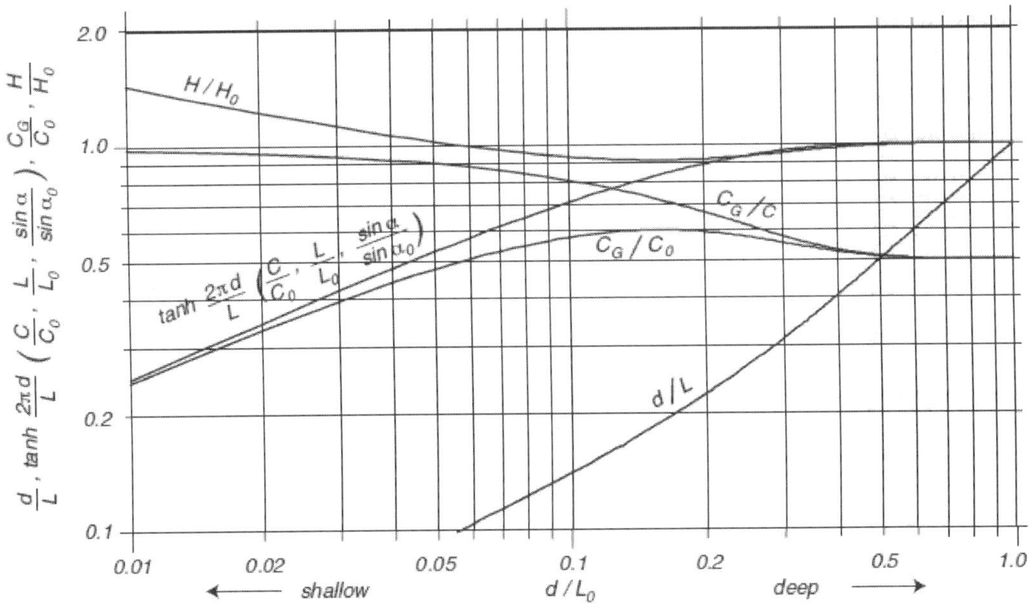

Figure 17.1 Properties of waves in intermediate depths between deep and shallow water. d **= depth,** L **= wavelength,** C **= phase velocity,** Cg **= group velocity,** α **= angle of crests relative to contours of constant depth,** H **= wave height. Subscript 0 refers to properties in deep water. From values in Wiegel (1964: A1).**

Because wave velocity is a function of depth in shallow water, variations in offshore water depth can focus or defocus wave energy reaching the shore. Consider the simple case of waves with deep-water crests almost parallel to astraight beach with two ridges each extending seaward from a headland (figure 17.2). Wave group velocity is faster in the deeper water between the ridges, and the wave crests become progressively deformed as the wave propagates toward the beach. Wave energy, which propagates perpendicular to wave crests, is refracted out of the region between the headland. As a result, wave energy is focused into the headlands, and breakers there are much larger than breakers in the bay. The difference in wave height can be surprisingly large. On a calm day, breakers can be knee high shoreward of a submarine canyon at La Jolla Shores, California, just south of the Scripps Institution of Oceanography. At the same time, waves just north of the canyon can be high enough to attract surfers.

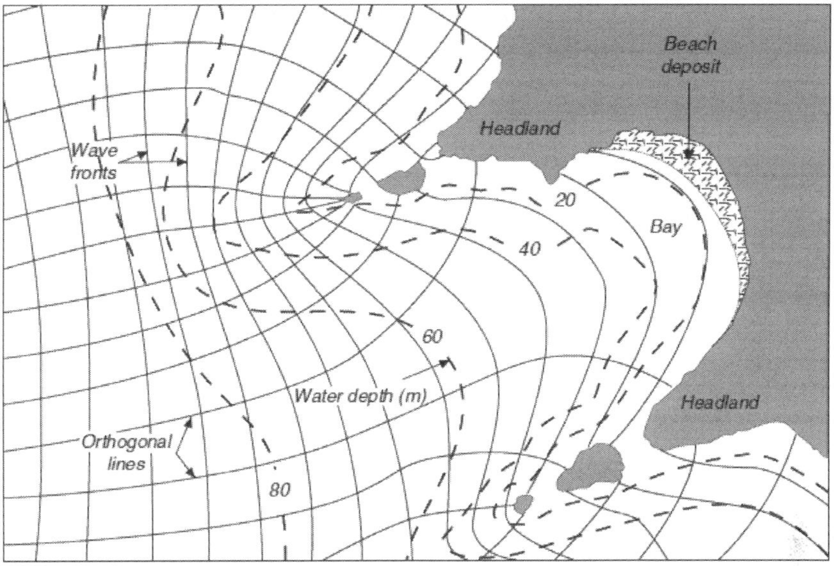

Figure 17.2 sub-sea features, such as submarine canyons and ridges, offshore of coasts can greatly influence the height of breakers inshore of the features. After Thurman (1985: 229).

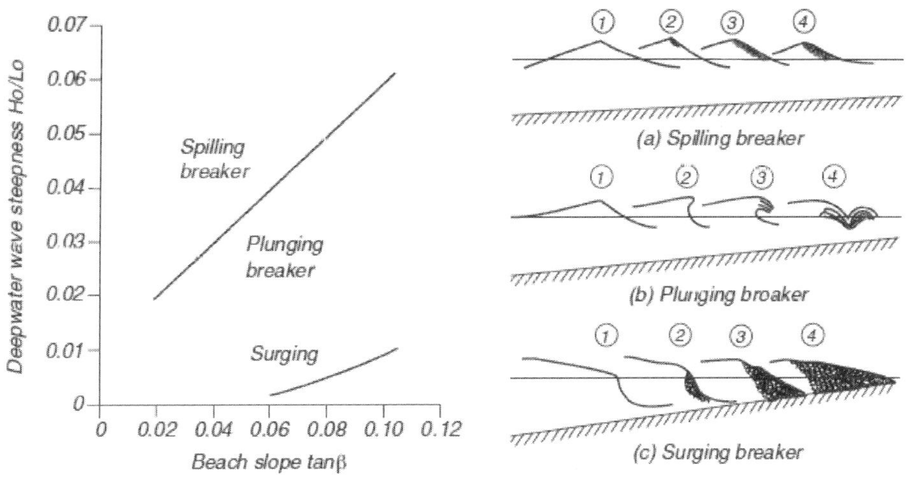

Figure 17.3 Left: Classification of breaking waves as a function of beach steepness and wave steepness offshore. Right: Sketch of types of breaking waves. After Horikawa (1988: 79, 81).

Breaking Waves: As waves move into shallow water, the group velocity becomes small, wave energy per square meter of sea surface increases, and non- linear terms in the wave equations become important. These processes cause waves to steepen, with short steep crests and broad shallow troughs. When wave slope at the crest becomes sufficiently steep, the wave breaks (figure 17.3 Right). The shape of the breaking wave depends on the slope of the bottom, and the steepness of waves offshore (figure 17.3 Left).

1. Steep waves tend to lose energy slowly as the waves moves into shallower water through water spilling down the front of the wave. These are spilling breakers.
2. Less steep waves on steep beaches tend to steepen so quickly that the crest of the wave moves much faster than the trough, and the crest, racing ahead of the trough, plunges into the trough (figure 17.4).
3. If the beach is sufficiently steep, the wave can surge up the face of the beach without breaking in the sense that white water is formed. Or if it is formed, it is at the leading edge of the water as it surges up the beach. An extreme example would be a wave incident on a vertical breakwater.

Figure 17.4 Steep, plunging breakers are the archetypical breaker. The edge of such breakers are ideal for surfing. From photo by Jeff Devine.

Wave-Driven Currents: Waves break in a narrow band of shallow water along the beach, the *surf zone*. After breaking, waves continues as a near-vertical wall of turbulent water called a *bore* which carries water to the beach. At first, the bore surges up the beach, then retreats. The water carried by the bore is left in the shallow waters inside the breaker zone.

Water dumped inside the breaker zone must return offshore. It does this by first moving parallel to the beach as an *along-shore current* . Then it turns and flows offshore perpendicular to the beach in a narrow, swift *rip current* . The rips are usually spaced hundreds of meters apart (figure 17.5). Usually there is a band of deeper water between the breaker zone and the beach, and the long- shore current runs in this channel. The strength of a rip current depends on the height and frequency of breaking waves and the strength of the onshore wind. Rips are a danger to unwary swimmers, especially poor swimmers bobbing along in the waves inside the breaker zone. They are carried along by the along-shore current until they are suddenly carried out to sea by the rip. Swimming against the rip is futile, but swimmers can escape by swimming parallel to the beach.

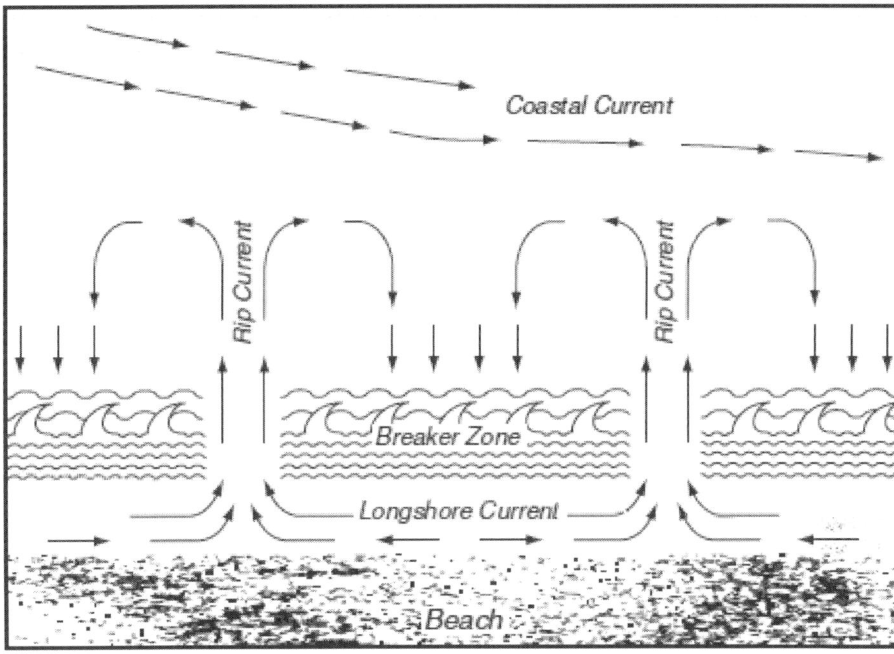

Figure 17.5 Sketch of rip currents generated by water carried to the beach by breaking waves.

Edge waves are produced by the variability of wave energy reaching shore. Waves tend to come in groups, especially when waves come from distant storms. For several minutes breakers may be smaller than average, then a few very large waves will break. The minute-to-minute variation in the height of breakers produces low-frequency variability in the along-shore current. This, in turn, drives a low-frequency wave attached to the beach, an edge wave. The waves have periods of a few minutes, a long-shore wave length of around a kilometer, and an amplitude that decays exponentially offshore (figure 17.6).

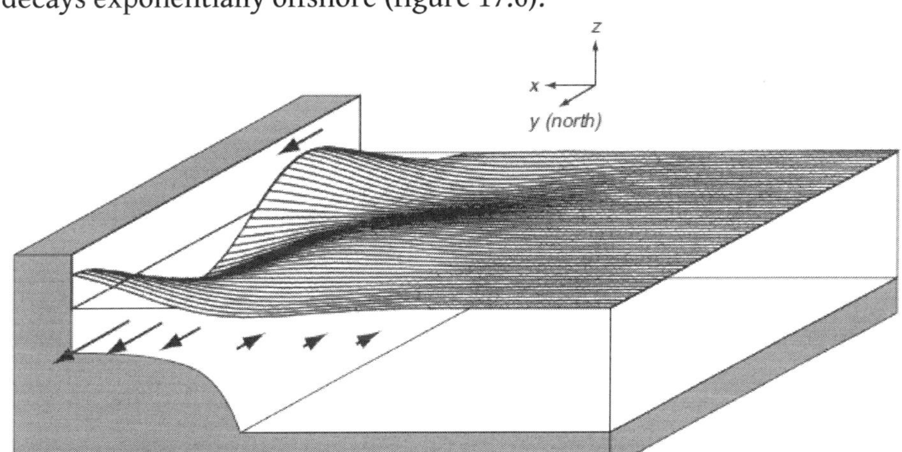

Figure 17.6 Computer-assisted sketch of an edge wave. Such waves exist in the breaker zone near the beach and on the continental shelf. After Cutch in and Smith (1973).

17.2 TSUNAMIS

Tsunamis are low-frequency ocean waves generated by submarine earth- quakes. The sudden motion of sea floor over distances of a hundred or more kilometers generates waves with periods of 15–40 minutes (figure 17.7). A quick calculation shows that such waves must be shallow-water waves, propagating at a speed of 180 m/s and having a wavelength of 130 km in water 3.6 km deep (figure 17.8). The waves are not noticeable at sea, but after slowing on approach to the coast, and after refraction by sub-sea features, they can come ashore and surge to heights ten or more meters above sea level. In an extreme example, the great 2004 Indian Ocean tsunami destroyed hundreds of villages, killing at least 200,000 people.

Shepard (1963, Chapter 4) summarized the influence of tsunamis based on his studies in the Pacific.

1. Tsunamis appear to be produced by movement (an earthquake) along a linear fault.

2. Tsunamis can travel thousands of kilometers and still do serious damage.

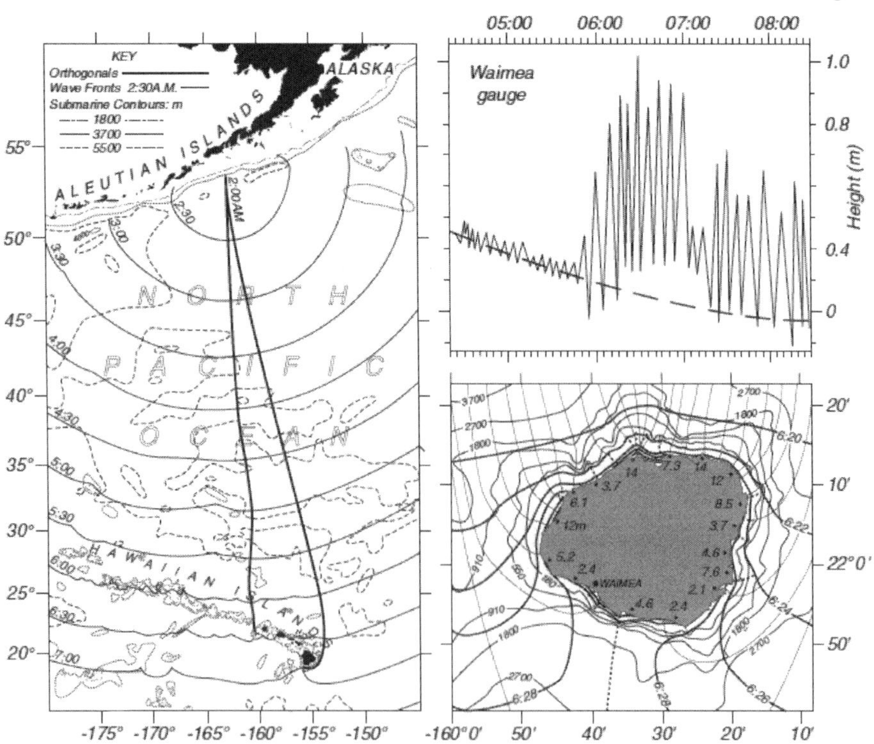

Figure 17.7 Left Hourly positions of leading edge of tsunami generated by the large earthquake in the Aleutian Trench on April 1, 1946 at 1:59 AM Hawaiian time (12:59 GMT). Right: top Sealevel recorded by a river gauge in the estuary of the Waimea River. Right: lower Map of Kauai showing the heights reached by the water (in meters above lower low water) during the tsunami, wave fronts, orthogonals, and submarine contours. Times refer to the computed arrival time of the first wave. After Shepard, MacDonald, and Cox (1950).

1. The first wave of a tsunami is not likely to be the biggest.
2. Wave amplitudes are relatively large shoreward of submarine ridges. They are relatively low shoreward of submarine valleys, provided the features extend into deep water.
3. Wave amplitudes are decreased by the presence of coral reefs bordering the coast.
4. Some bays have a funneling effect, but long estuaries attenuate waves.
5. Waves can bend around circular islands without great loss of energy, but they are considerably smaller on the backsides of elongated, angular is- lands.

Numerical models are used to forecast tsunami heights throughout ocean basins and the inundation of coasts. For example, NOAA's Center for Tsunami Research uses the Method of Splitting Tsunami most model (Titov and Gon- zalez, 1997). The model uses nested grids to resolve the tsunami wavelength, it propagates the wave across ocean basins, and then calculates run-up when the wave comes ashore. It is initialized from a ground deformation model that uses measured earthquake magnitude and location to calculate vertical displacement of the sea floor. The forcing is modified once waves are measured near the earthquake by sea floor observing stations.

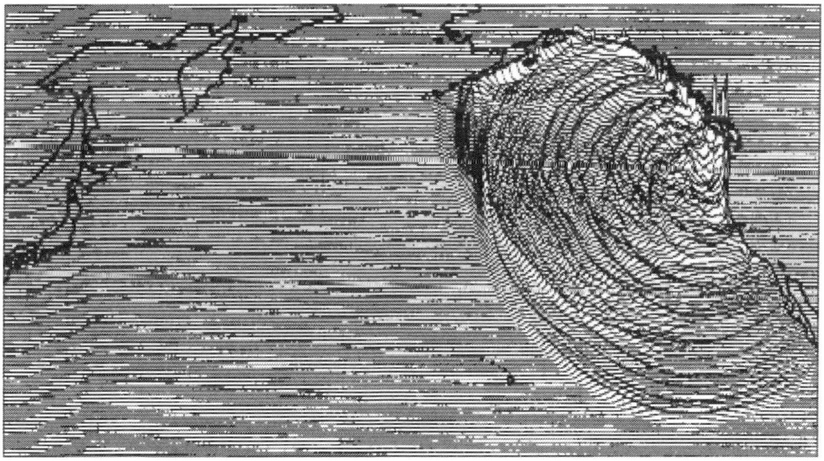

Figure 17.8 Tsunami waves four hours after the great M9 Cascadia earthquake off the coast of Washington on 26 January 1700 calculated by a finite-element, numerical model. Maximum open-ocean wave height, about one meter, is north of Hawaii. After Satake et al. (1996).

17.3 STORM SURGES

Storm winds blowing over shallow, continental shelves pile water against the coast. The increase in sea level is known as a storm surge. Several processes are important:

- Ekman transport by winds parallel to the coast transports water toward the coast causing a rise in sea level.

- Winds blowing toward the coast push water directly toward the coast.
- Wave run-up and other wave interactions transport water toward the coast adding to the first two processes.
- Edge waves generated by the wind travel along the coast.
- The low pressure inside the storm raises sea level by one centimeter for each millibar decrease in pressure through the inverted-barometer effect.
- Finally, the storm surge adds to the tides, and high tides can change a relative weak surge into a much more dangerous one.

See Graber et al (2006) and §15.5 for a description of Advanced Circulation Model used by the National Hurricane Center for predicting storm-surges.

To a crude first approximation, wind blowing over shallow water causes a slope in the sea surface proportional to wind stress.

$$\frac{\partial \zeta}{\partial x} = \frac{\tau_0}{\rho g H} \quad (17.4)$$

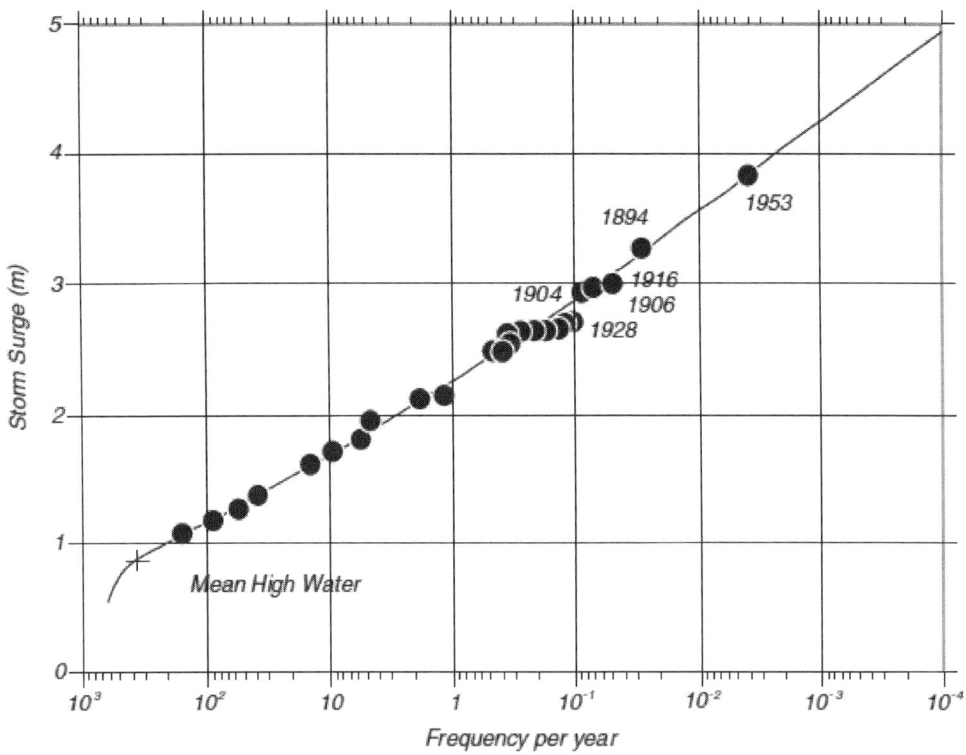

Figure 17.9 Probability (per year) density distribution of vertical height of storm surges in the Hook of Holland in the Netherlands. The distribution function is Rayleigh, and the probability of large surges is estimated by extrapolating the observed frequency of smaller, more common surges. After Wiegel (1964: 113).

where ζ is sea level, x is horizontal distance, H is water depth, $T0$ is wind stress at the sea surface, ρ is water density; and g is gravitational acceleration.

If $x = 100$ km, $U = 40$ m/s, and $H = 20$ m, values typical of a hurricane offshore of the Texas Gulf Coast, then $T = 2.7$ Pa, and $\zeta = 1.3$ m at the shore. Figure 17.9 shows the frequency of surges at the Netherlands and a graphical method for estimating the probability of extreme events using the probability of weaker events.

17.4 PRINCIPLES UNDERLYING OCEANIC TIDAL PHENOMENA

Tides have been so important for commerce and science for so many thou- sands of years that tides have entered our everyday language: *time and tide wait for no one, the ebb and flow of events, a high-water mark,* and *turn the tide of battle.*

1. Tides produce strong currents in many parts of the ocean. Tidal currents can have speeds of up to 5 m/s in coastal waters, impeding navigation and mixing coastal waters.
2. Tidal currents generate internal waves over seamounts, continental slopes, and mid-ocean ridges. The waves dissipate tidal energy. Breaking internal waves and tidal currents are a major force driving oceanic mixing.
3. Tidal mixing helps drive the deep circulation, and it influences climate and abrupt climate change.
4. Tidal currents can suspend bottom sediments, even in the deep ocean.
5. Earth's crust is elastic. It bends under the influence of the tidal potential. It also bends under the weight of oceanic tides. As a result, the sea floor, and the continents move up and down by about 10 cm in response to the tides. The deformation of the solid earth influence almost all precise geodetic measurements.
6. Oceanic tides lag behind the tide-generating potential. This produces forces that transfer angular momentum between earth and the tide producing body, especially the moon. As a result of tidal forces, earth's rotation about it's axis slows, increasing the length of day; the rotation of the moon about earth slows, causing the moon to move slowly away from earth; and moon's rotation about it's axis slows, causing the moon to keep the same side facing earth as the moon rotates about earth.
7. Tides influence the orbits of satellites. Accurate knowledge of tides is needed for computing the orbit of altimetric satellites and for correcting altimeter measurements of oceanic topography.
8. Tidal forces on other planets and stars are important for understanding many aspects of solar-system dynamics and even galactic dynamics. For example, the rotation rate of Mercury, Venus, and Io result from tidal forces.

Mariners have known for at least four thousand years that tides are related to the phase of the moon. The exact relationship, however, is hidden behind many complicating factors, and some of the greatest scientific minds of the last four centuries worked to understand, calculate, and predict tides. Galileo, Descartes, Kepler, Newton, Euler, Bernoulli, Kant, Laplace, Airy, Lord Kelvin, Jeffreys, Munk and many others contributed. Some of the first computers were developed to compute and predict tides. Ferrel built a tide-predicting machine in 1880 that was used by the U.S. Coast Survey to predict nineteen tidal constituents. In 1901, Harris extended the capacity to 37 constituents.

Despite all this work important questions remained: What is the amplitude and phase of the tides at any place on the ocean or along the coast? What is the speed and direction of tidal currents? What is the shape of the tides on the ocean? Where is tidal energy dissipated? Finding answers to these simple questions is difficult, and the first, accurate, global maps of deep-sea tides were only published in 1994 (LeProvost et al. 1994). The problem is hard because the tides are a self-gravitating, near-resonant, sloshing of water in a rotating, elastic, ocean basin with ridges, mountains, and submarine basins.

Predicting tides along coasts and at ports is much easier. Data from a tide gauge plus the theory of tidal forcing gives an accurate description of tides near the tide gauge.

Tidal Potential: Tides are calculated from the hydrodynamic equations for a self-gravitating ocean on a rotating, elastic earth. The driving force is the gradient of the gravity field of the moon and sun. If the earth were an ocean planet with no land, and if we ignore the influence of inertia and currents, the gravity gradient produces a pair of bulges of water on earth, one on the side facing the moon or sun, one on the side away from the moon or sun. A clear derivation of the forces is given by Pugh (1987) and by Dietrich, Kalle, Krauss, and Siedler (1980). Here I follow the discussion in Pugh (1987: §3.2).

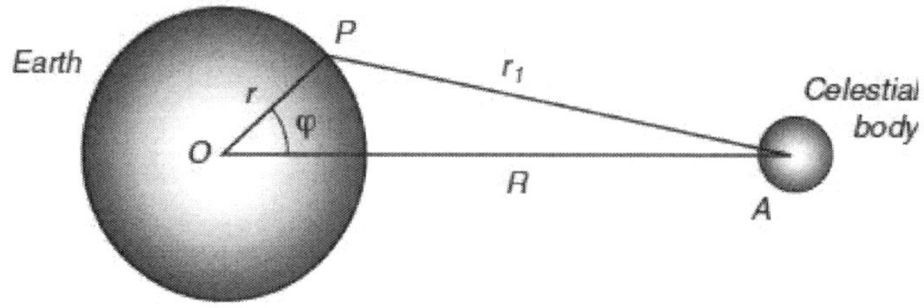

Figure 17.10 Sketch of coordinates for determining the tide-generating potential.

Note that many oceanographic books state that the tide is produced by two processes: i) the centripetal acceleration at earth's surface as the earth and moon circle around a common center of mass, and ii) the gravitational attraction of mass on earth and the moon. However, the derivation of the tidal potential does not involve centripetal acceleration, and the concept is not used by the astronomical or geodetic communities.

To calculate the amplitude and phase of the tide on an ocean planet, we begin by calculating the tide-generating potential. This is much easier than calculating the forces. Ignoring for now earth's rotation, the rotation of moon about earth produces a potential *VM* at any point on earth's surface

$$V_M = -\frac{\gamma M}{r_1} \quad (17.5)$$

where the geometry is sketched in figure 17.10, γ is the gravitational constant, and *M* is moon's mass. From the triangle *OPA* in the figure,Using this in (17.5) gives

$$r_1^2 = r^2 + R^2 - 2rR\cos\varphi \quad (17.6)$$

$$V_M = -\frac{\gamma M}{R}\left\{1 - 2\left(\frac{r}{R}\right)\cos\varphi + \left(\frac{r}{R}\right)^2\right\}^{-1/2} \quad (17.7)$$

$r/R \approx 1/60$, and (17.7) may be expanded in powers of r/R using Legendre polynomials (Whittaker and Watson, 1963: §15.1):

$$V_M = -\frac{\gamma M}{R}\left\{1 + \left(\frac{r}{R}\right)\cos\varphi + \left(\frac{r}{R}\right)^2\left(\frac{1}{2}\right)(3\cos^2\varphi - 1) + \cdots\right\} \quad (17.8)$$

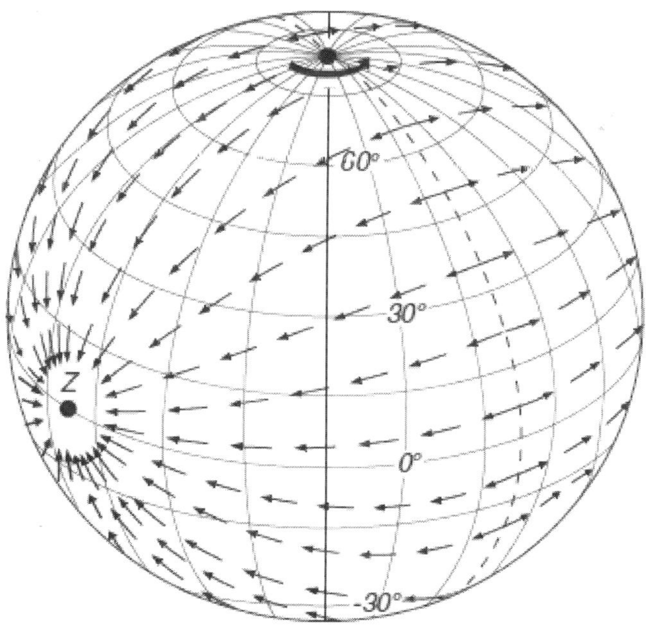

Figure 17.11 The horizontal component of the tidal force on earth when the tide-generating body is above the Equator at *Z*. After Dietrich et al. (1980: 413).

The tidal forces are calculated from the spatial gradient of the potential. The first term in (17.8) produces no force. The second term, when differentiated with respect to (r cos ϕ) produces a constant force $\gamma M/R^2$ parallel to OA that keeps earth in orbit around the center of mass of the earth-moon system. The third term produces the tides, assuming the higher-order terms can be ignored. The tide-generating potential is therefore:

$$V = -\frac{\gamma M r^2}{2R^3}(3\cos^2\varphi - 1) \quad (17.9)$$

The tide-generating force can be decomposed into components perpendicular P and parallel H to the sea surface. Tides are produced by the horizontal component. "The vertical component is balanced by pressure on the sea bed, but the ratio of the horizontal force per unit mass to vertical gravity has to be balanced by an opposing slope of the sea surface, as well as by possible changes in current momentum" (Cartwright, 1999: 39, 45). The horizontal component, shown in figure 17.11, is:

$$H = -\frac{1}{r}\frac{\partial V}{\partial \varphi} = \frac{2G}{r}\sin 2\varphi \quad (17.10)$$

where

$$G = \frac{3}{4}\gamma M\left(\frac{r^2}{R^3}\right) \quad (17.11)$$

The tidal potential is symmetric about the earth-moon line, and it produces symmetric bulges.

If we allow our ocean-covered earth to rotate, an observer in space sees the two bulges fixed relative to the earth-moon line as earth rotates. To an observer on earth, the two tidal bulges seems to rotate around earth because moon appears to move around the sky at nearly one cycle per day. Moon produces high tides every 12 hours and 25.23 minutes on the equator if the moon is above the equator. Notice that high tides are not exactly twice per day because the moon is also rotating around earth. Of course, the moon is above the equator only twice per lunar month, and this complicates our simple picture of the tides on an ideal ocean-covered earth. Furthermore, moon's distance from earth R varies because moon's orbit is elliptical and because the elliptical orbit is not fixed.

Clearly, the calculation of tides is getting more complicated than we might have thought. Before continuing on, we note that the solar tidal forces are derived in a similar way. The relative importance of the sun and moon arenearly the same. Although the sun is much more massive than moon, it is much further away.

$$G_{sun} = G_S = \frac{3}{4}\gamma S\left(\frac{r^2}{R_{sun}^3}\right) \quad (17.12)$$

$$G_{moon} = G_M = \frac{3}{4}\gamma M\left(\frac{r^2}{R_{moon}^3}\right) \quad (17.13)$$

$$\frac{G_S}{G_M} = 0.46051 \quad (17.14)$$

where R_{sun} is the distance to the sun, S is the mass of the sun, R_{moon} is the distance to the moon, and M is the mass of the moon.

Coordinates of Sun and Moon: Before we can proceed further we need to know the position of moon and sun relative to earth. An accurate description of the positions in three dimensions is very difficult, and it involves learning arcane terms and concepts from celestial mechanics. Here, I paraphrase a simplified description from Pugh (1987). See also figure 4.1.

A natural reference system for an observer on earth is the equatorial system described at the start of Chapter 3. In this system, *declinations* δ of a celestial body are measured north and south of a plane which cuts the earth's equator.

Angular distances around the plane are measured relative to a point on this celestial equator which is fixed with respect to the stars. The point chosen for this system is the *vernal equinox*, also called the 'First Point of Aries'. . . The angle measured eastward, between Aries and the equatorial intersection of the meridian through a celestial object is called the *right ascension* of the object. The declination and the right ascension together define the position of the object on a celestial background. . .

[Another natural reference] system uses the plane of the earth's revolution around the sun as a reference. The celestial extension of this plane, which is traced by the sun's annual apparent movement, is called the *ecliptic*. Conveniently, the point on this plane which is chosen for a zero reference is also the vernal equinox, at which the sun crosses the equatorial plane from south to north around 21 March each year. Celestial objects are located by their ecliptic latitude and ecliptic longitude. The angle between the two planes, of 23.45°, is called the obliquity of the ecliptic. . . —Pugh (1987: 72).

Tidal Frequencies: Now, let's allow earth to spin about its polar axis. The changing potential at a fixed geographic coordinate on earth is:

$$\cos\varphi = \sin\varphi_p \sin\delta + \cos\varphi_p \cos\delta \cos(\tau_1 - 180^\circ) \quad (17.15)$$

where ϕp is latitude at which the tidal potential is calculated, δ is declination of moon or sun north of the equator, and τ_1 is the hour angle of moon or sun. The *hour angle* is the longitude where the imaginary plane containing the sun or moon and earth's rotation axis crosses the Equator.

The period of the solar hour angle is a solar day of 24 hr 0 m. The period of the lunar hour angle is a lunar day of 24 hr 50.47 m.

Earth's axis of rotation is inclined 23.45° with respect to the plane of earth's orbit about the sun. This defines the ecliptic, and the sun's declination varies between $\delta = \pm 23.45^\circ$ with a period of one solar year. The orientation of earth's rotation axis precesses with respect to the stars with a period of 26 000 years. The rotation of the ecliptic plane causes δ and the vernal equinox to change slowly, and the movement called the *precession of the equinoxes.*

Earth's orbit about the sun is elliptical, with the sun in one focus. That point in the orbit where the distance between the sun and earth is a minimum is called *perigee.* The orientation of the ellipse in the ecliptic plane changes slowly with time, causing perigee to rotate with a period of 20 942 years. There fore R_{sun} varies with this period.

Moon's orbit is also elliptical, but a description of moon's orbit is much more complicated than a description of earth's orbit. Here are the basics. The moon's orbit lies in a plane inclined at a mean angle of 5.15° relative to the plane of the ecliptic. And lunar declination varies between δ = 23.45 ± 5.15°with a period of one tropical month of 27.32 solar days. The actual inclination of moon's orbit varies between 4.97°, and 5.32°.

The shape of moon's orbit also varies. First, perigee rotates with a period of 8.85 years. The eccentricity of the orbit has a mean value of 0.0549, and it varies between 0.044 and 0.067. Second, the plane of moon's orbit rotates around earth's axis of rotation with a period of 18.613 years. Both processes cause variations in R_{moon}.

Note that I am a little imprecise in defining the position of the sun and moon. Lang (1980: § 5.1.2) gives much more precise definitions.

$$V = \frac{\gamma M r^2}{R^3} \frac{1}{4} \left[\left(3 \sin^2 \varphi_p - 1\right) \left(3 \sin^2 \delta - 1\right) + 3 \sin 2\varphi_p \; \sin 2\delta \; \cos \tau_1 + 3 \cos^2 \varphi_p \; \cos^2 \delta \; \cos 2\tau_1 \right] \quad (17.16)$$

Equation (17.16) separates the period of the lunar tidal potential into three terms with periods near 14 days, 24 hours, and 12 hours. Similarly the solar potential has periods near 180 days, 24 hours, and 12 hours. Thus there are three distinct groups of tidal frequencies: twice-daily, daily, and long period, having different latitudinal factors $\sin^2 \theta$, $\sin 2\theta$, and $(1 - 3 \cos^2 \theta)/2$, where θ is the co-latitude (90° - ϕ).

Doodson (1922) expanded (17.16) in a Fourier series using the cleverly chosen frequencies in table 17.1. Other choices of fundamental frequencies are possible, for example the local, mean, solar time can be used instead of the local, mean, lunar time. Doodson's expansion, however, leads to an elegant decomposition of tidal constituents into groups with similar frequencies and spatial variability.

Using Doodson's expansion, each constituent of the tide has a frequency

$$f = n_1 f_1 + n_2 f_2 + n_3 f_3 + n_4 f_4 + n_5 f_5 + n_6 f_6 \quad (17.17)$$

Table 17.1 Fundamental Tidal Frequencies

	Frequency °/hour		Period	Source
f_1	14.49205211	1	lunar day	Local mean lunar time
f_2	0.54901653	1	month	Moon's mean longitude
f_3	0.04106864	1	year	Sun's mean longitude
f_4	0.00464184	8.847	years	Longitude of moon's perigee
f_5	-0.00220641	18.613	years	Longitude of moon's ascending node
f_6	0.00000196	20,940	years	Longitude of sun's perigee

where the integers n_i are the *Doodson numbers*. n_1 = 1, 2, 3 and n_2–n_6 are between -5 and +5. To avoid negative numbers, Doodson added five to $n_{2\cdots6}$.

Each tidal wave having a particular frequency given by its Doodson number is called a *tidal constituent*, sometimes called a *partial tides*. For example, the principal, twice-per-day, lunar tide has the number 255.555. Because the very long-term modulation of the tides by the change in sun's perigee is so small, the last Doodson number *n*6 is usually ignored.

If the ocean surface is in equilibrium with the tidal potential, which means we ignore inertia and currents and assume no land (Cartwright 1999: 274), the largest tidal constituents would have amplitudes given in table 17.2. Notice that tides with frequencies near one or two cycles per day are split into closely spaced lines with spacing separated by a cycle per month. Each of these lines is further split into lines separated by a cycle per year (figure 17.12). Furthermore, each of these lines is split into lines separated by a cycle per 8.8 yr, and so on.

Table 17.2 Principal Tidal Constituents

Tidal							Equilibrium Amplitude†	Period
Species	Name	n_1	n_2	n_3	n_4	n_5	(*m*)	(hr)
Semidiurnal	n_1 = 2							
Principal lunar	M_2	2	0	0	0	0	0.242334	12.4206
Principal solar	S_2	2	2	-2	0	0	0.112841	12.0000
Lunar elliptic	N_2	2	-1	0	1	0	0.046398	12.6584
Lunisolar	K_2	2	2	0	0	0	0.030704	11.9673
Diurnal	n_1 = 1							
Lunisolar	K_1	1	1	0	0	0	0.141565	23.9344
Principal lunar	O_1	1	-1	0	0	0	0.100514	25.8194
Principal solar	P_1	1	1	-2	0	0	0.046843	24.0659
Elliptic lunar	Q_1	1	-2	0	1	0	0.019256	26.8684

Tidal							**Equilibrium Amplitude†**	**Period**
Long Period	*n*1 = 0							
Fortnightly	*Mf*	0	2	0	0	0	0.041742	327.85
Monthly	*Mm*	0	1	0	-1	0	0.022026	661.31
Semiannual	*Ssa*	0	0	2	0	0	0.019446	4383.05

†Amplitudes from Apel (1987)

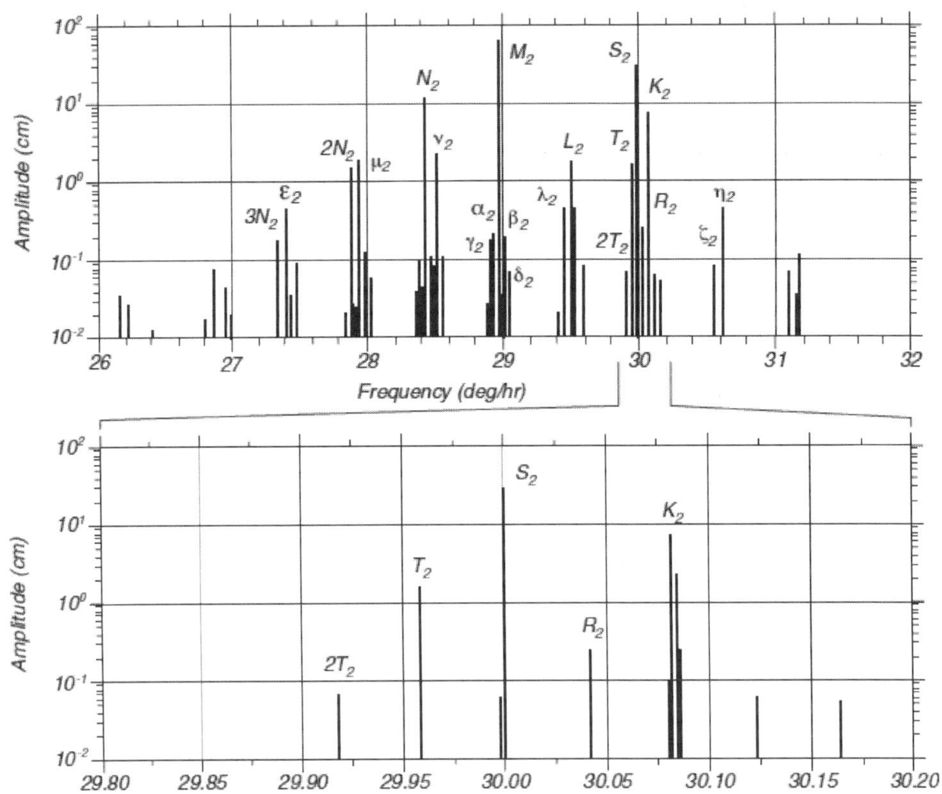

Figure 17.12 Upper: Spectrum of equilibrium tides with frequencies near twice per day. The spectrum is split into groups separated by a cycle per month (0.55 deg/hr). Lower: Expanded spectrum of the $S2$ **group, showing splitting at a cycle per year (0.04 deg/hr). The finest splitting in this figure is at a cycle per 8.847 years (0.0046 deg/hr). From Richard Eanes, Center for Space Research, University of Texas.**

Clearly, there are very many possible tidal constituents.

Why is the tide split into the many constituents shown in figure 17.12? To answer the question, suppose moon's elliptical orbit was in the equatorial plane of earth. Then $\delta = 0$. From (17.16), the tidal potential on the equator, where $\phi p = 0$, is:

$$V = \frac{\gamma M r^2}{R^3}\frac{1}{4}\cos\left(4\pi f_1\right) \quad (17.18)$$

If the ellipticity of the orbit is small, $R = R0(1+E)$, and (17.18) is approximately

$$V = a(1 - 3\epsilon)\cos(4\pi f_1) \quad (17.19)$$

where $a = (\gamma M r^2) / (4R^3)$ is a constant. E varies with a period of 27.32 days, and we can write $E = b\cos(2\pi f2)$ where b is a small constant. With these simplifications, (17.19) can be written:

$$V = a\cos(4\pi f_1) - 3ab\cos(2\pi f_2)\cos(4\pi f_1) \quad (17.20a)$$
$$V = a\cos(4\pi f_1) - 3ab\left[\cos 2\pi(2f_1 - f_2) + \cos 2\pi(2f_1 + f_2)\right] \quad (17.20b)$$

which has a spectrum with three lines at $2f1$ and $2f1 \pm f2$. Therefore, the slow modulation of the amplitude of the tidal potential at two cycles per lunar day causes the potential to be split into three frequencies. This is the way amplitude modulated AM radio works. If we add in the slow changes in the shape of the orbit, we get still more terms even in this very idealized case of a moon in an equatorial orbit.

If you are very observant, you will have noticed that the tidal spectrum in figure 17.12 does not look like the ocean-wave spectrum of ocean waves in figure 16.6. Ocean waves have all possible frequencies, and their spectrum is continuous. Tides have precise frequencies determined by the orbit of earth and moon, and their spectrum is not continuous. It consists of discrete lines.

Doodson's expansion included 399 constituents, of which 100 are long period, 160 are daily, 115 are twice per day, and 14 are thrice per day. Most have very small amplitudes, and only the largest are included in table 17.2. The largest tides were named by Sir George Darwin (1911) and the names are included in the table. Thus, for example, the principal, twice-per-day, lunar tide, which has Doodson number 255.555, is the *M2* tide, called the *M-two* tide.

17.5 FORECASTING TIDAL BEHAVIOR

If tides in the ocean were in equilibrium with the tidal potential, tidal pre- diction would be much easier. Unfortunately, tides are far from equilibrium. The shallow-water wave which is the tide cannot move fast enough to keep up with sun and moon. On the equator, the tide would need to propagate around the world in one day. This requires a wave speed of around 460 m/s, which is only possible in an ocean 22 km deep. In addition, the continents interrupt the propagation of the wave. How to proceed?

We can separate the problem of tidal prediction into two parts. The first deals with the prediction of tides in ports and shallow water where tides can be measured by tide gauges. The second deals with the prediction of tides in the deep ocean where tides are measured by satellite altimeters.

Tidal Prediction for Ports and Shallow Water: Two methods are used to predict future tides at a tide gauge station using past observations of sea level measured at the gauge.

The Harmonic Method This is the traditional method, and it is still widely used. The method typically uses 19 years of data from a coastal tide gauge from which the amplitude and phase of each tidal constituent (the tidal harmonics) in the tide-gage record are calculated. The frequencies used in the analysis are specified in advance from the basic frequencies given in table 17.1.

Despite its simplicity, the technique had disadvantages compared with the response method described below.

1. More than 18.6 years of data are needed to resolve the modulation of the lunar tides.
2. Amplitude accuracy of 10^{-3} of the largest term require that at least 39
3. frequencies be determined. Doodson found 400 frequencies were needed for an amplitude accuracy of 10^{-4} of the largest term.
4. Non-tidal variability introduces large errors into the calculated amplitudes and phases of the weaker tidal constituents. The weaker tides have amplitudes smaller than variability at the same frequency due to other processes such as wind set up and currents near the tide gauge.
5. At many ports, the tide is non-linear, and many more tidal constituents are important. For some ports, the number of frequencies is unmanageable. When tides propagate into very shallow water, especially river estuaries, they steepen and become non-linear. This generates harmonics of the original frequencies. In extreme cases, the incoming waves steepens so much the leading edge is nearly vertical, and the wave propagates as solitary wave. This is a *tidal bore.*

The Response Method This method, developed by Munk and Cartwright (1966), calculates the relationship between the observed tide at some point and the tidal potential. The relationship is the spectral admittance between the major tidal constituents and the tidal potential at each station. The admittance is assumed to be a slowly varying function of frequency so that the admittance of the major constituents can be used for determining the response at nearby frequencies. Future tides are calculated by multiplying the tidal potential by the admittance function.

1. The technique requires only a few months of data.
2. The tidal potential is easily calculated, and a knowledge of the tidal frequencies is not needed.
3. The admittance is $Z(f) = G(f)/H(f)$. $G(f)$ and $H(f)$ are the Fourier transforms of the potential and the tide gage data, and f is frequency.

4. The admittance is inverse transformed to obtain the admittance as a function of time.

5. The technique works only if the waves propagate as linear waves.

Tidal Prediction for Deep-Water: Prediction of deep-ocean tides has been much more difficult than prediction of shallow-water tides because tide gauges were seldom deployed in deep water. All this changed with the launch of Topex/Poseidon. The satellite was placed into an orbit especially designed for observing ocean tides (Parke et al. 1987), and the altimetric system was sufficiently accurate to measure many tidal constituents. Data from the satellite have now been used to determine deep-ocean tides with an accuracy of ± 2 cm. For most practical purposes, the tides are now known accurately for most of the ocean.

Two avenues led to the new knowledge of deep-water tides using altimetry. *Prediction Using Hydrodynamic Theory* Purely theoretical calculations of tides are not very accurate, especially because the dissipation of tidal energy is not well known. Nevertheless, theoretical calculations provided insight into processes influencing ocean tides. Several processes must be considered:

1. The tides in one ocean basin perturb earth's gravitational field, and the mass in the tidal bulge attracts water in other ocean basins. The self- gravitational attraction of the tides must be included.
2. The weight of the water in the tidal bulge is sufficiently great that it deforms the sea floor. The earth deforms as an elastic solid, and the deformation extends thousands of kilometers.
3. The ocean basins have a natural resonance close to the tidal frequencies. The tidal bulge is a shallow-water wave on a rotating ocean, and it propagates as a high tide rotating around the edge of the basin. Thus the tides are a nearly resonant sloshing of water in the ocean basin. The actual tide heights in deep water can be higher than the equilibrium values noted in table 17.2.
4. Tides are dissipated by bottom friction especially in shallow seas, by the flow over seamounts and mid-ocean ridges, and by the generation of in- ternal waves over seamounts and at the edges of continental shelves. If the tidal forcing stopped, the tides would continue sloshing in the ocean basins for several days.
5. Because the tide is a shallow-water wave everywhere, its velocity depends on depth. Tides propagate more slowly over mid-ocean ridges and shallow seas. Hence, the distance between grid points in numerical models must be proportional to depth with very close spacing on continental shelves (LeProvost et al. 1994).
6. Internal waves generated by the tides produce a small signal at the sea surface near the tidal frequencies, but not phase-locked to the potential. The noise near the frequency of the tides causes the spectral cusps in the spectrum of sea-surface elevation first seen by Munk and Cartwright (1966). The noise is due to deep-water, tidally generated, internal waves.

Altimetry Plus Response Method Several years of altimeter data from Topex/Poseidon have been used with the response method to calculate deep-sea tides almost everywhere equator ward of 66° (Ma et al. 1994). The altimeter measured sea-surface heights in geocentric coordinates at each point along the sub- satellite track every 9.97 days. The temporal sampling aliased the tides into long frequencies, but the aliased periods are precisely known and the tides can be recovered (Parke et al. 1987). Because the tidal record is shorter than 8 years, the altimeter data are used with the response method to obtain predictions for a much longer time.

Recent solutions by ten different groups, have accuracy of ± 2.8 cm in deep water (Andersen, Wood worth, and Flather, 1995). Work has begun to improve knowledge of tides in shallow water.

Altimetry Plus Numerical Models Altimeter data can be used directly with numerical models of the tides to calculate tides in all areas of the ocean from deep water all the way to the coast. Thus the technique is especially useful for determining tides near coasts and over sea-floor features where the altimeter ground track is too widely spaced to sample the tides well in space. Tide models use finite-element grids similar to the one shown in figure 15.3. Recent numerical calculations by (LeProvost et al. 1994; LeProvost, Bennett, and Cartwright, 1995) give global tides with ± 2–3 cm accuracy and full spatial resolution.

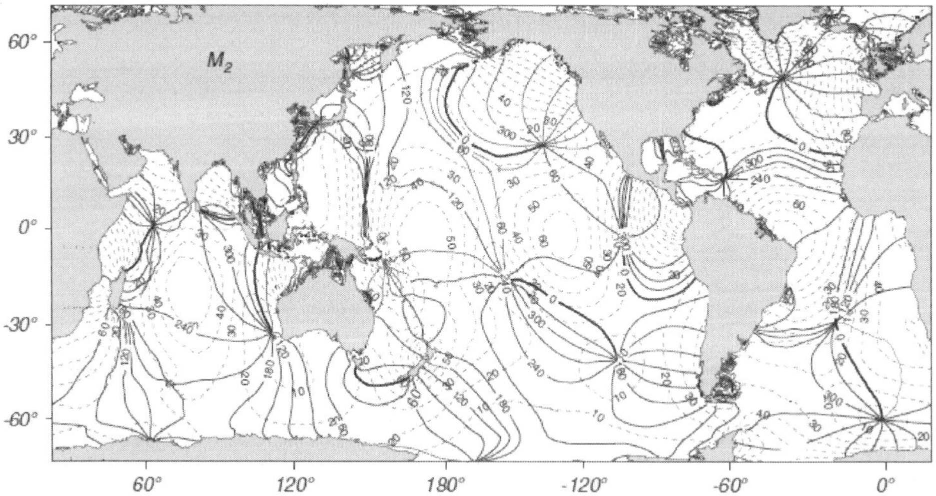

Figure 17.13 Global map of ***M2*** **tide calculated from Topex/Poseidon observations of the height of the sea surface combined with the response method for extracting tidal information. Full lines are contours of constant tidal phase, contour interval is 30°. Dashed lines are lines of constant amplitude, contour interval is 10 cm. From Richard Ray, NASA Goddard Space Flight Center.**

Maps produced by this method show the essential features of the deep-ocean tides (figure 17.13). The tide consists of a crest that rotates counterclockwise around the ocean basins in the northern hemisphere, and in the opposite direction in the southern hemisphere. Points of minimum amplitude are called *amphidromes.* Highest tides tend to be along the coast.

The maps also show the importance of the size of the ocean basins. The semi-diurnal (12 hr period) tides are relatively large in all ocean basins. But the diurnal (24 hr period) tides are small in the Atlantic and relatively large in the Pacific and Indian ocean. The Atlantic is too small to have a resonant sloshing with a period near 24 hr.

Tidal Dissipation: Tides dissipate 3.75 ± 0.08 TW of power (Kantha, 1998), of which 3.5 TW are dissipated in the ocean, and much smaller amounts in the atmosphere and solid earth. The dissipation increases the length of dayby about 2.07 milliseconds per century, it causes the semimajor axis of moon's orbit to increase by 3.86 cm/yr, and it mixes water masses in the ocean.

The calculations of dissipation from Topex/Poseidon observations of tides are remarkably close to estimates from lunar-laser ranging, astronomical observations, and ancient eclipse records. The calculations show that roughly two thirds of the M2 tidal energy is dissipated on shelves and in shallow seas, and one third is transferred to internal waves and dissipated in the deep ocean (Egbert and ray, 2000). 85 to 90% of the energy of the K1 tide is dissipated in shallow water, and only about 10–15% is transferred to internal waves in the deep ocean (LeProvost 2003, personal communication).

Overall, our knowledge of the tides is now sufficiently good that we can begin to use the information to study mixing in the ocean. Recent results show that "tides are perhaps responsible for a large portion of the vertical mixing in the ocean" (Jayne et al. 2004). Remember, mixing helps drive the abyssal circulation in the ocean as discussed in §13.2 (Munk and Wunsch, 1998). Who would have thought that an understanding of the influence of the ocean on climate would require accurate knowledge of tides?

17.6 CONCISE SUMMARY

- Waves propagating into shallow water are refracted by features of the sea floor, and they eventually break on the beach. Breaking waves drive near-shore currents including long-shore currents, rip currents, and edge waves.
- Storm surges are driven by strong winds in storms close to shore. The amplitude of the surge is a function of wind speed, the slope of the seafloor, and the propagation of the storm.
- Tides are important for navigation; they influence accurate geodetic measurements; and they change the orbits and rotation of planets, moons, and stars in galaxies.

- Tides are produced by a combination of time-varying gravitational potential of the moon and sun and the centrifugal forces generated as earth rotates about the common center of mass of the earth-moon-sun system.
- Tides have six fundamental frequencies. The tide is the superposition of hundreds of tidal constituents, each having a frequency that is the sum and difference of five fundamental frequencies.
- Shallow water tides are predicted using tide measurements made in ports and other locations along the coast. Tidal records of just a few months duration can be used to predict tides many years into the future.
- Tides in deep water are calculated from altimetric measurements, especially Topex/Poseidon measurements. As a result, deep water tides areknown almost everywhere with an accuracy approaching ±2 cm.
- The dissipation of tidal energy in the ocean transfers angular momentum from moon to earth, causing the day to become longer.
- Tidal dissipation mixes water masses, and it is a major driver of the deep, meridional-overturning circulation. Tides, abyssal circulation, and climate are closely linked.

REFERENCES

Alley R.B. 2000. Ice-core evidence of abrupt climate change. *Proceedings of National Academy of Sciences* 97 (4): 1331–1334.

Apel J.R. 1987. *Principles of Ocean Physics*. New York: Academic Press.

Anderson J.D. 2005. Ludwig Prandtl's Boundary Layer. *Physics Today* 58 (12): 42–48.

Andersen O.B., P.L. Woodworth, and R.A. Flather. 1995. Inter comparison of recent ocean tide models. *J. Geophysical Research* 100 (C12): 25,262–25,282.

Arthur R.S. 1960. A review of the calculation of ocean currents at the equator. *Deep-Sea Research* 6 (4): 287–297.

Atlas R., R.N. Hoffman, and S.C. Bloom. 1993. Surface wind velocity over the oceans. In: *Atlas of satellite observations related to global change*. Edited by R. J. Gurney, J. L. Foster and C. L. Parkinson. 129–140. Cambridge: University Press.

Baker D.J. 1981. Ocean instruments and experiment design. In *Evolution of Physical Oceanography: Scientific Surveys in Honor of Henry Stommel*. Edited by B. A. Warren and C. Wunsch. 396–433. Cambridge: Massachusetts Institute of Technology Press.

Barnett T.P., M. Latif, N.E. Graham, M. Flugel, S. Pazan, and W. White. 1993. enso and enso-related predictability. Part I: Prediction of equatorial Pacific sea surface temperature in a hybrid coupled ocean-atmosphere model. *Journal of Climate* 6: 1545– 1566.

Barnier B., L. Siefridt, and P. Marchesiello. 1995. Thermal forcing for a global ocean circulation model using a three-year climatology of ECMWF analyses. *Journal of Marine Systems* 6: 393–380.

Barnston A.G., Y. Hea, and M.H. Glantz. 1999. Predictive Skill of Statistical and Dynami- cal Climate Models in SST Forecasts during the 1997–98 El Niño Episode and the 1998 La Nin˜a Onset. *Bulletin of the American Meteorological Society* 80 (2): 217–243.

Batchelor G.K. 1967. *An Introduction to Fluid Dynamics*. Cambridge: University Press.

Baumgartner A., and E. Reichel. 1975. *The World Water Balance*. New York: Elsevier.

Beardsley R.C., A.G. Enriquez, C.A. Friehe, and C.A. Alessi. 1997. Intercomparison of aircraft and buoy measurements of wind speed and wind stress during smile. *Journal of Atmospheric and Oceanic Technology* 14: 969–977.

Behringer D.W., M. Ji and A. Leetmaa. 1998. An improved coupled model for enso pre- diction and implications for ocean initialization: Part 1: The ocean data assimilation. *Monthly Weather Review* 126(4): 1013–1021.

Bennett A.F. 1992. *Inverse Methods in Physical Oceanography*. Cambridge: University Press.

Berlinski D. 1996. The end of materialist science. *Forbes ASAP* (December 2, 1996): 146– 160.

Binder R.C. 1949. *Fluid Mechanics*. Second ed. New York: Prentice–Hall.

Bjerknes J. 1966. The possible response of the atmospheric Hadley circulation to equatorial anomalies of ocean temperature. *Tellus* 4: 820-929.

Bjerknes J. 1972. Large-scale atmospheric response to the 1964–65 Pacific equatorial warm- ing. *Journal of Physical Oceanography* 2 (3): 212–217.

Bjerknes V. and J.W. Sandstro¨m. 1910. Dynamic Meteorology and Hydrography, Part I. Statics. Carnegie Institution of Washington DC, Publication No. 88.

Bleck R. 2002. An oceanic general circulation model framed in hybrid isopycnic-Cartesian coordinates. *Ocean Modeling* 4: 55–88.

Blumberg A.F., and G.L. Mellor. 1987. A description of a three-dimensional ocean circu- lation model. In: *Three-Dimensional Coastal Ocean Models*. Edited by N. S. Heaps. 1–16. Washington, DC: American Geophysical Union.

Bond G., H. Heinrich, W. Broecker, L. Labeyrie, J. McManus, J. Andrews, S. Huon, R. Jantschik, S. Clasen, C. Simet, K. Tedesco, M. Klas, G. Bonani, and S. Ivy. 1992. Evidence for massive discharges of icebergs into the North Atlantic ocean during the last glacial period. *Nature* 360, 245–.

Boville B.A., and P.R. Gent. 1998. The NCAR Climate System Model, Version One. *Journal of Climate* 11 (6): 1115–1130.

Bowden K.F. 1962. Turbulence. In: *The Sea Volume 1*. Edited by M. N. Hill. 802–825. New York: Interscience Publishers, John Wiley and Sons.

Bracewell R.N. 1986. *The Fourier Transform and Its Applications*. Second, revised ed. New York: McGraw-Hill Publishing Company.

Brauer A., G. H. Haug, et al. 2008. An abrupt wind shift in western Europe at the onset of the Younger Dryas cold period. *Nature Geoscience* 1 (8): 520–523.

Broecker W.S. 1987. Unpleasant surprises in the greenhouse? *Nature* 328: 123–126.

Broecker W.S. 1997. Thermohaline circulation, the Achilles heel of our climate system: Will man-made CO2 upset the current balance? *Science* 278 (5343): 1582–1588.

Bryan K. 1969. A numerical method for the study of the world ocean. *Journal of Compu- tational Physics* 4: 347–376.

Bryden H.L. and T.H. Kinder. 1991. Steady two-layer exchange through the Strait of Gibraltar. *Deep Sea Research* 38 Supplement 1A: S445–S463.

Carritt D.E., and J.H. Carpenter. 1959. The composition of sea water and the salinity- chlorinity-density problems. National Academy of Sciences-National Research Council, Report 600: 67–86.

Cartwright D.E. 1999. *Tides: A Scientific History*. Cambridge, University Press.

Cazenave A., and J.Y. Royer. 2001. Applications to marine Geophysics. In *Satellite Al- timetry and earth Sciences*. 407–439. San Diego: Academic Press.

Cess R.D., M.H. Zhang, P.Minnis, L.Corsetti, E.G. Dutton, B.W. Forgan, D.P. Garber, W.L. Gates, J.J. Hack, E.F. Harrison, X. Jing, J.T. Kiehl, C.N. Long, J.-J. Morcrette, G.L. Potter, V. Ramanathan, B. Subasilar, C.H. Whitlock, D.F. Young, and Y. Zhou. 1995. Absorption of solar radiation by clouds: Observations versus models. *Science* 267 (5197): 496–499.

Chambers D.P., B.D. Tapley, and Stewart, R.H. 1998. Measuring heat storage changes in the equatorial Pacific: A comparison between Topex altimetry and Tropical Atmo- sphere-Ocean buoys. *Journal of Geophysical Research* 103(C9): 18,591–18,597.

Charnock H. 1955. Wind stress on a water surface. *Quarterly Journal Royal Meteorological Society* 81: 639–640.

Chelton D.B., J.C. Ries, B.J. Haines, L.L. Fu, and P.S. Callahan. 2001. Satellite Altimetry. In: *Satellite Altimetry and Earth Sciences: A handbook of techniques and applications*. Editors: L.-L. Fu and A. Cazenave. Academic Press: 1–131.

Chen D., S.E. Zebiak, A.J. Busalacchi, and M.A. Cane. 1995. An improved proceedure for El Niño forecasting: Implications for predictability. *Science* 269 (5231): 1699–1702.

Chereskin T.K., and D. Roemmich. 1991. A comparison of measured and wind-derived Ekman transport at 11°N in the Atlantic Ocean. *Journal of Physical Oceanography* 21 (6): 869–878.

Chen D., S.E. Zebiak, A.J. Busalacchi, and M.A. Cane. 1995. An improved proceedure for El Niño forecasting: Implications for predictability. *Science* 269 (5231): 1699–1702.

Chou S.-H., E. Nelkin, et al. 2004. A comparison of latent heat fluxes over global oceans for four flux products. *Journal of Climate* 17 (20): 3973–3989.

Church J. A. 2007. Oceans: A Change in Circulation? *Science* 317 (5840): 908–909.

Clarke G.L., G.C. Ewing, and C.J. Lorenzen. 1970. Spectra of backscattered light from the sea obtained from aircraft as a measure of chlorophyll concentration. *Science* 167: 1119–1121.

Coakley J.A., and F.P. Bretherton. 1982. Cloud cover from high resolution scanner data: Detecting and allowing for partially filled fields of view. *Journal of Geophysical. Re- search* 87 (C7): 4917–4932.

Cooley J.W., P.A. Lewis, and P.D. Welch. 1970. The fast Fourier transform algorithm: Programming considerations in the calculation of sine, cosine and Laplace transforms. *Journal of Sound and Vibration* 12: 315–337.

Couper A. Editor. 1983. *The Times Atlas of the Oceans*. New York: Van Nostrand Reinhold Company.

Cox M.D. 1975. A baroclinic model of the world ocean: Preliminary results. In: *Numer- ical Models of Ocean Circulation*: 107–120. Washington, DC: National Academy of Sciences.

Cromwell T., R.B. Montgomery, and E.D. Stroup. 1954. Equatorial undercurrent in Pacific Ocean revealed by new methods. *Science* 119 (3097): 648–649.

Cushman-Roisin B. 1994. *Introduction to Geophysical Fluid Dynamics*. Englewood Cliffs: Prentice Hall.

Cutchin D.L., and R.L. Smith. 1973. Continental shelf waves: Low-frequency variations in sea level and currents over the Oregon continental shelf. *Journal of Physical Oceanog- raphy* 3 (1): 73–82.

Daley R. 1991. *Atmospheric Data Analysis*. Cambridge: University Press.

Danabasoglu G., J.C. McWilliams, and P.R. Gent. 1994. The role of mesoscale tracer transports in the global ocean circulation. *Science* 264 (5162): 1123–1126.

Dansgaard W., S.J. Johnsen, H.B. Clausen, D. Dahl-Johnsen, N. Gunderstrup, C.U. Ham- mer, C. Hvidberg, J. Steffensen, A. Sveinbjrnsobttir, J. Jouze and G. Bond. 1993. Evidence for general instability of past climate from a 250-kyr ice core record. *Nature* 364: 218-220.

Darnell W.L., F. Staylor, S.K. Gupta, and F,M. Denn. 1988. Estimation of surface insola- tion using sun-synchronous satellite data. *Journal of Climate* 1 (8): 820–835.

Darnell W.L., W.F. Staylor, S.K. Gupta, N.A. Ritchey, and A.C. Wilbur. 1992. Seasonal variation of surface radiation budget derived from International Satellite Cloud Clima- tology Project C1 data. *Journal of Geophysical Research* 97: 15,741–15,760.

Darwin Sir G.H. 1911. *The Tides and Kindred Phenomena in the Solar System*. 3rd ed. London: John Murray.

DaSilva A., C.C. Young, and S. Levitus. 1995. Atlas of surface marine data 1994. Vol. 1: Algorithms and procedures. National Oceanic and Atmospheric Administration Report.

Davis R.A. 1987. *Oceanography: An Introduction to the Marine Environment*. Dubuque: Wm. C. Brown Publishers.

Davis R.E., R. DeSzoeke, and P. Niiler. 1981. Variability in the upper ocean during mile. Part II: Modeling the mixed layer response. *Deep-Sea Research* 28A (12): 1453–1475.

Davis R.E., D.C. Webb, L.A. Regier, and J. Dufour. 1992. The Autonomous Lagrangian Circulation Explorer (alace). *Journal of Atmospheric and Oceanic Technology* 9: 264–285.

Defant A. 1961. *Physical Oceanography*. New York: Macmillan Company.

Dietrich G., K. Kalle, W. Krauss, and G. Siedler. 1980. *General Oceanography*. 2nd ed. Translated by Susanne and Hans Ulrich Roll. New York: John Wiley and Sons (Wiley- Interscience).

Dittmar W. 1884. Report on researches into the composition of ocean water, collected by the HMS Challenger. *Challenger Reports, Physics and Chemistry* 1.

Dobson G.M.B. 1914. Pilot balloon ascents at the central flying school, Upavon, during the year 1913. *Quarterly Journal of the Royal Meteorological Society* 40: 123–135.

Domingues C. M., J. A. Church, et al 2008. Improved estimates of upper-ocean warming and multi-decadal sea-level rise. *Nature* 453 (7198): 1090–1093.

Doodson A.T. 1922. Harmonic development of the tide-generating potential. *Proceedings of the Royal Society of London* A 100: 305–329.

Dorman C.E. and R.H. Bourke. 1981. Precipitation over the Atlantic ocean, 30°S to 70°N. *Monthly Weather Review* 109: 554–563.

Dritschel D.G., M. de la T. Juarez and M.H.P. Ambaum. 1999. The three-dimensional vortical nature of atmospheric and oceanic turbulent flows. *Physics of Fluids* 11(6): 1512–1520.

Dushaw B.D., P.F. Worcester, B.D. Cornuelle, and B.M. Howe. 1993. On equations for the speed of sound in sea water. *Journal of the Acoustical Society of America* 93: 255–275.

Ebbesmeyer C.C., and W.J. Ingraham. 1992. Shoe spill in the North Pacific. *EOS, Trans- actions of the American Geophysical Union* 73 (34): 361, 365.

Ebbesmeyer C.C., and W.J. Ingraham. 1994. Pacific toy spill fuels ocean current pathways research. *EOS Transactions of the American Geophysical Union* 75 (37): 425, 427,430.

Eden C., and J. Willebrand. 1999. Neutral density revisited. *Deep-Sea Research* Part II: Topical Studies in Oceanography. 46: 33–54.

Egbert G.B. and R.D. Ray 2000. Significant dissipation of tidal energy in the deep ocean inferred from satellit altimeter data. *Nature* 405: 775–778.

Ekman V.W. 1905. On the influence of the Earth's rotation on ocean currents. *Arkiv for Matematik, Astronomi, ochFysik* : 2 (11).

Emery W., and P. Schussel. 1989. Global difference between skin and bulk sea surface temperatures. *EOS: Transactions of the American Geophysical Union* 70 (14): 210– 211.

Feynman R.P., R.B. Leighton, and M. Sands. 1964. *The Feynman Lectures on Physics.* Addison-Wesley Publishing Company.

Fofonoff N.P., and R.C. Millard. 1983. Algorithms for computation of fundamental proper- ties of sea water. unesco Technical Papers in Marine Science 44.

Friedman R.M. 1989. *Appropriating the Weather. Vilhelm Bjerknes and the Construction of a Modern Meteorology.* Ithaca and London: Cornell University Press.

Friedrichs M.A.M., and M.M. Hall. 1993. Deep circulation in the tropical North Atlantic.*Journal of Marine Research* 51 (4): 697–736.

Freilich M.H., and R.S. Dunbar. 1999. The accuracy of the NSCAT 1 vector winds: Com- parisons with National Data Buoy Center buoys. *Journal of Geophysical Research* submitted

Garabato A.C.N., K.L. Polzin, B.A. King, K.J. Heywood, and M. Visbeck. 2004. Widespread Intense Turbulent Mixing in the Southern Ocean. *Science* 303 (5655): 210–213.

Garabato A.C.N., D.P. Stevens, A.J. Watson, and W. Roether. 2007. Short-circuiting of the overturning circulation in the Antarctic Circumpolar Current. *Nature* 447 (7141): 194–197.

Gargett A.E., and G. Holloway. 1992. Sensitivity of the gfdl ocean model to different diffusivities of heat and salt. *Journal of Physical Oceanography* 22 (10): 1158–1177.

Garrett C. 2006. Turbulent dispersion in the ocean. *Progress in Oceanography* 70 (2–4): 113–125.

Gates W.L., A. Henderson-Sellers, G.J. Boer, C.K. Folland, A. Kitoh, B.J. McAvaney, F. Semazzi, N. Smith, A.J. Weaver, and Q.-C. Zeng. 1996. *Climate Models–Evaluation.* In: *Climate Change 1995*. Edited by J.T. Houghton, L.G.M. Filho, B.A. Callander, N. Harris, A. Kattenberg and K. Maskell. 229–284. Cambridge: University Press.

Gates W.L. 1992. amip: The Atmospheric Model Intercomparison Project. *Bulletin American Meteorological Society* 73: 1962–1970.

Gent P.R., and J.C. McWilliams. 1990. Isopycnal mixing in ocean circulation models. *Journal of Physical Oceanography* 20: 150–155.

Gleckler P.J., and B. Weare. 1997. Uncertainties in global ocean surface heat flux climatologies derived from ship observance. *Journal of Climate* 10: 2763–2781.

Gill A.E. 1982. *Atmosphere-Ocean Dynamics*. New York: Academic Press.

Gnadadesikan A. 1999. A simple predictive model for the structure of the oceanic pycno- cline. *Science* 283 (5410): 2077–2079.

Goldenberg S.B., and J.J. O'Brien. 1981. *Monthly Weather Review* 109: 1190.

Goldstein S. 1965. *Modern Developments in Fluid Dynamics*: Two Volumes. New York: Dover Publications.

Gordon H.R., D.K. Clark, J.W. Brown, O.B. Brown, R.H. Evans, and W.W. Broenkow. 1983. Phytoplankton pigment concentrations in the Middle Atlantic Bight: comparison of ship determinations and czcs estimates. *Applied Physics* 22 (1): 20-36.

Gouretski V., and K. Jancke. 1995. A consistent prewoce hydrographic data set for the south Atlantic: Station data and gridded fields. woce Report No. 127/95.

Graber H.C., V.J. Cardone, R.E. Jensen, D.N. Slinn, S.C. Hagen, A.T. Cox, M.D. Powell, and C. Grassl. 2006. Coastal Forecasts and Storm Surge Predictions for Tropical Cyclones: A Timely Partnership Program. *Oceanography* 19 (1): 131–141.

Grassl H. 2000. Status and improvements of coupled general circulation models. *Science*288 (5473): 1991–1997.

Gregg M.C. 1987. Diapycnal mixing in the thermocline: A review. *Journal of Geophysical Research* 92 (C5): 5,249–5,289.

Gregg M.C. 1991. The study of mixing in the ocean: a brief history. *Oceanography* 4 (1): 39–45.

Hackett B., L.P. Roed, B. Gjevik, E.A. Martinsen, and L.I. Eide. 1995. A review of the MetoceanModeling Project (mompop) Part 2: Model validation study. In: *Quantitative Skill Assessment for Coastal Ocean Models*. Edited by D. R. Lynch and A. M. Davies. 307–327. Washington DC: American Geophysical Union.

Haidvogel D.B., and A. Beckmann. 1998. Numerical models of the coastal ocean. In: *The Sea, Volume 10*. Edited by K. H. Brink and A. R. Robinson. 457–482. New York: John Wiley and Sons.

Haidvogel D.B. and A. Beckmann. 1999. *Numerical Ocean Circulation Modeling*. London, Imperial College Press.

Hall M.M., and H.L. Bryden. 1982. Direct estimates and mechanisms of ocean heat trans- port. *Deep-Sea Research* 29: 339–359.

Harrison D.E. 1989. On climatological monthly mean wind stress and wind stress curl fields over the world ocean. *Journal of Climate* 2: 57.

Harrison D.E., and N.K. Larkin. 1996. The coads sea level pressure signal: A near-global elNiño composite and time series view, 1946–1993. *Journal of Climate* 9 (12): 3025– 3055.

Harrison D.E. and N.K. Larkin 1998. El Niño-Southern Oscillation sea surface temperature and wind anomalies, 1946–1993. *Reviews of Geophysics* 36 (3): 353–399.

Hartmann D.L. 1994. *Global Physical Climatology*. Academic Press.

Hasselmann K. 1961. On the non-linear energy transfer in a gravity-wave spectrum Part 1.General theory. *Journal of Fluid Mechanics* 12 (4): 481–500.

Hasselmann K. 1963a. On the non-linear energy transfer in a gravity wave spectrum. PartConservation theorems; wave-particle analogy; irreversibility. *Journal of Fluid Mechanics* 15 (2): 273–281.

Hasselmann K. 1963b. On the non-linear energy transfer in a gravity wave spectrum. PartEvaluation of the energy flux and swell-sea interaction for a Neumann spectrum.*Journal of Fluid Mechanics.* 15 (3): 385–398.

Hasselmann K. 1966. Feynman diagrams and interaction rules of wave-wave scattering processes. *Reviews of Geophysical.* 4 (1): 1–32.

Hasselmann K. 1970. Wind–driven inertial oscillations. *Geophysical Fluid Dynamics* 1: 1 463–502.

Hasselmann K., T.P. Barnett, E. Bouws, H. Carlson, D.E. Cartwright, K. Enke, J.A. Ewing,H. Gienapp, D.E. Hasselmann, P. Kruseman, A. Meerburg, P. Mller, D.J. Olbers, K. Richter, W. Sell, and H. Walden. 1973. Measurements of wind-wave growth and swell decay during the Joint North Sea Wave Project (jonswap). *Ergnzungsheft zur Deutschen Hydrographischen Zeitschrift Reihe* A(8) (Nr. 12): 95.

Hasselmann K., and S. Hasselmann. 1991. On the nonlinear mapping of an ocean wave spectrum into a synthetic aperture radar image spectrum and its inversion. *Journal of Geophysical Research* C96 10,713–10,729.

Heaps N.S., Editor. 1987. *Three-Dimensional Coastal Ocean Models.* Washington DC: American Geophysical Union.

Hinze J.O. 1975. *Turbulence*. 2nd ed. New York: McGraw-Hill.

Hirst A.C., S.P. O'Farrell, and H.B. Gordon. 2000. Comparison of a coupled ocean-atmosphere model with and without oceanic eddy-induced advection. Part I: Oceanic spinup and control integration. *Journal of Climate* 13 (1): 139–163.

Hoffman D., and O.J. Karst. 1975. The theory of the Rayleigh distribution and some of its applications. *Journal of Ship Research* 19 (3): 172–191.

Hogg N., J. McWilliams, P. Niiler and J. Price. 2001. Objective 8—To determine the important processes and balances for the dynamics of the general circulation. In: *2001 U.S. WOCE Report*. College Station, Texas, U.S. woce Office: 50–59.

Holloway G. 1986. Eddies, waves, circulation, and mixing: statistical geofluid mechanics. *Annual Reviews of Fluid Mechanics* 18: 91–147.

Holloway G. 1986. Estimation of oceanic eddy transports from satellite altimetry. *Nature*323 (6085): 243–244.

Holloway G. 1994. Representing eddy forcing in models. woce Notes 6 (3): 7–9.

Horikawa K. 1988. *Nearshore Dynamics and Coastal Processes*. Tokyo: University of Tokyo Press.

Houghton J.T. 1977. *The Physics of Atmospheres*. Cambridge: University Press.

Houghton J.T., L.G.M. Filho, B.A. Callander, N. Harris, A. Kattenberg, and K. Maskell. 1996. *Climate Change 1995: The Science of Climate Change*. Cambridge: University Press.

Hoyt D.V., and K.H. Schatten. 1997. *The Role of the Sun in Climate Change*. Oxford: Oxford University Press.

Huffman G.J., R.F. Adler, B. Rudolf, U. Schneider, and P.R. Keehm. 1995. Global pre- cipitation estimates based on a technique for combining satellite-based estimates, rain gauge analysis, and nwp model precipitation information. *Journal of Climate* 8: 1284– 1295.

Huffman G.J., R.F. Adler, P.A. Arkin, A. Chang, R. Ferraro, A. Gruber, J. Janowiak, A. McNab, B. Rudolf, and U. Schneider. 1997. The Global Precipitation Climatology Project (gpcp) combined precipitation data set. *Bulletin of the American Meteorological Society* 78 (1): 5–20.

Ichiye T., and J. Petersen. 1963. The anomalous rainfall of the 1957–1958 winter in the equatorial central Pacific arid area. *Journal of the Meteorological Society of Japan* Series II, 41: 172–182.

International Hydrographic Bureau 1953. Limits of oceans and seas, 3rd ed. Special Report No. 53, Monte Carlo.

IPCC Intergovernmental Panel on Climate Change. 2007. *Climate Change 2007: The Physical Science Basis. Contribution of Working Group I to the Fourth Assessment Report of the Intergovernmental Panel on Climate Change*. [Solomon, S., D. Qin, M. Manning, Z. Chen, M. Marquis, K.B. Averyt, M. Tignor and H.L. Miller (eds.)]. Cambridge University Press, Cambridge, United Kingdom and New York, NY, USA, 996 pp.

Iselin C. 1936. A study of the circulation of the Western North Atlantic. *Physical Oceanography and Meteorology*. 6 (4): 101.

Isemer H.J., and L. Hasse. 1987. *The Bunker Climate Atlas of the North Atlantic*. Volume2. Berlin: Springer-Verlag.

Jackett D.R., and T.J. McDougall. 1997. A neutral density variable for the world's oceans. *Journal of Physical Oceanography* 27: 237–263.

Jan van Oldenborgh G., M.A. Balmaseda, L. Ferranti, T.N. Stockdale, and D.L.T. An- derson. 2005. Did the ECMWF Seasonal Forecast Model Outperform Statistical enso Forecast Models over the Last 15 Years? *Journal of Climate* 18 (16): 3240–3249.

Jarosz E., D.A. Mitchell, D.W. Wang, and W.J. Teague. 2007. Bottom-Up Determination of Air-Sea Momentum Exchange Under a Major Tropical Cyclone. *Science* 315 (5819): 1707–1709.

Jayne S.R., L.C.S. Laurent and S.T. Gille. 2004. Connections between ocean bottom topography and Earth's climate. *Oceanography* 17 (1): 65–74.

Jelesnianski C.P.J., P.C. Chen, and W.A. Shaffer. 1992. slosh: Sea, lake, and overland surges from hurricanes. NOAA Technical Report nws 48.

Jerlov N.G. 1976. *Marine Optics*. Amsterdam: Elsevier Scientific Publishing Company.

Ji M., A. Leetmaa, and V. Kousky. 1996. Coupled model predictions of enso during the 1980s and the 1990s at the National Centers for Environmental Prediction. *Journal of Climate* 9 (12): 3105–3120.

Ji M., D.W. Behringer, and A. Leetmaa. 1998. An improved coupled model for enso prediction and implications for ocean initialization. Part II: The coupled model. *Bulletin of the American Meteorological Society* 126 (4): 1022–1034.

Johns E., D.R. Watts, and H.T. Rossby. 1989. A test of geostrophy in the Gulf Stream. *Journal of Geophysical Research* 94 (C3): 3211–3222.

Johns T.C., R.E. Carnell, J.F. Crossley, J.M. Gregory, J.F.B. Mitchell, C.A. Senior, S.F.B. Trett, and R.A. Wood. 1997. The second Hadley Centre coupled ocean-atmosphere GCM: model description, spinup and validation. *Climate Dynamics* 13 (2): 103–134.

Joseph J., and H. Sender. 1958. Uber die horizontale diffusion immeere. *DeutschesHydrographiches Zeitung* 11: 49–77.

Josey S.A., E.C. Kent, and P.K. Taylor. 1999. New insights into the ocean heat budget closure problem from analysis of the soc air-sea flux climatology. *Journal of Climate* 12: 2856–2880.

JPOTS Joint Panel on Oceanographic Tables and Standards. 1981. *The practical salinity scale 1978 and the international equation of state of seawater 1980*. Paris: unesco Technical Papers in Marine Science 36: 25.

JPOTS Joint Panel on Oceanographic Tables and Standards. 1991. *Processing of Oceano- graphic Station Data*. Paris: unesco.

Kalnay E., M. Kanamitsu, R. Kistler, W. Collins, D. Deaven, L. Gandin, M. Iredell, S. Saha,G. White, J. Woollen, Y. Zhu, M. Chelliah, W. Ebisuzaki, W. Higgins, J. Janowiak,K.C. Mo, C. Ropelewski, J. Wang, A. Leetmaa, R. Reynolds, R. Jenne, and D. Joseph. 1996. The NCEP/NCAR 40-year reanalysis project. *Bulletin American Meteorological Society* 77: 437–471.

Kantha L.H. 1998. Tides–A modern perspective. *Marine Geodesy* 21: 275–297.

Kent E.C., and P.K. Taylor. 1997. Choice of a Beaufort Scale. *Journal of Atmospheric and Oceanic Technology* 14 (2): 228–242.

Kerr R.A. 1998. Models win big in forecasting El Niño. *Science* 280 (5363): 522–523.

Kerr R.A. 2002. Salt fingers mix the sea. *Science* 295 (5561): 1821.

Kiehl J.T., and K.E. Trenberth. 1996. Earth's annual global mean energy budget. *Bulletin of the American Meteorological Society* 78 (2): 197–208.

Kilpatrick K.A., G.P. Podesta, and R. Evans. 2001. Overview of the NOAA/nasa advanced very high resolution radiometer Pathfinder algorithm for sea surface temperature and associated matchup database. *Journal of Geophysical Research* 106: 9179–9198.

Kistler R.E., E. Kalnay, W. Collins, S. Saha, G. White, J. Woolen, M. Chelliah, and W. Ebisuzaki. 2000. The NCEP/NCAR 50-year reanalysis. *Bulletin of the American Mete- orological Society* 82: 247–267.

Komen G.J., L. Cavaleri, M. Donelan, K. Hasselmann, S. Hasselmann, and P.A.E.M. Jans- sen. 1996. *Dynamics and Modelling of Ocean Waves*. 1st paperback ed. Cambridge: University Press.

Kullenberg B. 1954. Vagn Walfrid Ekman 1874–1954. *Journal du Conseil international pour l'exploration de la mer* 20 (2): 140–143.

Kumar A., A. Leetmaa, and M. Ji. 1994. Simulations of atmospheric variability induced by sea surface temperatures and implications for global warming. *Science* 266 (5185): 632–634.

Kundu P.K. 1990. *Fluid Mechanics*. San Diego: Academic Press.

Kunze E., and J.M. Toole. 1997. Tidally driven vorticity, diurnal shear, and turbulence atop Fieberling Seamount. *Journal of Physical Oceanography* 27 (2): 2,663–2,693.

Lagerloef G.S.E., G. Mitchum, R. Lukas, and P. Niiler. 1999. Tropical Pacific near-surface current estimates from altimeter, wind and drifter data. *Journal of Geophysical Re- search* 104 (C10): 23,313–23,326.

Lamb H. 1945. *Hydrodynamics.* 6th, first American edition. New York: Dover Publications.

Lang K.R. 1980. *Astrophysical Formulae: A Compendium for the Physicist and Astrophysicist.* 2nd ed. Berlin: Springer-Verlag.

Langer J. 1999. Computing in physics: Are we taking it too seriously? Or not seriously enough? *Physics Today* 52 (7): 11–13.

Langmuir I. 1938. Surface motion of water induced by wind. *Science* 87: 119–123.

Larson R. 2002. E-Enabled textbooks: Lower cost, higher functionality. *Syllabus* 15(10): 44.

Latif M., A. Sterl, E. Maier-Reimer, and M.M. Junge. 1993. Structure and predictability of the El Niño/Southern Oscillation phenomenon in a coupled ocean-atmosphere model. *Journal of Climate* 6: 700–708.

Lawrence M.G., J. Landgraf, P. Jckel, and B. Eaton. 1999. Artifacts in global atmospheric modeling: Two recent examples. *EOS Transactions American Geophysical Union* 80 (11): 123, 128.

Lean J., J. Beer, and R. Bradley. 1995. Reconstruction of solar irradiance since 1610: Implications for climate change. *Geophysical Research Letters* 22 (23): 3195–3198.

Ledwell J.R., A.J. Watson and C.S. Law 1998. Mixing of a tracer in the pycnocline. *Journal of Geophysical Research* 103(C10): 21,499–421,529.

Leetmaa A., and A.F. Bunker. 1978. Updated charts of the mean annual wind stress, con- vergences in the Ekman layers and Sverdrup transports in the North Atlantic. *Journal of Marine Research* 36: 311–322.

Leetmaa A., J.P. McCreary, and D.W. Moore. 1981. Equatorial currents; observation and theory. In: *Evolution of Physical Oceanography.* Edited by B. A. Warren and C. Wunsch. 184–196. Cambridge: Massachusetts Institute of Technology Press.

LeProvost C., M.L. Genco, F. Lyard, P. Vincent, and P. Canceil. 1994. Spectroscopy of the world ocean tides from a finite element hydrodynamic model. *Journal Geophysical Research* 99 (C12): 24,777–24,797.

LeProvost C., A. F. Bennett and D. E. Cartwright. 1995. Ocean tides for and from Topex/ Poseidon. *Science* 267 (5198): 639–647.

Levitus S. 1982. *Climatological Atlas of the World Ocean.* NOAA Professional Paper 13.

Levitus S. 1994. *World Ocean Atlas 1994* cd-rom *Data Set*. NOAA National Oceanographic Data Center.

Lewis E.L. 1980. The Practical Salinity Scale 1978 and its antecedents. *IEEE Journal of Oceanic Engineering* oe-5: 3–8.

List R.J. 1966. *Smithsonian Meteorological Tables*. 6th ed. Washington DC: The Smithsonian Institution.

Liu W.T. 2002. Progress in scatterometer application. *Journal of Oceanography* 58: 121– 136.

Longuet-Higgins M.S., and O.M. Phillips. 1962. Phase velocity effects in tertiary wave interactions. *Journal of Fluid Mechanics.* 12 (3): 333–336.

Lynch D.R., J.T.C. Ip, C.E. Naimie, and F.E. Werner. 1996. Comprehensive coastal circu- lation model with application to the Gulf of Maine. *Continental Shelf Research* 16 (7): 875–906.

Lynn R.J., and J.L. Reid. 1968. Characteristics and circulation of deep and abyssal waters. *Deep-Sea Research* 15 (5): 577–598.

McAvaney B.J., C. Covey, S. Joussaume, V. Kattsov, A. Kitoh, W. Ogana, A.J. Pitman, A.J. Weaver, R. A. Wood and Z.-C. Zhao. 2001. Model Evaluation. In: *Climate Change 2001: The Scientific Basis. Contribution of Working Group 1 to the Third Assessment Report of the Intergovernmental Panel on Climate Change*. Editeb by J.T. Houghton, Y. Ding, D.J. Griggs, N. Noguer, P.J. v. d. Linden, X. Dai, K. Maskell and C.A. Johnson. Cambridge, University Press: 881.

MacKenzie K.V. 1981. Nine-term equation for sound speed in the ocean. *Journal of the Acoustical Society of America* 70: 807–812.

McDougall T.J. 1987. Neutral surfaces. *Journal of Physical Oceanography* 17 (11): 1950–1964.

McDougall T. J. and R. Feistel 2003. What causes the adiabatic lapse rate? *Deep Sea Research Part I: Oceanographic Research Papers* 50 (12): 1523–1535.

McIntyre M.E. 1981. On the 'wave momentum' myth. *Journal of Fluid Mechanics* 106: 331–347.

McNally G.J., W.C. Patzert, A.D. Kirwan, and A.C. Vastano. 1983. The near-surface circu- lation of the North Pacific using satellite tracked drifting buoys. Journal of Geophysical Research 88 (C12): 7,507–7,518.

McPhaden M.J. 1986. The equatorial undercurrent: 100 years of discovery. *EOS Transans- actions of the Amererican Geophysical Union* 67 (40): 762–765.

McPhaden M.J., A.J. Busalacchi, R. Cheney, J.R. Donguy, K.S. Gage, D. Halpern, M. Ji, P. Julian, G. Meyers, G.T. Mitchum, P.P. Niiler, J. Picaut, R.W. Reynolds, N. Smith K. Takeuchi. 1998. The Tropical Ocean-Global Atmosphere (textsctoa) observing system: A decade of progress. *Journal of Geophysical Research* 103: 14,169–14,240.

McPhaden M.J. 1999. Genesis and evolution of the 1997-1998 El Niño. *Science* 283 (5404): 950–954.

McPhaden M.J., S.E. Zebiak, and M.H. Glantz. 2006. enso as an Integrating Concept in Earth Science. *Science* 314 (5806): 1740–1745.

Ma X.C., C.K. Shum, R.J. Eanes, and B.D. Tapley. 1994. Determination of ocean tides from the first year of Topex/Poseidon altimeter measurements. *Journal of Geophysical Research* 99 (C12): 24,809–24,820.

Malanotte-Rizzoli P., Ed. 1996. *Modern Approaches to Data Assimilation in Ocean Modeling*. Amsterdam: Elsevier.

Maltrud M.E., R.D. Smith, A.J. Semtner, and R.C. Malone. 1998. Global eddy-resolving ocean simulations driven by 1985–1995 atmospheric winds. *Journal of Geophysical Research* 103 (C13): 30,825–30,852.

Margules M. 1906. Uber Temperaturschichtung in stationarbewegter und ruhender Luft. *MeteorologischeZeitschrift* 241–244.

Marotzke J., and J.R. Scott. 1999. Convective mixing and the thermohaline circulation. *Journal of Physical Oceanography* 29 (11): 2962–2970.

Marotzke J. 2000. Abrupt climate change and thermohaline circulation: Mechanisms and predictability. *Proceedings National Academy of Sciences* 97 (4): 1347–1350.

Martrat B., J.O. Grimalt, N.J. Shackleton, L. de Abreu, M.A. Hutterli, and T.F. Stocker. 2007. Four Climate Cycles of Recurring Deep and Surface Water Destabilizations on the Iberian Margin. *Science* 317 (5837): 502–507.

Matth¨aus W. 1969. Zur entdeckungsgeschichte des A¨ quatorialen Unterstroms im Atkantis- chen Ozean. *Beitrage Meereskunde* 23: 37–70.

Maury M.F. 1855. *Physical Geography of the Sea*. Harper.

May D.A., M.M. Parmenter, D.S. Olszewski, and B.D. McKenzie. 1998. Operational pro- cessing of satellite sea surface temperature retrievals at the Naval Oceanographic Office. *Bulletin of the American Meteorological Society* 79 (3): 397–407.

Mellor G.L., and T. Yamada. 1982. Development of a turbulence closure model for geo- physical fluid problems. *Reviews of Geophysics and Space Physics* 20 (4): 851–875.

Mellor G.L. 1998. *User's Guide for a Three-dimensional, Primitive equation, Numerical Ocean Model Version 1998*. Princeton, Princeton University: 41.

Menard H.W., and S.M. Smith. 1966. Hypsometry of ocean basin provinces. *Journal of Geophysical Research* 71: 4305–4325.

Mercier H., M. Arhan and J.R.E. Lutjeharm. 2003. Upper-layer circulation in the eastern Equatorial and South Atlantic Ocean in January–March 1995. *Deep-Sea Research* 50 (7): 863–887.

Merryfield W.J., G. Holloway, and A.E. Gargett. 1999. A global ocean model with double- diffusion mixing. *Journal of Physical Oceanography* 29 (6): 1124–1142.

Miles J.W. 1957. On the generation of surface waves by shear flows. *Journal of Fluid Mechanics*. 3 (2) 185–204.

Millero F.J., G. Perron, and J.F. Desnoyers. 1973. Heat capacity of seawater solutions from 5° to °35C and 0.05 to 22°/oo chlorinity. *Journal of Geophysical Research* 78 (21): 4499–4506.

Millero F.J., C.-T. Chen, A. Bradshaw, and K. Schleicher. 1980. A new high pressure equation of state for seawater. *Deep-Sea Research* 27A: 255–264.

Millero F.J., and A. Poisson. 1981. International one-atmosphere equation of state of seawater.*Deep-Sea Research* 28A (6): 625–629.

Millero F.J. 1996. *Chemical Oceanography* (2nd ed). New York, CRC Press.

Millero F. J., R. Feistel, et al. 2008. The composition of Standard Seawater and the definition of the Reference-Composition Salinity Scale. *Deep Sea Research Part I: Oceanographic Research Papers* 55 (1): 50–72.

Montgomery R.B., and E.D. Stroup. 1962. Equatorial Waters and Currents at 150°W in July–August, 1952. Baltimore: The Johns Hopkins Press.

Morel A. 1974. Optical porperties of pure water and pure seawater. In: *Optical Aspects of Oceanography*. Edited by N. G. Jerlov and E. S. Nielson. 1–24. Academic Press.

Moskowitz L. 1964. Estimates of the power spectrums for fully developed seas for wind speeds of 20 to 40 knots. *Journal of Geophysical Research* 69 (24): 5161–5179.

Moum J.N., and D.R. Caldwell. 1985. Local influences on shear-flow turbulence in the equatorial ocean. *Science* 230: 315–316.

Munk W.H. 1950. On the wind-driven ocean circulation. *Journal of Meteorology* 7 (2): 79–93.

Munk W.H., and E. Palmen. 1951. A note on the dynamics of the Antarctic Circumpolar Current. *Tellus* 3: 53–55.

Munk W.H. 1966. Abyssal recipes. *Deep-Sea Research* 13: 707–730.

Munk W.H., G.R. Miller, F.E. Snodgrass, and N.F. Barber. 1963. Directional recording of swell from distant storms. *Philosophical Transactions Royal Society of London* 255 (1062): 505–584.

Munk W.H., and D.E. Cartwright. 1966. Tidal spectroscopy and prediction. *Philosophical Transactions Royal Society London* Series A. 259 (1105): 533–581.

Munk W.H., R.C. Spindel, A. Baggeroer, and T.G. Birdsall. 1994. The Heard Island feasi- bility test. *Journal of the Acoustical Society of America* 96 (4): 2330–2342.

Munk W., P. Worcester, and C. Wunsch. 1995. *Ocean Acoustic Tomography*. Cambridge: University Press.

Munk W. and C. Wunsch 1998. Abyssal recipes II. *Deep-Sea Research* 45: 1976–2009.

NAS National Academy of Sciences. 1963. *Ocean Wave Spectra: Proceedings of a Conference*. Englewood Cliffs, New Jersey: Prentice-Hall.

Neal V.T., S. Neshyba, and W. Denner. 1969. Thermal stratification in the Arctic Ocean. *Science* 166 (3903): 373–374.

Nerem R.S., E. Leuliette, and A. Cazenave. 2006. Present-day sea-level change: A review. *Comptes Rendus Geosciences* 338 (14–15): 1077–1083.

Neumann G., and W.J. Pierson. 1966. *Principles of Physical Oceanography*. New Jersey: Prentice-Hall.

Newton P. 1999. A manual for planetary management. *Nature* 400 (6743): 399.

Niiler P.P., R.E. Davis, and H.J. White. 1987. Water-following characteristics of a mixed layer drifter. *Deep-Sea Research* 34 (11): 1867–1881.

Niiler P.P., A.S. Sybrandy, K. Bi, P.M. Poulain, and D. Bitterman. 1995. Measurement of the water following capability of holey-sock and tristar drifters. *Deep-Sea Research* 42 (11/12): 1951–1964.

North G.R. and S. Nakamoto. 1989. Formalism for comparing rain estimation designs. *Journal of Atmospheric and Oceanic Technology*. 6 (6): 985–992.

Oberhuber J.M. 1988. An atlas based on the coads data set: The budgets of heat, buoyancy and turbulent kinetic energy at the surface of the global ocean. *Max-Planck-Institutfu ̈rMeteorologie*: Report 15.

Open University 1989a. *Ocean Circulation*. Oxford: Pergamon Press.

Open University 1989b. *Seawater: Its Composition, Properties and Behaviour*. Oxford: Pergamon Press.

Open University 1989c. *Waves, Tides and Shallow Water-Processes*. Oxford: Pergamon Press.

Oppenheim A.V. and R.W. Schafer. 1975. *Digital Signal Processing*. Englewood Cliffs, New Jersey: Prentice-Hall.

Orsi A.H., T. Whitworth and W.D. Nowlin. 1995. On the meridional extent and fronts of the Antarctic Circumpolar Current. *Deep-Sea Research* 42(5): 641–673.

Orsi A.H., G.C. Johnson, and J.L. Bullister. 1999. Circulation, mixing, and production of Antarctic Bottom Water. *Progress in Oceanography* 43 55-109.

Pacanowski R., and S.G.H. Philander. 1981. Parameterization of vertical mixing in numer- ical models of tropical oceans. *Journal of Physical Oceanography* 11: 1443–1451.

Pacanowski R.C., and S.M. Griffies. 1999 *MOM 3.0 Manual*. NOAA/Geophysical Fluid Dynamics Laboratory, Princeton.

Parke M.E., R.H. Stewart, D.L. Farless, and D.E. Cartwright. 1987. On the choice of orbits for an altimetric satellite to study ocean circulation and tides. *Journal of Geophysical Research* 92: 11,693–11,707.

Parker R.L. 1994. *Geophysical Inverse Theory*. Princeton: Princeton University Press.

Pedlosky J. 1987. *Geophysical Fluid Dynamics*. 2nd ed. Berlin: Springer Verlag.

Pedlosky J. 1996. *Ocean Circulation Theory*. Berlin: Springer–Verlag.

Percival D.B., and A.T. Walden. 1993. *Spectral Analysis for Physical Applications: Multi-taper and Conventional Univariate Techniques*. Cambridge: University Press.

Philander S.G.H., T. Yamagata, and R.C. Pacanowski. 1984. Unstable air-sea interactions in the tropics. *Journal of Atmospheric Research* 41: 604–613.

Philander S.G. 1990. *El Niño, La Nin˜a, and the Southern Oscillation*. Academic Press.

Phillips O.M. 1957. On the generation of waves by turbulent wind. *Journal of Fluid Me- chanics*. 2 (5): 417–445.

Phillips O.M. 1960. On the dynamics of unsteady gravity waves of finite amplitude. Part I. The elementary interactions. *Journal of Fluid Mechanics* 9 (2): 193–217.

Picaut J., F. Masia, and Y.d. Penhoat. 1997. An advective-reflective coNCEPtual model for the oscillatory nature of the ENSO. *Science* 277 (5326): 663–666.

Pickard G.L., and W.J. Emery. 1990. *Descriptive Physical Oceanography: An Introduction*. 5th enlarged ed. Oxford: Pergamon Press.

Pierson W.J., and L. Moskowitz. 1964. A proposed spectral form for fully developed wind seas based on the similarity theory of S.A. Kitaigordskii. *Journal of Geophysical. Research* 69: 5181–5190.

Pinet P.R. 2000. *Invitation to oceanography*. 2nd Edition. Sudbury, Massachusetts: Jones and Bartlett Publishers.

Polzin K.L., J.M. Toole, J.R. Ledwell, and R.W. Schmitt. 1997. Spatial variability of turbulent mixing in the abyssal ocean. *Science* 276 (5309): 93–96.

Post D.E. and L.G. Votta. 2005. Computational science demands a new paradigm. *Physics Today* 58 (1): 35–41.

Powell M. D., P.J. Vickery and T.A. Reinhold. 2003. Reduced drag coefficient for high wind speeds in tropical cyclones. *Nature* 422 (6929): 279–283.

Press W.H., S.A. Teukolsky, W.T. Vetterling, and B.P. Flannery. 1992. *Numerical Recipes in FORTRAN*. 2nd ed. Cambridge: University Press.

Preston-Thomas H. 1990. The International Temperature Scale of 1990 (its-90). *Metrolo- gia* 27 (1): 3–10.

Proudman J. 1916. On the motion of solids in a liquid possessing vorticity. *Proceedings Royal Society (London)* A 92: 408–424

Pugh D.T. 1987. *Tides, Surges, and Mean Sea-Level*. Chichester: John Wiley.

Rahmstorf S. 1995. Bifurcations of the Atlantic thermohaline circulation in response to changes in the hydrological cycle. *Nature* 378: 145–149.

Ralph E.A., and P.P. Niiler. 2000. Wind-driven currents in the tropical Pacific. *Journal of Physical Oceanography* 29 (9): 2121–2129.

Ramanathan V., B. Subasilar, G.J. Zhang, W. Conant, R.D. Cess, J.T. Kiehl, H. Grassl, and L. Shi. 1995. Warm pool heat budget shortwave cloud forcing: A missing physics? *Science* 267 (5197): 499–503.

Rasmusson E.M., and T.H. Carpenter. 1982. Variations in tropical sea surface temperature and surface wind fields associated with the Southern Oscillation/ El Niño. *Monthly Weather Review* 110: 354–384.

Rasmusson E.M., and J.M. Wallace. 1983. Meteorological Aspects of the El Niño/ Southern Oscillation. *Science* 222: 1195–1202.

Reid R.O. 1948. The equatorial currents of the eastern Pacific as maintained by the stress of the wind. *Journal of Marine Research* 7 (2): 75–99.

Reynolds O. 1883. An experimental investigation of the circumstances which determine whether the motion of water will be direct or sinuous, and the law of resistance in parallel channels. *Philosophical Transactions, Royal Society London* 174: 935.

Reynolds R.W., and D.C. Marsico. 1993. An improved real-time global sea surface temper- ature analysis. *Journal of Climate* 6: 114–119.

Reynolds R.W., and T.M. Smith. 1994. Improved global sea surface temperature analysis using optimum interpolation. *Journal of Climate* 7: 929–948.

Reynolds R.W., and T.M. Smith. 1995. A high-resolution global sea surface climatology.

Journal of Climate 8 (6): 1571–1583.

Reynolds R.W., N.A. Rayner, T. M. Smith, D. C. Stokes, W. Wang. 2002. An improved in situ and satellite SST analysis for climate. *Journal of Climate* 15: 1609–1625.

Rhines P.B. 1984. Notes on the general circulation of the ocean. In: *A Celebration in Geophysics and Oceanography-1982*. 83–86. Scripps Institution of Oceanography.

Richardson E.G. 1961. *Dynamics of Real Fluids*. 2nd ed. London: Edward Arnolds.

Richardson P.L. 1981. Eddy kinetic energy in the North Atlantic from surface drifters. *Journal of Geophysical Research*. 88 (C7): 4355–4367.

Richardson P.L. 1993. Tracking Ocean Eddies. *American Scientist* 81: 261–271.

Richardson P. L. 2008. On the history of meridional overturning circulation schematic diagrams. *Progress In Oceanography* 76 (4): 466–486.

Ring Group R.H. Backus, G.R. Flierl, D.R. Kester, D.B. Olson, P.L. Richardson, A.C. Vastano, P.H. Wiebe, and J.H. Wormuth. 1981. Gulf Stream cold-core rings: Their physics, chemistry, and biology. *Science* 212 (4499): 1091–1100.

Rintoul S.R., and C. Wunsch. 1991. Mass, heat, oxygen and nutrient fluxes and budgets in the North Atlantic Ocean. *Deep-Sea Research* 38 (Supplement 1): S355–S377.

Riser S. C., L. Ren, et al. 2008. Salinity in ARGO. *Oceanography* 21 (1): 56–67.

Robinson A.R., S.M. Glenn, M.A. Spall, L.J. Walstad, G.M. Gardner, and W.G. Leslie. 1989. Forecasting Gulf Stream meanders and rings. *EOS Transactions American Geophysical Union* 70: (45).

Roed L.P., B. Hackett, B. Gjevik, and L.I. Eide. 1995. A review of the MetoceanModeling Project (mompop) Part 1: Model comparison study. In: *Quantitative Skill Assessment for Coastal Ocean Models*. Edited by D. R. Lynch and A. M. Davies. 285–305. Washington DC: American Geophysical Union.

Ropelewski C.F., and M.S. Halpert. 1987. Global and regional precipitation associated with El Niño/Southern Oscillation. *Monthly Weather Review* 115: 1606–1626.

Rossby C.C. 1936. Dynamics of steady ocean currents in the light of experimental fluid mechanics. *Papers in Physical Oceanography and Meteorology, Massachusetts Institute of Technology and Woods Hole Oceanographic Institution*. 5(1): 43.

Rossow W.B. and R.A. Schiffer. 1991. Isccp Cloud Data Products. *Bulletin of the American Meteorology Society*, 72 (1): 2–20.

Rudnick , D.L., T.J. Boyd, R.E. Brainard, G.S. Carter, G.D. Egbert, M.C. Gregg, P.E. Holloway, J.M. Klymak, E. Kunze, C.M. Lee, M.D. Levine, D.S. Luther, J.P. Martin, M.A. Merrifield, J.N. Moum, J.D. Nash, R. Pinkel, L. Rainville, T.B. Sanford. 2003. From Tides to Mixing Along the Hawaiian Ridge. *Science* 301 (5631): 355–357.

Sabine C.L., R.A. Feely, N. Gruber, R.M. Key, K. Lee, J.L. Bullister, R. Wanninkhof, C.S. Wong, D.W.R. Wallace, B. Tilbrook, F.J. Millero, T.-H. Peng, A. Kozyr, T. Ono and A.F. Rios. 2004. The Oceanic Sink for Anthropogenic CO2. *Science* 305 (5682): 367–371.

Sandwell D.T., and W.H.F. Smith. 2001. Bathymetric Estimation. In: *Satellite Altimetry and Earth Sciences*. Edited by L.-L. Fu and A. Cazanave. 441–457. San Diego: Academic Press.

Satake K., K. Shimazaki, Y. Tsuji, and K. Ueda. 1996. Time and size of a giant earthquake in Cascadia inferred from Japanese tsunami records of January 1700. *Nature* 379 (6562): 246–249.

Saunders P.M. and N. P. Fofonoff. 1976. Conversion of pressure to depth in the ocean. *Deep-Sea Research* 23: 109–111.

Saunders P.M. 1986. The accuracy of measurements of salinity, oxygen and temperature in the deep ocean. *Journal of Physical Oceanography* 16 (1): 189–195.

Schmitt R.W., H. Perkins, J.D. Boyd, and M.C. Stalcup. 1987. C-SALT: An investigation of the thermohaline staircase in the western tropical North Atlantic. *Deep-Sea Research* 34 (10): 1655–1665.

Schmitt R.W., P.S. Bogden, and C.E. Dorman. 1989. Evaporation minus precipitation and density fluxes for the North Atlantic. *Journal of Physical Oceanography* 19: 1208–1221.

Schmitt R.W. 1994. The ocean freshwater cycle. jsc Ocean Observing System Development Panel, Texas A&M University, College Station, Texas: 40.

Schmitz W.J. 1996. On the World Ocean Circulation: Volume I. Some Global Features/ North Atlantic Circulation. Woods Hole Oceanographic Institution, Technical Report whoi–96–03.

Schubert S.D., R.B. Rood, and J. Pfaendtner. 1993. An assimilated dataset for Earth science applications. *Bulletin American Meteorological Society* 74 (12): 2331–2342.

Selby J.E.A., and R.A. McClatchey. 1975. Atmospheric transmittance from 0.25 to 28.5 μm: Computer code lowtran 3. Air Force Cambridge Research Laboratories, Optical Physics Laboratory Technical Report TR–75–0255.

Service R.F. 1996. Rock chemistry traces ancient traders. *Science* 274 (5295): 2012–2013.

Sette O.E., and J.D. Isaacs. 1960. Symposium on "The Changing Pacific Ocean in 1957 and 1958". California Cooperative Oceanic Fisheries Investigations Reports VII: 13–217.

Shamos M.H. 1995. *The Myth of Scientific Literacy*. New Brunswick: Rutger University Press.

Shchepetkin A. F. and J.C. McWilliams. 2004. The Regional Oceanic Modeling System: A split-explicit, free-surface, topography-following-coordinate ocean model. *Ocean Mod- elling* 9: 347–404.

Shepard F.P., G.A. MacDonald and D.C. Cox. 1950. The tsunami of April 1, 1946. *Bulletin of the Scripps Institution of Oceanography* 5(6): 391–528.

Shepard F.P. 1963. *Submarine Geology*. 2nd ed. New York: Harper and Row.

Slingo J.M., K.R. Sperber, J.S. Boyle, J.-P. Ceron, M. Dix, B. Dugas, W. Ebisuzaki, J. Fyfe, D. Gregory, J.-F. Gueremy, J. Hack, A. Harzallah, P. Inness, A. Kitoh, W.K.-M. Lau, B. McAvaney, R. Madden, A. Matthews, T.N. Palmer, C.-K. Park, D. Randall, and N. Renno. 1995. Atmospheric Model Inter comparison Project (amip): Intraseasonal Oscillations in 15 Atmospheric General Circulation Models (Results From an aimp Diagnostics Subproject). World Meteorological Organization/World Climate Research Programme, WCRP–88 (WMO/TD No. 661).

Slutz R.J., S.J. Lubker, J.D. Hiscox, S.D. Woodruff, R.L. Jenne, D.H. Joseph, P.M. Steurer, and J.D. Elms. 1985: Comprehensive Ocean-Atmosphere Data Set; Release 1. NOAA Environmental Research Laboratories, Climate Research Program, Boulder, Colorado:, 268. (NTIS PB86–105723).

Smagorinski J. 1963. General circulation experiments with primitive equations I. The basic experiment. *Monthly Weather Review* 91: 99–164.

Smith R.D., M.E. Maltrud, F.O. Bryan and M.W. Hecht. 2000. Numerical simulation of the North Atlantic ocean at 1/10°. *Journal of Physical Oceanography* 30 (7): 1532–1561.

Smith S.D. 1980. Wind stress and heat flux over the ocean in gale force winds. *Journal of Physical Oceanography* 10: 709-726.

Smith S.D. 1988. Coefficients for sea surface wind stress, heat flux and wind profiles as a function of wind speed and temperature. *Journal of Geophysical Research* 93: 15,467– 15,472.

Smith T.M., T.R. Karl and R.W. Reynolds. 2002. How accurate are climate simulations? *Science* 296 (5567): 483–484.

Smith T.M., and R.W. Reynolds. 2004. Improved extended reconstruction of sst (1854– 1997). *Journal of Climate* 17 (6): 2466–2477.

Smith W.H.F., and D.T. Sandwell. 1994. Bathymetric prediction from dense satellite al- timetry and sparse shipboard bathymetry. *Journal of Geophysical Research* 99 (B11): 21,803–21,824.

Smith W.H.F. and D.T. Sandwell 1997. Global sea floor topography from satellite altimetry and ship depth soundings. *Science* 277 (5334): 1956–1962.

Snodgrass F.E. 1964. Precision digital tide gauge. *Science* 146 (3641): 198–208.

Soulen R.J., and W.E. Fogle. 1997. Temperature scales below 1 kelvin. *Physics Today* 50 (8 Part 1): 36–42.

Stammer D., R. Tokmakian, A. Semtner, and C. Wunsch. 1996. How well does a 1/4°global circulation model simulate large-scale oceanic observations? *Journal of Geophysical Research* 101 (C10): 25,779–25,811.

Starr V.P. 1968. *Physics of Negative Viscosity Phenomena*. New York: McGraw-Hill.

Steig E.J. 2006. Climate change: The south-north connection. *Nature* 444 (7116): 152–153.

Stern M.E. 1960. The 'salt fountain' and thermohaline convection. *Tellus* 12: 172–175.

Stewart I. 1992. Warning—handle with care! *Nature* **355**: 16–17.

Stewart R.H. 1980. Ocean wave measurement techniques. In: *Air Sea Interaction, Instruments and Methods*. Edited by L. H. F. Dobson and R. Davis. 447–470. New York: Plenum Press.

Stewart R.H. 1985. *Methods of Satellite Oceanography*. University of California Press.

Stewart R.H. 1995. Predictability of Texas Rainfall Patterns on Time Scales of Six to Twelve Months: A Review. In: *The Changing Climate of Texas: Predictability and Implications for the Future* Edited by J. Norwine, J.R. Giardino, G.R. North and J.B. Valdes. College Station, Texas: Texas A&M University. 38–47

Stocker T.F., and O. Marchal. 2000. Abrupt climate change in the computer: Is it real? *Proceedings National Academy of Sciences* 97 (4): 1362–1365.

Stokes G.G. 1847. On the theory of oscillatory waves. *Cambridge Transactions* 8: 441–473.

Stommel H. 1948. The westward intensification of wind-driven ocean currents. *Transac- tions, American Geophysical Union* 29 (2): 202–206.

Stommel H. 1958. The abyssal circulation. *Deep-Sea Research* 5 (1): 80–82.

Stommel H., A.B. Arons, and A.J. Faller. 1958. Some examples of stationary flow patterns in bounded basins. *Tellus* 10 (2): 179–187.

Stommel H., and A.B. Arons. 1960. On the abyssal circulation of the world ocean—II. An idealized model of the circulation pattern and amplitude in oceanic basins. *Deep-Sea Research* 6: 217–233.

Stommel H.M., and D.W. Moore. 1989. *An Introduction to the Coriolis Force*. Cambridge: University Press.

Stott P.A., S.F.B. Tett, G.S. Jones, M.R. Allen, J.F.B. Mitchell and G.J. Jenkins 2000. External Control of 20th Century Temperature by Natural and Anthropogenic Forcings. *Science* 290 (5499): 2133-2137.

Strub P.T., T.K. Chereskin, P.P. Niiler, C. James, and M.D. Levine. 1997. Altimeter-derived variability of surface velocities in the California Current System 1. Evaluation of Topex altimeter velocity resolution. *Journal of Geophysical Research* 102 (C6): 12,727–12,748.

SUN Working Group on Symbols, Units and Nomenclature in Physical Oceanography. 1985. *The International System of units (SI) in oceanography*. iapso Paris: Unesco Techni- cal Papers in Marine Science 45: 124.

Sverdrup H.U. 1947. Wind-driven currents in a baroclinic ocean: with application to the equatorial currents of the eastern Pacific. *Proceedings of the National Academy of Sciences* 33 (11): 318–326.

Sverdrup H.U., M.W. Johnson, and R.H. Fleming. 1942. *The Oceans: Their Physics, Chemistry, and General Biology*. Englewood Cliffs, New Jersey: Prentice-Hall.

SWAMP Group Sea Wave Modeling Project. 1985. *Ocean Wave Modeling*. New York: Plenum Press.

Swenson K.R., and A.E. Shaw. 1990. The Argos system: Monitoring the world's environ- ment. *Oceanography* 3 (1): 60–61.

Takayabu Y.N., T. Ihuchi, M. Kachi, A. Shibata, and H. Kanzawa. 1999. Abrupt termination of the 1997–98 El Nino in response to a Madden-Julian oscillation. *Nature* 402 (6759): 279–282.

Tapley B.D., and M.-C. Kim. 2001. Applications to geodesy. In *Satellite Altimetry and Earth Sciences*. Edited by L.-L. Fu and A. Cazenave. 371–406. San Diego: Academic Press.

Taylor G.I. 1921. Experiments with rotating fluids. *Proceedings Royal Society (London)* A 100: 114–121.

Taylor P.K. Editor. 2000. Intercomparison and validation of ocean-atmosphere energy flux fields: Final Report of the Joint World Climate Research Program and Scientific Com- mittee on Ocean Research Working Group on Air-sea Fluxes, *World Climate Research Program Report* wcrp-112: 303.

Taylor P. K., E.F. Bradley, C.W. Fairall, D. Legler, J. Schultz, R.A. Weller and G.H. White. 2001. Surface fluxes and surface reference sites. In: *Observing the Oceans in the 21st Century*. Edited by C.J. Koblinsky and N.R. Smith. Melbourne, Bureau of Meteorology: 177–197.

Tchernia P. 1980. *Descriptive Regional Oceanography*. Oxford: Pergamon Press.

Tennekes H., and J.L. Lumley. 1990. *A First Course in Turbulence*. Boston: MIT Press.

Thurman H.V. 1985. *Introductory Oceanography*. Fourth ed. Columbus: Charles E. Merrill Publishing Company.

Titov V.V. and F.I. Gonzalez. 1997. Implementation and testing of the Method of Splitting Tsunami (most) model. NOAA Pacific Marine Environmental Laboratory Contribution 1927: 14.

Toggweiler J.R. 1994. The ocean's overturning circulation. *Physics Today* 47 (11): 45–50.

Toggweiler J. R. and J. Russell 2008. Ocean circulation in a warming climate. *Nature* 451 (7176): 286–288.

Tolmazin D. 1985. *Elements of Dynamic Oceanography*. Boston: Allen and Unwin.

Tomczak M., and J.S. Godfrey. 1994. *Regional Oceanography: An Introduction*. London: Pergamon.

Tomczak M. 1999. Some historical, theoretical and applied aspects of quantitative water mass analysis. *Journal of Marine Research* 57 (2): 275–303.

Trenberth K.E., and D.J. Shea. 1987. On the evolution of the Southern Oscillation.

Monthly Weather Review 115: 3078–3096.

Trenberth K.E., W.G. Large, and J.G. Olson. 1989. The effective drag coefficient for evaluating wind stress over the oceans. *Journal of Climate*, 2: 1507–1516.

Trenberth K.E., W.G. Large, and J.G. Olson. 1990. The mean annual cycle in global ocean wind stress. *Journal of Physical Oceanography* 20 (11): 1742–1760.

Trenberth K.E., and A. Solomon. 1994. The global heat balance: heat transports in the atmosphere and ocean. *Climate Dynamics* 10 (3): 107–134.

Trenberth K.E. 1997. The use and abuse of climate models. *Nature* 386 (6621): 131–133.

Trenberth K.E. 1997. The definition of El Nin˜no. *Bulletin of the American Meteorological Society* 78 (12): 2771–2777.

Trenberth K.E. and J.M. Caron. 2001. Estimates of meridional atmospheric and oceanic heat transports. *Journal of Climate* 14 (16): 3433–3443.

Uchida H., S. Imawaki, and J.-H. Hu. 1998. Comparisons of Kuroshio surface velocities derived from satellite altimeter and drifting buoy data. *Journal of Oceanography* 54: 115–122.

Uppala S.M., and P.W. K°allberg, A.J. Simmons, U. Andrae, V. Da Costa Bechtold, M. Fiorino, J. K. Gibson, J. Haseler, A. Hernandez, G. A. Kelly, X. Li, K. Onogi, S. Saarinen, N. Sokka, R. P. Allan, E. Andersson, K. Arpe, M. A. Balmaseda, A. C. M. Beljaars, L. Van De Berg, J. Bidlot, N. Bormann, S. Caires, F. Chevallier, A. Dethof, Dragosavac, M. Fisher, M. Fuentes, S. Hagemann, E. Ho´lm, B. J. Hoskins, L.Isaksen, P.A.E.M. Janssen, R. Jenne, A. P. McNally, J.-F. Mahfouf, J.-J. Morcrette,A. Rayner, R. W. Saunders, P. Simon, A. Sterl, K. E. Trenberth, A. Untch, D. Vasiljevic, P. Viterbo, J. Woollen. 2005. The era–40 re-analysis. *Quarterly Journal of the Royal Meteorological Society* 131(612): 2961–3012.

Ursell F. 1950. On the theoretical form of ocean swell on a rotating earth. *Monthly Notices Royal Astronomical Society, Geophysical Supplement* 6 (1): G1–G8.

van Meurs P. 1998. Interactions between near-inertial mixed layer currents and the meso- scale: The importance of spatial variability in the vorticity field. *Journal of Physical Oceanography* 28 (7): 1363–1388.

Vesecky J.F., and R.H. Stewart. 1982. The observation of ocean surface phenomena using imagery from the Seasat synthetic aperture radar: An assessment. *Journal of Geo- physical Research* 87 (C5): 3397–3430.

von Arx W.S. 1962. *An Introduction to Physical Oceanography*. Reading, Massachusetts: Addison-Wesley.

WAMDI Group S. Hasselmann, K. Hasselmann, E. Bauer, P.A.E.M. Janssen, G.J. Komen,L. Bertotti, P. Lionello, A. Guillaume, V.C. Cardone, J.A. Greenwood, M. Reistad,L. Zambresky, and J.A. Ewing. 1988. The WAM model—A third generation wave prediction model. *Journal of Physical Oceanography* 18: 1775–1810.

Warren B.A. 1973. Transpacific hydrographic sections at Latitudes 43° S and 28° S: Thescorpio Expedition–II. Deep Water. *Deep-Sea Research* 20: 9–38.

WCRP World Climate Research Program. 1995. *Proceedings of the Workshop on Global Coupled General Circulation Models*. World Meteorological Organization/World Cli- mate Research Program, WCRP-87 (WMO/TD Number 655).

Weaver A.J. and C. Hillaire-Marcel 2004. Global Warming and the Next Ice Age. *Science*304 (5669): 400–402.

Webb D.J., and N. Suginohara. 2001. Vertical mixing in the ocean. *Nature* 409 (6816): 37.

Webster F. 1968. Observations of inertial-period motions in the deep sea. Reviews of Geophysics 6 (4): 473–490.

Webster P.J., and R. Lukas. 1992. toga coare: The Coupled Ocean-Atmosphere Response Experiment. *Bulletin American Meteorological Society* 73 (9): 1377–1416.

Weller R.A., J.P. Dean, J. Marra, J.F. Price, E.A. Francis, and D.C. Boardman. 1985. Three–dimensional flow in the upper ocean. *Science* 227: 1552–1556.

Weller R.A., and A.J. Plueddmann. 1996. Observations of the vertical structure of the oceanic boundary layer. *Journal of Geophysical Research* 101 (C4): 8,789–8,806.

Wentz P.J., S. Peteherych, and L.A. Thomas. 1984. A model function for ocean radar cross sections at 14.6 GHz. *Journal of Geophysical Research* 89 (C3): 3689–3704.

West G.B. 1982. Mean Earth ellipsoid determined from Seasat 1 altimetric observations. *Journal of Geophysical Research* 87 (B7): 5538–5540.

White G. Editor. 1996. wcrp Workshop on Air-Sea Flux Fields for Forcing Ocean Models and Validating GCMs. Geneva: World Meteorological Organization Report WCRP–95 (WMO/TD Number 762).

White W.B., and D.R. Cayan. 1998. Quasi-periodicity and global symmetries in inter-decadal upper ocean temperature variability. *Journal of Geophysical Research* 103 (C10): 21,335–21,354.

Whitham G.B. 1974. *Linear and Nonlinear Waves.* New York: John Wiley.

Whittaker E.T., and G.N. Watson. 1963. *A Course of Modern Analysis.* 4th ed. Cambridge: University Press.

Whitworth T., and R.G. Peterson. 1985. Volume transport of the Antarctic Circumpolar Current from bottom pressure measurements. *Journal of Physical Oceanography* 15 (6): 810–816.

Wiegel R.L. 1964. *Oceanographical Engineering.* Englewood Cliffs, New Jersey: Prentice Hall.

Wilson W.D. 1960. Equation for the speed of sound in sea water. *Journal of the Acoustical Society of America* 32 (10): 1357.

Woodruff S.D., R.J. Slutz, R.L. Jenne, and P.M. Steurer. 1987. A comprehensive ocean-atmosphere data set. *Bulletin American Meteorological Society* 68: 1239–1250.

Wooster W.S. 1960. Investigations of equatorial undercurrents. *Deep-Sea Research* 6 (4): 263–264.

Wooster W.S., A.J. Lee, and G. Dietrich. 1969. Redefinition of salinity. *Deep-Sea Research* 16 (3): 321–322.

Worthington L.V. 1981. The water masses of the World Ocean: Some results of a fine-scale census. In: *Evolution of Physical Oceanography: Scientific surveys in honor of Henry Stommel.* Edited by B. A. Warren and C. Wunsch. 42–69. Cambridge: Massachusetts Institute of Technology.

Wunsch C. 1996. *The Ocean Circulation Inverse Problem.* Cambridge: University Press.

Wunsch C. 2002. Ocean observations and the climate forecast problem. In: *Meteorology at the Millennium*, Edited by R.P. Pearce. London: Royal Meteorological Society: 233– 245.

Wunsch C. 2002b. What is the thermohaline circulation? *Science* 298 (5596): 1179–1180.

Wust G. 1964. The major deep–sea expeditions and research vessels 1873–1960. *Progress in Oceanography* 2: 3–52.

Wyrtki K. 1975. El Niño—The dynamic response of the equatorial Pacific Ocean to atmospheric forcing. *Journal of Physical Oceanography* 5 (4): 572–584.

Wyrtki K. 1979. Sea level variations: monitoring the breath of the Pacific. *EOS Transactions of the American Geophysical Union* 60 (3): 25-27.

Wyrtki K. 1985. Water displacements in the Pacific and the genesis of El Niño cycles. *Journal of Geophysical Research* 90 (C4): 7129-7132.

Xie P., and P.A. Arkin. 1997. Global precipitation: A 17-year monthly analysis based on gauge observations, satellite estimate, and numerical model outputs. *Bulletin of the American Meteorological Society* 78 (1): 2539–2558.

Yelland M., and P.K. Taylor. 1996. Wind stress measurements from the open ocean. *Journal of Physical Oceanography* 26 (4): 541–558.

Yelland M.J., B.I. Moat, P.K. Taylor, R.W. Pascal, J. Hutchings and V.C. Cornell. 1998. Wind stress measurements from the open ocean corrected for airflow distortion by the ship. *Journal of Physical Oceanography* 28 (7): 1511–1526.

Yu X., and M.J. McPhaden. 1999. Dynamical analysis of seasonal and inter annual variability in the equatorial Pacific. *Journal of Physical Oceanography* 29 (9): 2350–2369.

Zahn R. 1994. Core correlations. *Nature* 371 (6495): 289–290.

Index

A

Absorption 36, 55, 60, 100-101

Aerosols 54, 60, 94, 103

Agreement 198

Air-sea 338

Altimeters 32, 303

Analogy 124

Analysis 47-48, 262, 278, 346, 355

Animals 221

Anomaly 86, 168, 170, 176, 252

Antarctic 224, 226, 233, 235-236, 238-239, 241-242, 282, 345

Antarctica 23, 226, 235-236, 240

Atlantic 7, 10, 22-24, 80-81, 83, 96, 129-130, 135, 183, 199-201, 211, 221-227, 229, 231, 233-236, 238, 240, 242, 245, 268, 271, 298, 327, 335, 338, 347-348

Atmosphere v, 2, 8, 39, 41, 46, 49-51, 53-56, 60-62, 66, 69-71, 73, 83, 103, 119, 146, 152, 155, 183, 203, 205, 220-221, 231, 242-243, 253, 258, 260-263, 268, 280-282

B

Bandwidth 294

Baroclinic 169-170, 188, 203

Barotropic 169, 207, 210

Barrel 209

Bermuda 37, 84-85

Bottom 37, 69-70, 83, 129, 142, 146, 170, 187, 207, 219, 221-223, 225, 228, 235, 238, 242, 270, 272, 274

Bulletin 339, 354

C

Carbon 220-221

Climate 1-2, 4, 51, 73, 139, 220, 224-225, 227, 238, 272, 280-283, 315, 329-330, 332, 334, 336-338, 340-342, 346-347, 349-352

Cloud-free 55

Clouds 56, 60, 64-65, 69, 93

Components 111, 150, 281, 290, 301

Convection 136, 152, 227-228, 230, 258, 282

Coriolis 106-107, 111, 113, 126, 140, 142, 145-146, 157-158, 189, 196, 204, 214, 249, 255

Coupling 262

Crystal 97

Darwin 251
Deep-sea 158, 224, 343-344, 348, 354
Deep-water 286-287, 325
Density 53, 75, 77, 84-90, 104, 109, 115-117, 127, 131-137, 143, 165, 169, 177-179, 182, 203, 217, 219, 231, 233-236, 248, 269, 271, 274, 279, 301
Deuterium 224
Digital 34
Dispersion 286-287, 304
Doldrums 245
Dominant 39, 157, 268
Dominated 69
Doodson 320-321, 323
Doppler 180-181
Double 135-136, 303
Downward 36, 84, 128, 215
Downwind 142
Drifter 180, 182
Drifters 7, 147, 180
Drifting 200
Driven 204
Drives 39, 155, 215, 263
Drogue 180-181
Drought 258
During 16, 53, 154, 223-225, 258, 329, 333
Dynamic 13, 134, 166
Dynamics v, 3-4, 8, 105, 119, 257, 263, 265, 269, 315, 335, 337, 345, 347

E

Earth-moon 318
Earthquake 312-313
Ecliptic 319-320
Ecuador 244, 253, 255
Ellipsoid 32
Elliptical 318, 320
Energy 51-52, 66, 86-87, 100, 200, 203, 223, 230, 263, 288-289, 296, 300, 308, 327, 336
Equator 21, 40, 229, 244, 246-248, 253, 255-257, 318-319
Equinox 319

F

Floats 99-100, 181-182
Fluids 119
Flushing 109
Forecasts 47-48, 261, 263-264, 278-279, 301
Freezing 79
Friction 106, 111, 117, 119, 125-126, 136, 159, 197-198, 209
Frozen 292

G

Geophysics 8
Gradients 166, 208, 216
Greenhouse 55
Greenland 201, 221, 224-227
Greenwich 21
Gridded 46-47

H

Harmonics 324
Heating 41, 69, 243
Heights 289, 303, 312
Hemisphere 40, 47, 63, 144, 150, 155, 169, 222, 225, 227, 242, 245

Histogram 26

Horizontal 24-25, 33, 36-38, 48, 106, 127, 130-131, 140, 143, 153-154, 157-159, 169, 187, 189, 216, 268, 271, 274, 281, 318

Humidity 57-59

Hurricanes 276

Hybrid 271

I

Inflow 109

Infrared 51-52, 55-56, 59, 61, 65-66, 69, 71, 92-94, 204

Injection 92

Inland 275

Intensity 36

Interface 132, 135-136, 178-179

Ionosphere 164

Irradiance 100-101

Islands 246

J

Joules 52

Journal 329-350, 352-355

K

Kelvin 253, 255-257

Kinetic 200

L

Langmuir 154

Latent 57, 62, 66, 69

Long-wave 61

M

Magnetic 97

Marine 62, 139, 221, 237, 259-260, 333-334, 339, 352

Mechanics 105, 117, 336-337, 341

Mercury 90-91

Meridian 21

Mesoscale 163

Meteor 9

Microwave 45-46, 61

Mid-ocean 26, 130, 230, 325

Milestones 12

Models x, 3-4, 47-48, 61-62, 65-66, 73, 131, 230, 261-263, 265-269, 271-272, 274-275, 277-278, 280-284, 301, 326, 329, 335

Modular 262

Moisture 259

Molecular 87, 120, 124-125, 135

Molecules 103, 120, 182

Moment 209

Momentum 48, 105, 111, 113, 116, 120, 122-125, 158, 208-209, 242, 269, 290

Motion 105-106, 116-117, 119, 139, 141, 171, 177, 179, 192, 212, 215, 265, 279, 283, 285, 346

N

Nature 330, 333-334, 346, 350, 352

Nonlinear 19

Nutrients 152

O

Oceans v, 1, 51, 331, 352

Offshore 46, 48, 69, 152, 189, 285, 288, 304, 309-310

Open-ocean 129, 276

Organisms 237

Oxygen 220, 232

P

Perigee 320

Petawatts 223

Physics 329, 333, 340, 346, 350

Pivotal v

Platinum 95

Population 17-18

Pumping 153, 214-215, 217

R

Radars 304

Radiance 92-93, 100-101, 103

Radiation 51-52, 54-56, 59, 61, 64, 66-67, 71, 92, 94

Radiometer 92

Radius 21

Rainfall 61, 253, 259-260

Resistance 79

Resolution 33-34, 45, 48, 50, 65, 92, 262, 268, 270, 272, 274, 276, 279, 281

Rowboat 157

Rubber 184-185

Rudolf 337

Running 252

S

Salinity 35-36, 75-78, 80, 82-83, 85-86, 89-90, 95-96, 98-100, 104, 109, 219, 223, 225, 231-232, 234, 236, 281-282

Sampling 17-18, 20, 164, 293-294

Sandwell 31, 34, 350

Satellite 14, 32-33, 37, 45-46, 55, 60, 62, 64, 68, 93, 103, 162, 164, 180-181, 260, 277, 303, 325, 331-332, 350

Scanner 103

Seafloor 34

Sediments 224-225

Sensors 304

Sigma-t 86

Signal 37, 181, 294

Simulation 268

Solutions 2, 117, 144, 158, 192-193, 277

Sounds 37

Spacing 18-19, 274

Sparse 63, 265

Spatial 48, 155, 304

Spectra 288, 292, 297-298, 305

Spectrum 291, 295-301, 305, 322-323, 336

Stability 131-134, 136-137, 147, 149

Station 84-85, 170, 324

Storms 61, 200, 275, 285, 288-289, 301

Stream 13, 170, 175, 179, 194-204, 206, 222, 232, 277-279, 347

Strength 134, 230, 310

Submarine 312-313

Sunlight 39, 54-55, 60, 84, 100, 104

Systems 7, 32, 106, 109, 162-164, 266

T

Telescope 92

Thermal 98

Thermistor 92, 98

Titration 95

Topography 32-33, 161-164, 187, 207, 248, 277

Torque 196

Transform 295-296

Transition 121

Tritium 183, 237

Tsunamis 3, 312

V

Variables 19, 63, 90

Variance 296, 305

Variations 32-33, 35, 73

Vector 110, 113, 153, 194, 206

Velocity 25, 71, 100, 110, 113, 115, 119-120, 125-126, 134, 142, 144-146, 153-154, 158, 169, 171, 179-181, 186-188, 194, 206, 212-215, 228, 246, 255-257, 274, 278, 286-287, 304, 308

Viscosity 119-120, 124-126, 157, 189, 203, 268

Viscous 124, 266

Volcanic 224

W

Warming 226

Waters 79-81, 84, 102-103, 236, 244, 315

Wavelength 55, 101, 103, 286, 304

Weather 1, 46-48, 62, 65-66, 73, 152, 243, 258, 261-262, 275-276, 298, 301

Winter 18, 53, 84, 219, 259